BRITISH GEOLOGICAL SURVEY

Ordovician (Caradoc) marginal basin volcanism in Snowdonia (north-west Wales)

M F Howells, A J Reedman and S D G Campbell

Northamptonshire DISCARDED Libraries

Reference only Not to be taken away.

LONDON: HMSO 1991

ii

© *NERC copyright 1991*

First published 1991

ISBN 0 11 884465 2

Bibliographic reference

HOWELLS, M F, REEDMAN, A J, and CAMPBELL, S D G.
1991. *Ordovician (Caradoc) marginal basin volcanism in
Snowdonia (north-west Wales).* (London : HMSO for the
British Geological Survey.)

Authors

M F Howells, BSc, PhD
A J Reedman, BSc, PhD, DIC
S D G Campbell, BSc, PhD
British Geological Survey

Notes

Numbers preceded by the letters A or L refer to the British
Geological Survey Photograph Collection.

Maps used in this book are based on Ordnance Survey
mapping.

Printed in the UK for HMSO

Dd 291125 C10 8/91

CONTENTS

vi

PLATES

TABLES

PREFACE

North-west Wales reserves a special place in the history of geology. It was the exposures of rocks in Snowdonia which caused the great controversy between Sedgwick and Murchison in the nineteenth century that led Lapworth to devise the Ordovician System to accommodate them. Volcanic rocks are a major component of the Ordovician succession in Snowdonia, even though in the central and northern parts they occur within just two stages of the Caradoc. The volcanic rocks are interstratified with mainly marine sedimentary rocks and they have been folded and metamorphosed. After an absence of 100 years, the work of remapping this region was begun by the Geological Survey in the late 1960s and during this resurvey, special attention was given to the volcanic rocks.

This book owes much to the effort of many members of staff who were involved in the slow process of unravelling the complex volcanic stratigraphy of the Bangor sheet (106) published in 1985, but the main impetus for it came after a multidisciplinary study, specifically of the Caradoc volcanic rocks, was begun in April 1982.

The book differs from the standard Survey 1:50 000 sheet memoir in that it does not systematically present comprehensive details of localities. It is a process-orientated account of the volcanic evolution during Caradoc times of part of the Lower Palaeozoic Welsh Basin. It draws on the results of specific, detailed studies, many of which have been published elsewhere, so that they can be considered together in the overall context of the evolution of this sector of the basin.

Using the stratigraphy established by the detailed 1:10 000 mapping as a basis, a major geochemical study of the volcanic rocks has been accomplished. This has been brought together with studies of sedimentology, physical volcanology, palaeontology, structure and metamorphism. The result, a detailed account of the development of this volcanic province, shows the value of taking such a broadly multidisciplinary approach in studies of ancient volcanic successions. Moreover, the great age of this volcanic province, subsequent tectonic events, uplift and erosion have allowed the study to encompass all stages from its birth, through its maturity into its eventual death in a way that is not usually possible with younger volcanoes.

The book represents a major advance in our understanding of volcanic processes in marine environments, acidic ash-flow tuffs in subaerial and sub-aqueous settings, tectonic controls on volcanism and sedimentation and petrogenetic modelling. Its publication is therefore of importance to both national and international geoscience. It also draws attention to this classical area in one of the most beautiful and easily accessible regions of southern Britain, which is enjoyed by a great many people. That enjoyment will be enhanced by this eloquent and informative text and the insights it provides into the rocks that make up this area and of the manner in which they were deposited.

Dr Peter Cook
Director
British Geological Survey
Keyworth
Nottingham NG12 5GG

January 1991

ACKNOWLEDGEMENTS

This book results from a multidisciplinary study involving close collaboration with specialists from both within the British Geological Survey (BGS) and the universities. For his initial encouragement we are deeply indebted to Sir Malcolm Brown (Director BGS, 1979–85) who recognised the benefits of a concerted enquiry by individuals with widely differing specialist interests.

The study has provided the first comprehensive set of geochemical data from this classic area of Ordovician volcanism. In this work, we received considerable contributions from Dr T K Ball (BGS), Dr R E Bevins (National Museum of Wales), Dr B P Kokelaar (University of Ulster), Dr P Leat (Durham University), Dr A C Mann (Goldsmith's College), Mr R J Merriman (BGS) and Dr R S Thorpe (Open University). Major and trace element XRF analyses were carried out by Drs P K H Harvey and B P Atkin (University of Nottingham) and Miss A Hughes (University of Leeds). Rare earth element analyses were carried out by Dr J N Walsh (King's College, University of London), Dr A C Mann, Dr R S Thorpe and Miss A Hughes. Mr A T Kearsley (Oxford Polytechnic) made rare earth mineral determination using back scattered electron imagery. Preparation of thin sections and sample crushing was organised by Mr I Chaplin at the Open University.

Full chemical analytical and sample data are lodged with the National Geosciences Data Centre (BGS, Keyworth) in the Wales Survey Database developed by Mr K A Holmes (BGS).

The nature of base-metal sulphide mineralisation and its relationships to the volcanic rocks have been closely examined in collaboration with Dr T B Colman (BGS). Much information has been provided by Dr J A Evans (NERC Isotope Geology Unit) on isotope geochemistry, and by Dr B Roberts (Birkbeck College), Mr R J Merriman (BGS) and Dr R E Bevins (National Museum of Wales) on the conditions of low-grade regional metamorphism.

The environmental context of the volcanic rocks in the stratigraphic sequence has been the subject of detailed studies by Dr B P Kokelaar (University of Ulster) and Professor W J Fritz (Georgia State University, USA), and the associated sediments have been considered by Dr G J Orton (University of Oxford). Palaeoecological studies of macrofaunas by Mr S P Tunnicliff and Dr A W A Rushton (BGS) have provided additional environmental data and the results of microfloral analysis by Dr S G Molyneux (BGS) indicate the possibility of establishing a more comprehensive biostratigraphical framework in the area.

Geophysical analysis, in particular of the aeromagnetic and gravity anomalies, by Mr R B Evans and Mr B C Chacksfield (BGS) has been aided by an examination of the physical properties of the major rock types. This work has developed into a consideration of Snowdonia in its local setting in north-west Wales and in its context within the whole Lower Palaeozoic basin of Wales.

Structural interpretation has been an integral part of the study, concentrating in particular on the influence of contemporaneous structures on volcanism and sedimentation, the effects of the Caledonian deformation on these structures and the variation in this deformation resulting from the disposition of the volcanic rocks and associated intrusions. These aspects have benefited from the work of Dr I Wilkinson (University College of Wales, Aberystwyth) and Dr M Smith (BGS). During fieldwork, valuable assistance was provided by Mr K H Park and Mr J H Hwang (Korea Institute of Energy and Resources) and Mr T Lewis (University College of Wales, Aberystwyth). The onerous task of processing the text was accomplished by Mrs L M Ellis.

The contributions of all of these collaborators, in publications and personal communications, facilitated the production of this book and they are gratefully acknowledged. In addition, we recognise the contributions made by all of our colleagues who surveyed parts of the Bangor and Snowdon 1:50 000 geological sheets, in particular Professor E H Francis, who initiated the mapping programme, Drs B E Leveridge, C D R Evans, R Addison, B D T Lynas, B C Webb and M Smith.

M F Howells
A J Reedman
S D G Campbell

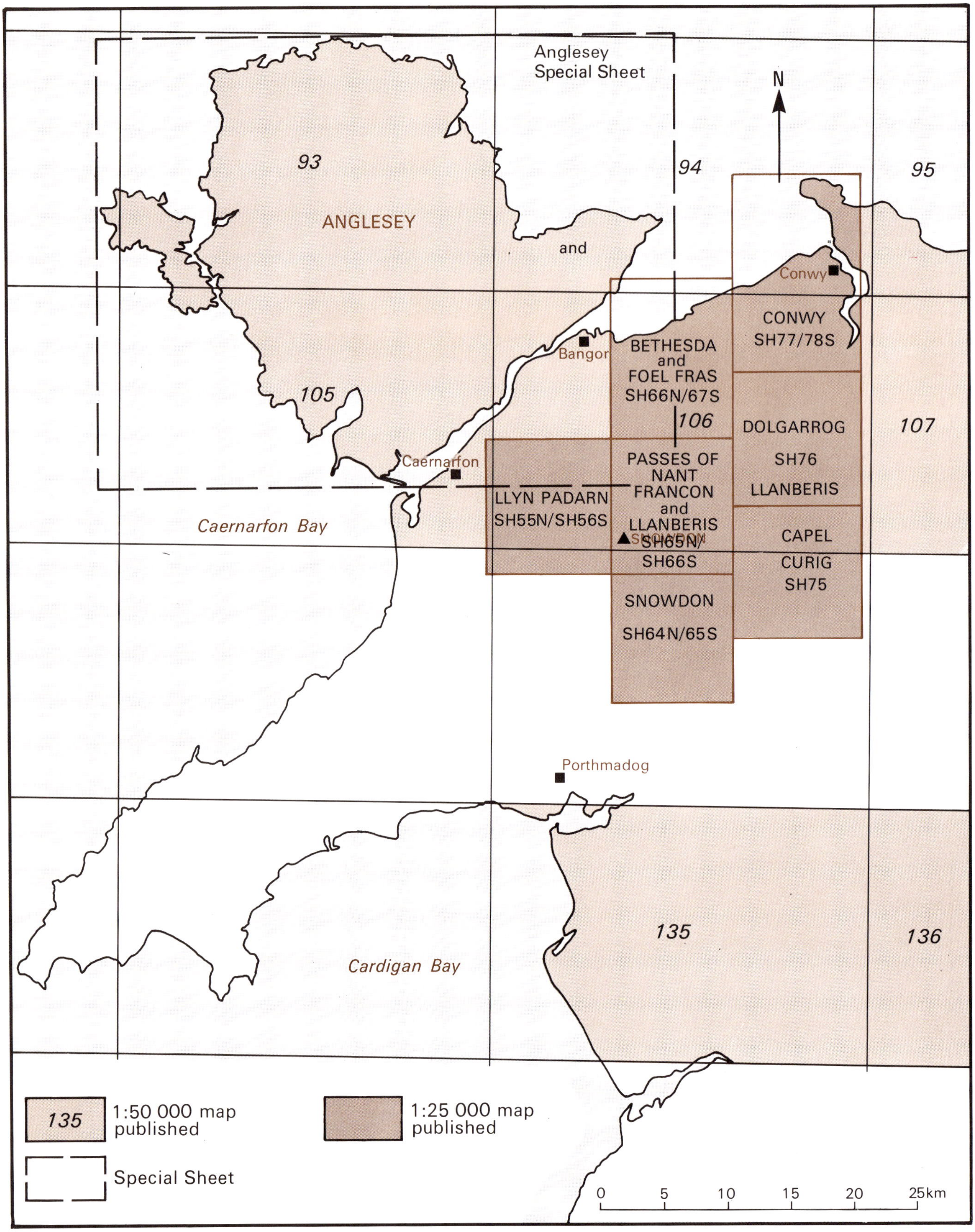

Anglesey
Special Sheet

N

94

95

93

ANGLESEY

and

Conwy ■

105

Bangor ■

BETHESDA
and
FOEL FRAS
SH66N/67S

CONWY
SH77/78S

106

DOLGARROG

SH76

107

Caernarfon

PASSES OF
NANT
FRANCON
and
LLANBERIS
▲ SNOWDON
SH65N/
SH66S

LLANBERIS

LLYN PADARN
SH55N/SH56S

CAPEL
CURIG
SH75

Caernarfon Bay

SNOWDON

SH64N/65S

Porthmadog ■

135

136

Cardigan Bay

	1:50 000 map published
135	

1:25 000 map
published

Special Sheet

0   5   10   15   20   25km

Key to the published geological maps of north-west Wales

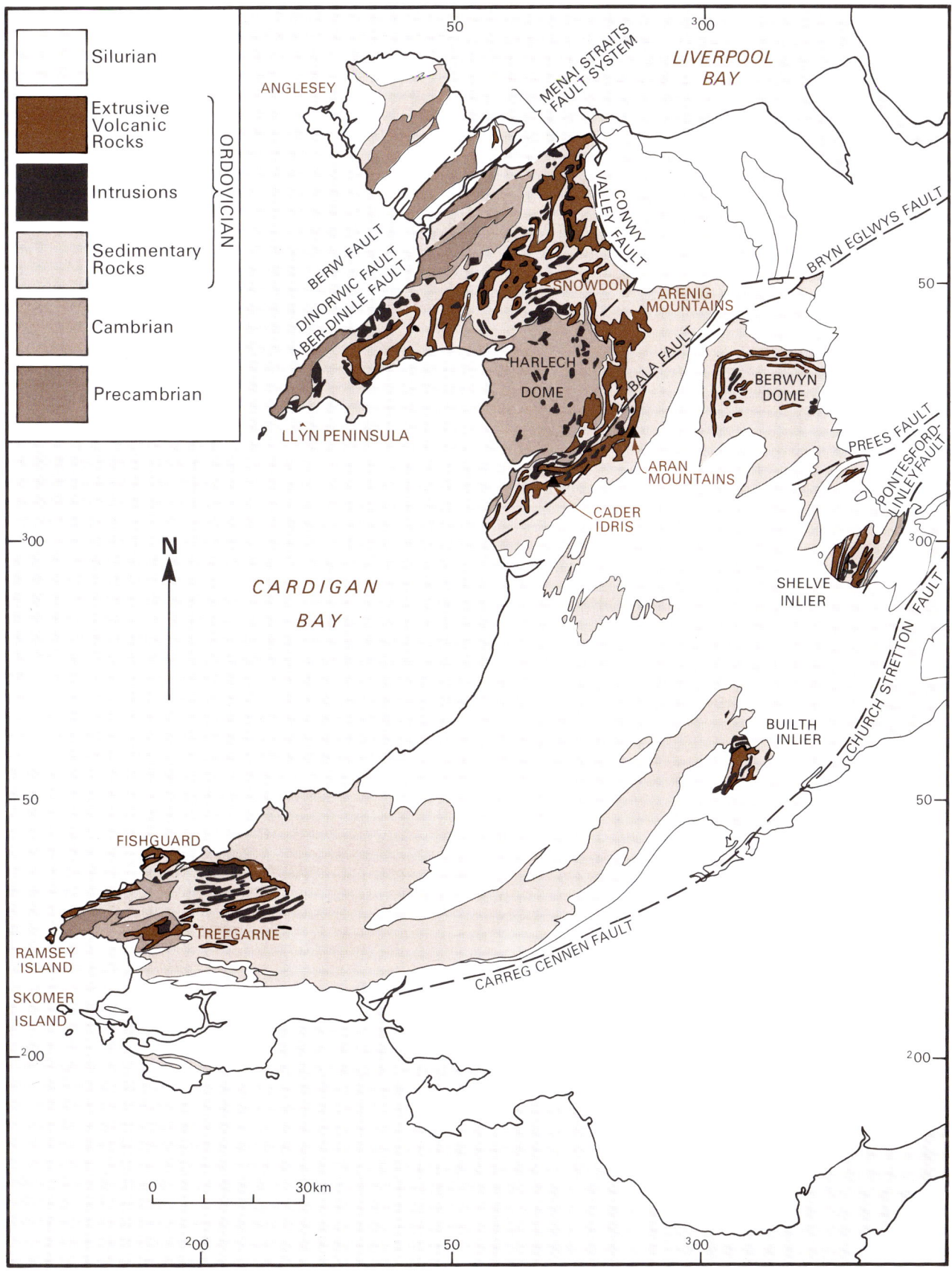

Figure 1 Simplified geological map of Wales.

Plate 1 The peaks of Y Lliwedd, Snowdon and Crib Goch (left to right) viewed south-westwards from Llynnau Mymbyr [SH 713 577] in the core of the Capel Curig Anticline.

CHAPTER 1

Introduction

The main outcrop of Ordovician (Caradoc) rocks in north-west Wales lies in the area of the Snowdonia National Park, in the county of Gwynedd. The area is renowned for its spectacular scenery and includes the fourteen peaks of Wales of 3000 ft (914 m) or more. It is one of the United Kingdom's most popular natural classrooms for the study of environmental science.

Wales holds an important place in the history of geology, particularly with respect to the Lower Palaeozoic rocks which occupy approximately two thirds of its area (Figure 1). It was here that the component Cambrian, Ordovician and Silurian systems were originally named (Murchison, 1835; Sedgwick, 1843; Lapworth, 1879) and many of the biostratigraphical divisions of these systems are themselves distinguished by Welsh place names.

In 1846 the Geological Survey began the first systematic mapping of North Wales on the scale of 1 inch to 1 mile, and this was completed in 1852. A sheet explanation (Ramsay, 1866) was followed by the classic North Wales memoir (Ramsay, 1881) which was the first comprehensive account of the geology. This account described the basic elements of the stratigraphy and structure over a wide area, and highlighted the occurrence of extrusive and intrusive volcanic rocks within the sequences.

The first detailed petrographical examination and interpretation of the volcanic rocks was accomplished by Harker (1889) in his distinguished essay on 'The Bala Volcanic Series, Caernarvonshire' which was facilitated by 'the admirable maps of the Geological Survey, supplemented by Sir Andrew Ramsay's Memoir'. In eastern Caernarvonshire, Harker (1889) related the extrusive volcanic rocks to separate centres of eruption which were marked by later intrusion (Figure 2). In the final summary of his essay, Harker expressed the dynamic character of the sequence of events encapsulated within the stratigraphical column in a manner that is more akin to current fashion than that which prevailed for much of the intervening century. His belief in such an approach was clearly stated in his final sentence:

'The history of vulcanicity in North Wales is the history of the pre-Cambrian and the Ordovician periods; and when this area comes to be studied in detail by abler geologists, we may hope to learn from it at least as much of the mechanism of igneous action and the internal economy of volcanoes as any other district, Palaeozoic or Neozoic, is able to teach us.'

Much of the work on the Lower Palaeozoic rocks of North Wales in the subsequent years was directed, however, to refining local stratigraphy and biostratigraphical correlation (for references see Bassett, 1969). In this general context three massive contributions were made, by E H Greenly on Anglesey, P G H Boswell on the Silurian strata, and O T Jones. Jones, having systematically published the stratigraphical and structural detail of Lower Palaeozoic successions in many parts of Wales, much of the Ordovician area in col-

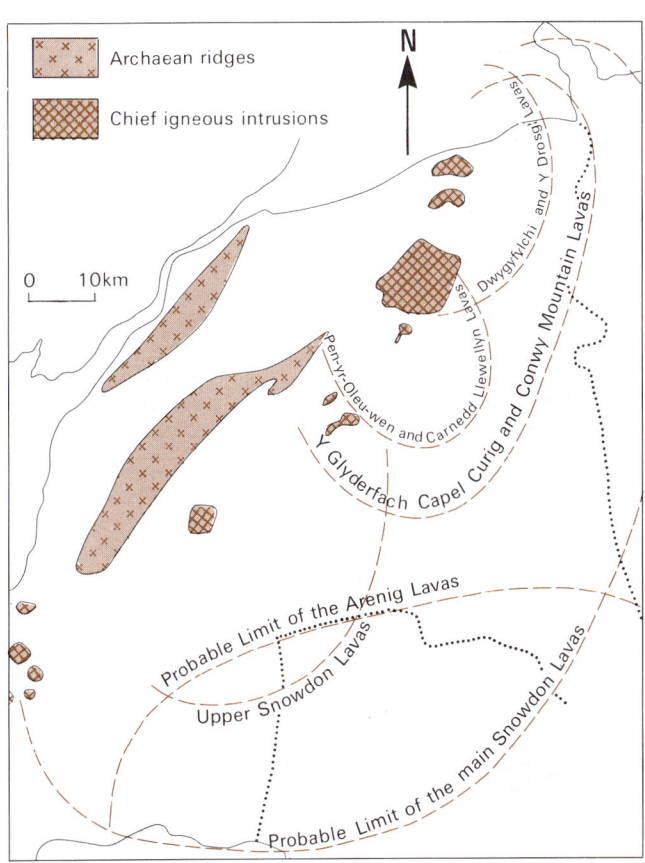

Figure 2 Representation of Harker's diagram, NW Wales (1889, fig. 6).

laboration with W J Pugh, eventually interpreted its development in terms of a geosynclinal model in his Presidential address to the Geological Society of London (Jones, 1938).

During this period, the interpretation of the volcanic sequences was rarely considered. An exception was the work of Howel Williams with the publication in 1927 of his classic map and paper on the geology of Snowdon. He recognised a wide variation in the lithologies and form of the volcanic rocks, and the close comparison of some of the fabrics in the acidic volcanic rocks with Peléan deposits (Anderson and Flett, 1903), substantiating an observation made earlier by Greenly (in Dakyns and Greenly, 1905). From Sedgwick (1843) onwards there had been agreement that, as the extrusive volcanic rocks were interbedded with sedimentary rocks containing marine fossils, the volcanic rocks were emplaced in a submarine environment. This assumption pertained until the independent recognition by Oliver (1954) and Rast et al. (1958; and see Fitch, 1967 for references) that many of the 'rhyolites' in Snowdonia were welded ash-flow tuffs and that, as such, they must have been erupted and emplaced subaerially. The repercussions from this observation initiated the latest phase of investigation and interpretation.

ORDOVICIAN VOLCANISM IN THE WELSH BASIN

Regional tectonic setting

The Lower Palaeozoic rocks of Wales, a thick sequence of predominantly marine sedimentary and intercalated volcanic rocks, accumulated in a basin sited on continental crust comprising accreted volcanic arcs (Thorpe, 1979). This Precambrian crust, probably no older than about 900 Ma (Thorpe et al., 1984a), is exposed on Anglesey, on the north-west side of the basin, and in smaller inliers along the Welsh Borderland, on the south-east side. It formed part of a microcontinent, Eastern Avalonia (Soper and Hutton, 1984) derived from continental Gondwana, and was separated from Baltica to the north-east by an ocean, Tornquist's Sea (Cocks and Fortey, 1982) (Figure 3). Both Eastern Avalonia and Baltica were separated from the North American continent (Laurentia) to the north-west by an ocean, Iapetus.

During the Lower Palaeozoic, volcanism in the Welsh Basin was mainly restricted to Ordovician times. Its relationship to volcanism in other sectors of the southern British Caledonides has been incorporated into numerous plate-tectonic models (e.g. Phillips et al., 1976; Stillman and Francis, 1979; Dewey, 1969, 1982; Fitton and Hughes, 1970; Fitton et al., 1982). All models envisage the Ordovician Welsh Basin as situated close to a convergent plate margin. In the vicinity of Wales, this destructive plate margin was probably active from late Tremadoc to Caradoc times. Kokelaar et al. (1984) proposed that the Tremadoc arc volcanism marked the initiation of south-east subduction of Iapetus oceanic lithosphere beneath Wales. Subsequent Arenig to Caradoc tholeiitic volcanism was related (Kokelaar et al., 1984) to back-arc extension with the arc sited further to the north in the Lake District–Leinster zone of the Caledonides. More recently, however, Kokelaar (1988) has argued for east–west extension in Wales from Tremadoc to Caradoc times, which was produced by north–south plate convergence. He speculates that the Lake District–Leinster zone could have arrived in its present position by sinistral strike-slip and that the Ordovician marginal basin of Wales developed either behind some other arc which had been displaced, or that there was no arc north of Anglesey, and that the marginal basin was produced by strike-slip induced pull-apart of the Tremadoc arc.

Recently, Leat and Thorpe (1989) have recognised ocean-island basalt magmas, together with magmatic arc basalts, in the early Caradoc volcanic rocks of central Snowdonia and in the early Silurian (Llandovery) Skomer Volcanic Group in south-west Wales (Thorpe et al., 1989). They argue that the initial eruption (early Caradoc) of ocean-island basalt indicates the end of subduction of Iapetus oceanic lithosphere below the Welsh Basin.

In early Silurian (Llandovery and Wenlock) times, sedimentation was dominated by the development of submarine fan complexes, which gave way to widespread shallow-marine progradation during the Ludlow and, finally, the infilling of the basin prior to Devonian molasse deposition. Throughout this period, the area was dominated by sedimentation in sub-basins which were probably defined by major basement fractures. Localised unconformities, particularly those separating Caradoc and Ashgill strata and within the Llandovery, probably reflect phases of uplift generated by these fractures. Woodcock (1984a,b) has proposed that the Silurian evolution of the basin was significantly influenced by dextral strike-slip faulting.

Similarities of post-Ashgill faunas between southern Britain and Baltica suggest closure of Tornquist's Sea during the Caradoc (Cocks and Fortey, 1982; Soper and Hutton, 1984). From Caradoc times onwards, faunal provincialism on either side of Iapetus diminished (Williams, 1969, 1976; McKerrow and Cocks, 1976), heralding its closure. The timing of the collision of the two continents is currently a matter of debate (Soper et al., 1988). The proposed Silurian collision (Leggett et al., 1983; Soper and Hutton, 1984) has been questioned by Murphy and Hutton (1986), Hutton (1987) and Pickering (1987) who argue that closure was largely accomplished by the late Caradoc.

Evolution and distribution

The Ordovician volcanic rocks in Wales have recently been reviewed by Kokelaar et al., (1984) following detailed examination of localities in north Dyfed (Bevins, 1979, 1982; Bevins and Roach, 1979a,b; Kokelaar et al., 1985) and Snowdonia (see Howells and Leveridge, 1980, and Howells et

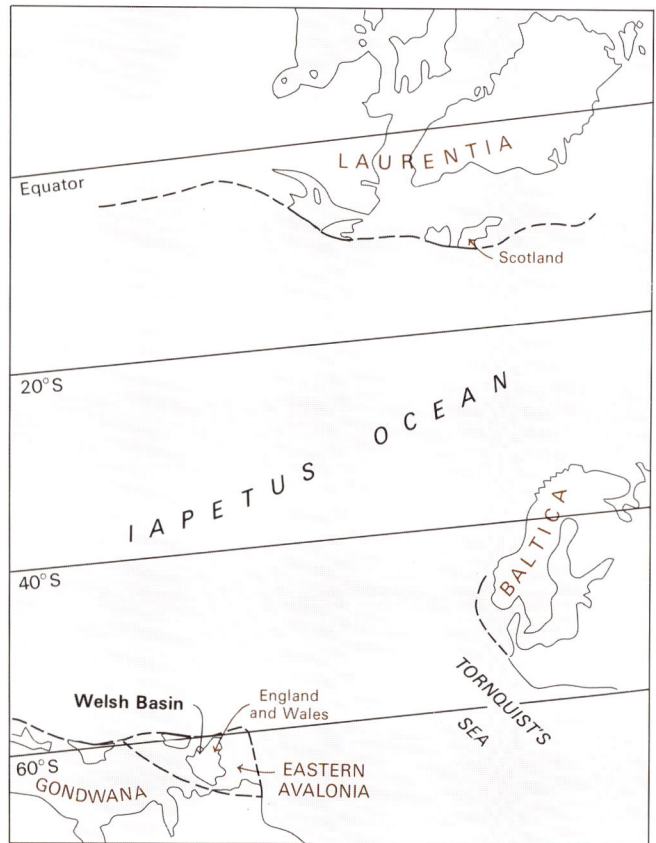

Figure 3 Disposition of England and Wales, and Scotland in Arenig times (after Cocks and Fortey, 1982).

al., 1986, for references). The physical character of the volcanic sequences was interpreted in the context of current knowledge of eruption and emplacement processes, and depositional environments.

The volcanism shifted in time and space. At the end of Tremadoc times, island-arc volcanism in South Wales (Trefgarn Volcanic Group; Kokelaar et al., 1984) and in southern Snowdonia (Rhobell Volcanic Complex; Kokelaar, 1979) (Figure 4) was associated with widespread tectonism. In South Wales, the main volcanism of mid-Arenig to Llanvirn age developed at three main centres (Kokelaar et al., 1984). It was associated with graptolitic black mudstone deposition, although periodic temporary emergence resulted in the restricted accumulation of littoral conglomerates. The volcanic sequences preserved are of bimodal, basalt/rhyolite composition with thick accumulations of acidic ash-flow tuff. However, the sequences commonly indicate subaqueous accumulation with the occurrence of acid and basic pillow lavas, breccias, hyalotuffs, and the intrusion of magma, with fluidisation at contacts, into wet unlithified sediment.

At Builth and Shelve (Figure 4), close to the south-east edge of the Ordovician outcrop, Llanvirn basic lavas and acidic pyroclastics were emplaced in relatively deep to shallow marine and subaerial environments close to the basin margin (Jones and Pugh, 1949; Lynas, 1983). In this area there is evidence of continued restricted volcanic activity into Caradoc times.

In North Wales two main phases of volcanic activity can be distinguished, in two main outcrops, both of which are predominantly of bimodal basalt/rhyolite composition and associated with marine sedimentary rocks. The deposits of the earlier, pre-Caradoc phase crop out about the south, east and north sides of the Harlech Dome, southern Snowdonia (Ridgway, 1975, 1976; Allen, 1982; Allen and Jackson, 1985; Kokelaar et al., 1984; Institute of Geological Sciences, 1982a) (Figures 4 and 5). The deposits of the later, Caradoc phase crop out in the broad synclinal complex of central and northern Snowdonia (Rast, 1969; Roberts, 1969; Kokelaar et al., 1984; Institute of Geological Sciences, 1976, 1981b; British Geological Survey, 1985a, 1985b, 1986, 1988, 1989). Subsidiary centres developed in the Lleyn Peninsula and in the Berwyn Hills in north-east Wales (Figure 4) (Brenchley, 1964, 1972, 1978).

The stratigraphical detail and context of the Caradoc volcanic rocks in central and northern Snowdonia are shown on the Generalised Vertical Section of 1:50 000 Bangor (106) Sheet (British Geological Survey, 1985a), 1:25 000 Snowdon Sheet (SH 64N 65S) (British Geological Survey, 1989), and in cartoon form (Figure 6).

The volcanism can be divided into two eruptive cycles which are reflected in two volcanic groups: the lower (1st Eruptive Cycle) Llewelyn Volcanic Group and the upper (2nd Eruptive Cycle) Snowdon Volcanic Group. This activity was largely confined to just two chronostratigraphical stages, about 5 per cent of total (72 Ma) Ordovician time. However, locally, volcanic rocks account for about half the total thickness of the Ordovician sequences.

Tectonic controls on the distribution of Ordovician volcanism in Snowdonia

The distribution of Ordovician eruptive centres in Wales was controlled by deep-seated fractures cutting the ensialic basement (Rast, 1969; Campbell et al., 1988; Kokelaar, 1988). The eruption of basaltic magmas with little evidence of crustal contamination has led Kokelaar (1988) to suggest that the fractures penetrated the base of the crust and that they originated as major strike-slip faults or shear zones. The absence of evidence of significant lateral displacement in the Lower Palaeozoic cover suggests that the fractures formed, at the latest, by early Cambrian times. In north-west Wales, the volcanism occurred mainly in narrow grabens or rifts, in which repeated subsidence occurred between deep-seated fractures. The grabens lay in the area between the north-east–south-west-trending Menai Straits Fault System in the north-west and the Bala Fault in the south-east. Late Precambrian major strike-slip movements have been determined in the former (Gibbons, 1983, 1985; Gibbons and Gayer, 1985) and inferred for the latter (Arthur, 1982; Fitches and Campbell, 1987).

In southern Snowdonia, the pre-Caradoc volcanism was associated with east–west extension and was controlled by the north–south-trending fractures of the Rhobell Fracture Zone and the Barmouth Fracture Zone (Kokelaar, 1988). Kokelaar (1988) argues that the north-east–south-west-orientated Snowdon graben, which controlled Caradoc volcanism in central and northern Snowdonia, also resulted from east–west extension and is a continuation of the north–south structures in southern Snowdonia which were rotated clockwise during later (end-Silurian to early Devonian) compressional deformation.

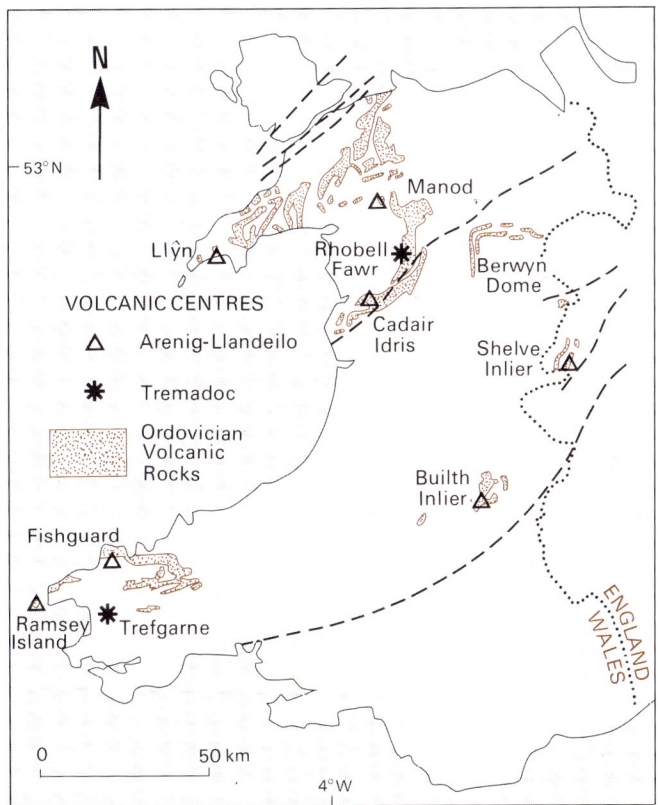

Figure 4 Ordovician, pre-Caradoc, volcanic centres in the Welsh Basin.

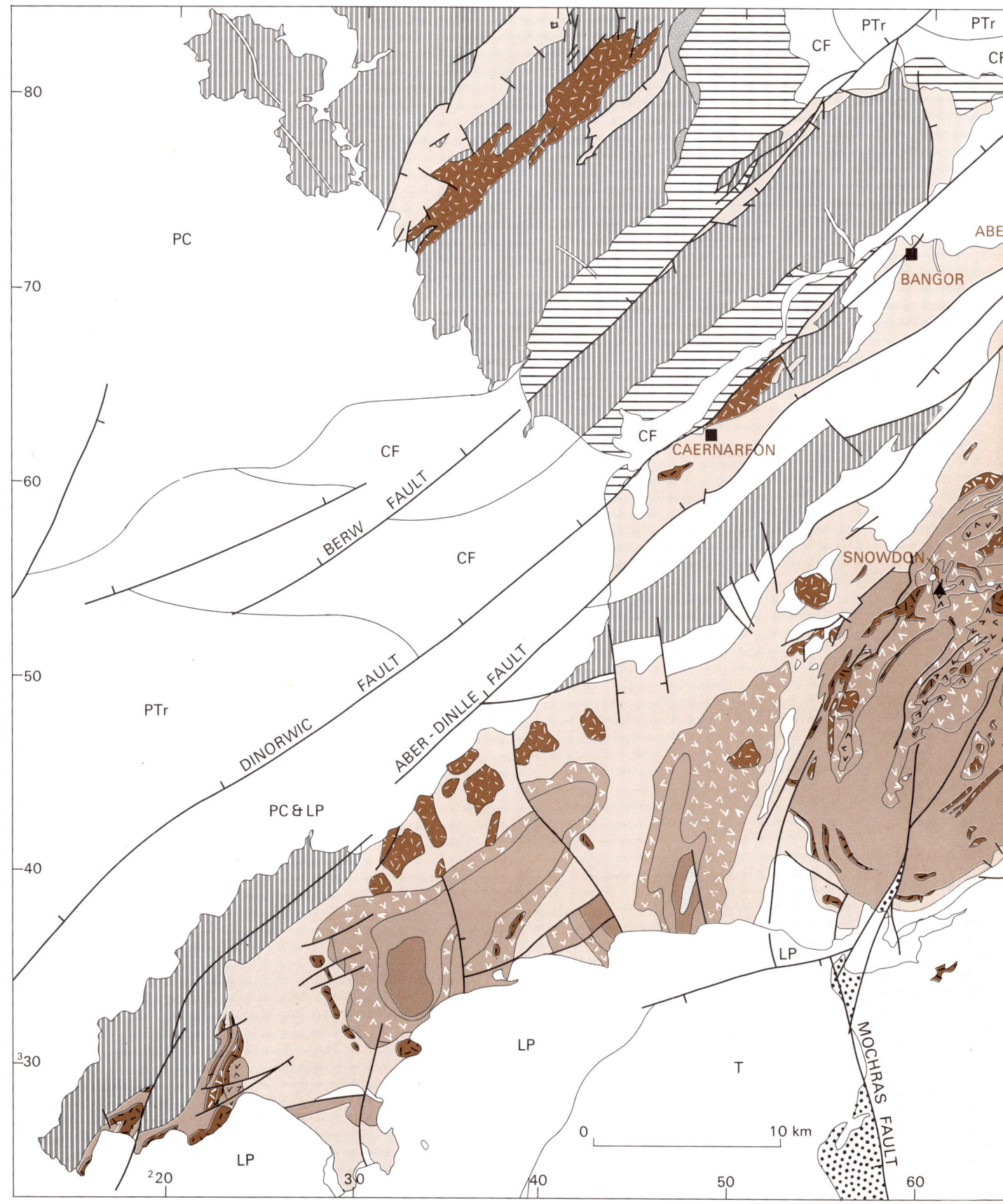

Figure 5 Geological map of North Wales and surrounding offshore region (based on British Geological Survey 1:250 000 Sheets Cardigan Bay (1982), Anglesey (1981), Liverpool Bay (1978) and Mid Wales and Marches (in press)).

See p.6
for key

CF

CF

PTr

P

CONWY

CONWY VALLEY FAULT

BRYN EGLWYS FAULT

BALA FAULT

70

80

90

300

10

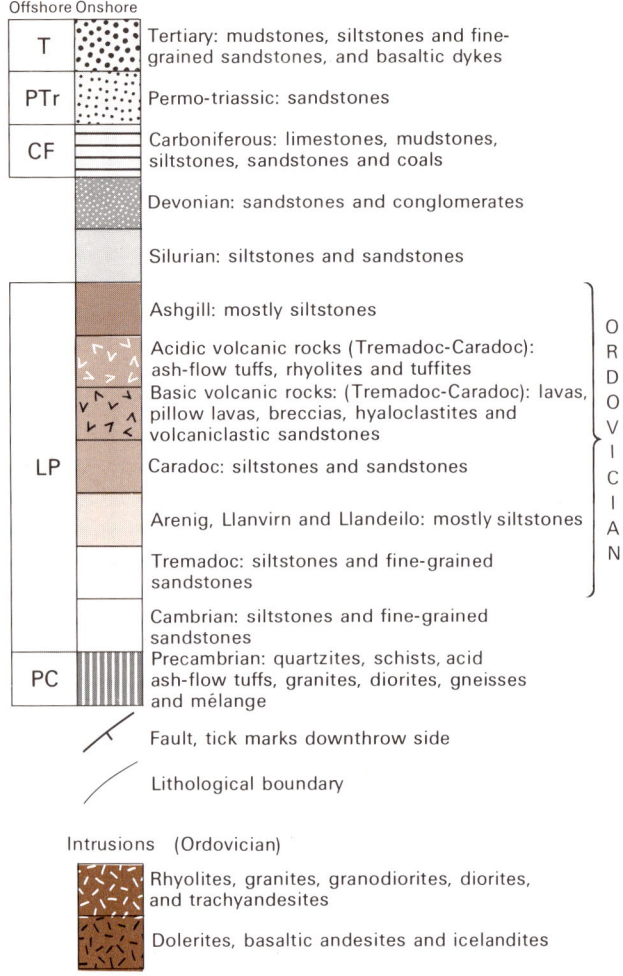

Offshore Onshore

T		Tertiary: mudstones, siltstones and fine-grained sandstones, and basaltic dykes
PTr		Permo-triassic: sandstones
CF		Carboniferous: limestones, mudstones, siltstones, sandstones and coals
		Devonian: sandstones and conglomerates
		Silurian: siltstones and sandstones
LP		Ashgill: mostly siltstones
		Acidic volcanic rocks (Tremadoc-Caradoc): ash-flow tuffs, rhyolites and tuffites
		Basic volcanic rocks: (Tremadoc-Caradoc): lavas, pillow lavas, breccias, hyaloclastites and volcaniclastic sandstones
		Caradoc: siltstones and sandstones
		Arenig, Llanvirn and Llandeilo: mostly siltstones
		Tremadoc: siltstones and fine-grained sandstones
		Cambrian: siltstones and fine-grained sandstones
PC		Precambrian: quartzites, schists, acid ash-flow tuffs, granites, diorites, gneisses and mélange

ORDOVICIAN

Fault, tick marks downthrow side

Lithological boundary

Intrusions (Ordovician)

Rhyolites, granites, granodiorites, diorites, and trachyandesites

Dolerites, basaltic andesites and icelandites

Campbell et al. (1988) and Smith (1988) consider that the Caradoc Snowdon graben transects the earlier north–south structure in southern Snowdonia. They recognise, however, that the dominantly north-east–south-west fractures are associated with a conjugate north–south set which suggests that the rift development was not solely the product of north-west–south-east extension but was influenced, to some extent, by transtensional stress.

Recognition of ignimbrites/ash-flow tuffs in Snowdonia

A major factor in recent interpretations of the volcanic rocks of the Welsh Basin has been the recognition of widespread ignimbrites or ash-flow tuffs. Subsequent to their original recognition in Snowdonia by Oliver (1954), they have been described at localities throughout the basin; in south-west Wales (Bevins, 1979; Lowman and Bloxam, 1981; Kokelaar et al., 1984), in southern Snowdonia (Davies, 1959; Dunkley, 1979) and central and northern Snowdonia (Howells et al., 1973, 1979, 1985a,b, 1986, 1987; Francis and Howells, 1973; Howells and Leveridge, 1980; Reedman et al., 1987a).

Marshall (1935) coined the term ignimbrite (glowing cloud rocks) for deposits in North Island, New Zealand, which bore a great similarity to deposits of glowing avalanches (*nuées*

ardentes) previously described from Soufrière (Anderson and Flett, 1903), Alaska (Fenner, 1923), Martinique (Perret, 1935) and Java (Neumann van Padang, 1933) (see Macdonald, 1972). The term has been used variously for deposits of ash and block grade. It has been used synonymously with welded tuff, even though many ignimbrites are entirely nonwelded (see Cas and Wright, 1987) and, consequently, the term is surrounded by confusion. Sparks et al. (1973) defined ignimbrite as 'a deposit formed from pumiceous pyroclastic flows irrespective of the degree of welding' and its use in this way is still common. However, in this work the pyroclastic flow deposits are described by the dominant grade of their components, e.g. ash-flow tuff (Smith, 1960a,b; Ross and Smith, 1961), block and ash-flow tuff.

The term 'flow unit' is used for the deposit of a single pyroclastic flow (Sparks et al., 1973) comprising a basal layer, caused by the interaction of the flow with the substrate, a central zone, formed from the main body of the flow, and a fine-grained top, comprising fine ash elutriated from the top of the flow during transport and which settled back onto the flow deposit after its emplacement (Figure 7).

Welding is the fusion of hot glass fragments under a compactional load and is a common feature of pyroclastic flow deposits (Smith, 1960a,b; Ross and Smith, 1961). Its occurrence is dependent on the viscosity (dependent on temperature and composition) of the glass, and lithostatic load (see Fisher and Schmincke, 1984; Cas and Wright, 1987). Its clear recognition in ancient pyroclastic rocks can be difficult because devitrification and metamorphic recrystallisation obscure the evidence of fusion of the adjacent glass fragments. As a result, its determination is by recognition of a flattening of the fragments which is distinct from that which could be produced by tectonic distortion. From such evidence, nonwelded, partially welded and densely welded acidic ash-flow tuffs are recognised. Characteristically, the flattened shards and juvenile pumice clasts (fiamme) in the welded tuff define a planar foliation (eutaxitic) which locally is extreme (parataxitic). Folding of the foliation indicates secondary mass flowage (rheomorphism) of the tuff during and immediately after welding. In some instances, it is suggested that welding may be a primary flow feature (Chapin and Lowell, 1979; Reedman et al., 1987a).

The foundation of a clear understanding of acidic ash-flow tuff eruption and emplacement was established by Smith (1960a,b) and Ross and Smith (1961). After a slightly delayed response, there have been numerous subsequent publications (see Chapin and Elston, 1979) and they continue unabated. Most of these describe geologically recent eruptions and their deposits in subaerial environments. A major contribution has been made by applying granulometric analytical techniques to these recent deposits and showing that on these data differing pyroclastic flow, fall and surge deposits can be categorised (Fisher and Schmincke, 1984; Cas and Wright, 1987). However, such an approach is not feasible in ancient consolidated sequences.

Interpretation of the Ordovician pyroclastic rocks in Snowdonia has been founded upon detailed mapping, on aerial photographs at about 1:8000 to produce geological maps at 1:10 000 scale, which established the correlation of individual pyroclastic units across the outcrop. The interpretation of these units is based on mesoscopic and micro-

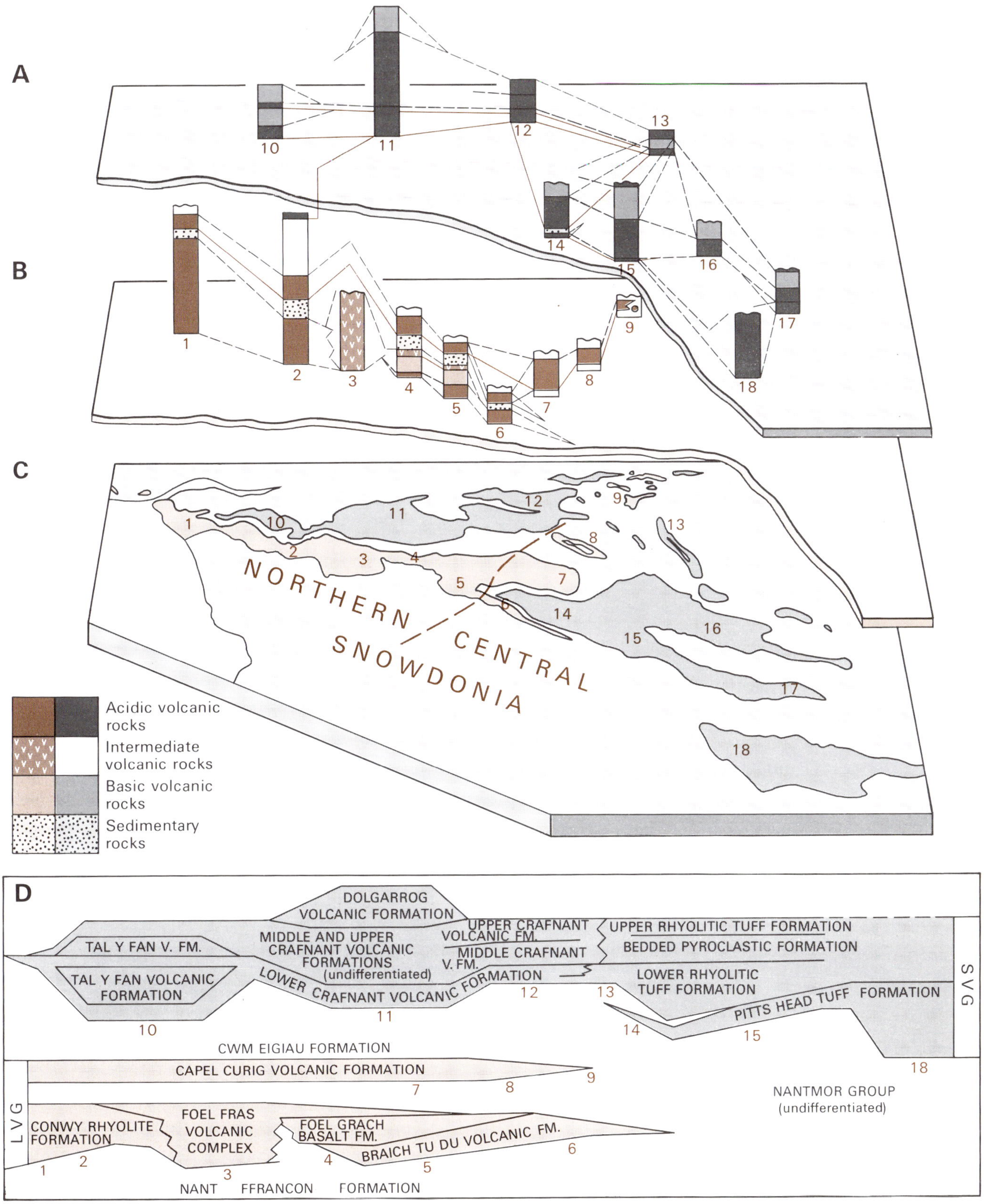

A

B

C

Acidic volcanic rocks

Intermediate volcanic rocks

Basic volcanic rocks

Sedimentary rocks

NORTHERN

CENTRAL

SNOWDONIA

D

DOLGARROG VOLCANIC FORMATION

TAL Y FAN V. FM.

MIDDLE AND UPPER CRAFNANT VOLCANIC FORMATIONS (undifferentiated)

UPPER CRAFNANT VOLCANIC FM.

UPPER RHYOLITIC TUFF FORMATION

BEDDED PYROCLASTIC FORMATION

MIDDLE CRAFNANT V. FM.

TAL Y FAN VOLCANIC FORMATION

LOWER CRAFNANT VOLCANIC FORMATION

LOWER RHYOLITIC TUFF FORMATION

SVG

PITTS HEAD TUFF FORMATION

CWM EIGIAU FORMATION

CAPEL CURIG VOLCANIC FORMATION

NANTMOR GROUP (undifferentiated)

LVG

CONWY RHYOLITE FORMATION

FOEL FRAS VOLCANIC COMPLEX

FOEL GRACH BASALT FM.

BRAICH TU DU VOLCANIC FM.

NANT FFRANCON FORMATION

Figure 6 Caradoc volcanic sequences in Snowdonia; cartoons of outcrop and stratigraphic detail.

A. Snowdon Volcanic Group (SVG). B. Llewelyn Volcanic Group (LVG). C. Outcrop, perspective grid of 6 km 'squares'. D. Stratigraphic terminology. In A, B and D, tie lines mark the bases of major ash-flow tuffs and approximate time planes.

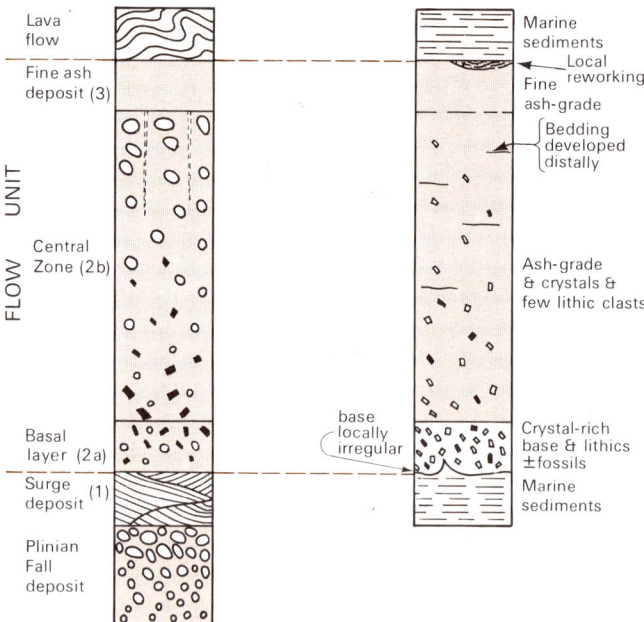

FLOW UNIT

Lava flow

Fine ash deposit (3)

Central Zone (2b)

Basal layer (2a)

Surge deposit (1)

Plinian Fall deposit

Marine sediments

Fine ash-grade Local reworking

Bedding developed distally

Ash-grade & crystals & few lithic clasts

base locally irregular

Crystal-rich base & lithics ±fossils

Marine sediments

Figure 7 Comparative sections, subaerial and subaqueous, of ash-flow tuff units.

A. Schematic section of subaerial flow unit (after Sparks et al., 1973 and Sheridan, 1979). B. Subaqueous flow unit in the Ordovician sequence of North Wales.

scopic examination of selected serial sections along their length and environmental interpretation of the subjacent and suprajacent sedimentary strata (e.g. Howells et al., 1973; Howells and Leveridge, 1980) (Figure 6).

Because of the concentration of research on geologically recent ash-flow eruptions and their deposits in subaerial environments, the assumption that ash-flow tuffs were diagnostic of a subaerial environment inevitably grew. As a result, following the recognition of acidic ash-flow tuffs in Snowdonia (Oliver, 1954; Rast et al., 1958) the palaeo-geographic interpretation of North Wales was modified to account for widespread and repeated emergence of land. Subsequently, a major subaerial volcano was proposed (Rast, 1969; Bromley, 1969; Beavon, 1980), complete with caldera, ring faults, rim syncline and resurgent cauldron.

Recently, theoretical considerations (Sparks et al., 1980) and field observations (Howells et al., 1973; Francis and Howells, 1973; Howells and Leveridge, 1980; Kokelaar et al., 1984, 1985; Howells et al., 1985a) have shown that subaqueous emplacement and subsequent welding of acidic ash-flow tuffs is possible and occurred commonly in the Welsh Basin, with some of the Ordovician examples described in this book being amongst the best known.

Geochemistry

A major factor, in the original proposal, in support of the multidisciplinary study was the dearth of geochemical data for the volcanic rocks in this classic Ordovician sequence. This was in marked contrast to the wealth of data on the physical characters of the volcanism that had emerged during the mapping of the Bangor 1:50 000 Sheet (106) (British

Geological Survey, 1985a). It was realised that the detailed stratigraphy which had been established would provide an ideal framework for a major study on the geochemical evolution of the volcanism. As a result, the geochemistry of the volcanic rocks is a major component of this book. Approximately 1400 samples have been analysed for major trace element concentrations and these form the Snowdon Geochemical Database of the British Geological Survey.

As proposed by Kokelaar et al. (1984), Bevins et al. (1984) and Leat et al. (1986) and substantiated by Merriman et al. (1986), interpretation has concentrated on those elements least likely to have been affected by hydrothermal alteration, metasomatism and low-grade regional metamorphism (Chapter 6), particularly the high field strength (HFS) and rare earth elements (REE). Accordingly, the geochemical characters of the volcanic rocks have been classified by using the Zr/TiO_2 vs. Nb/Y discriminant diagram of Winchester and Floyd (1979). Comparisons have been made both graphically and statistically with respect to element ratios (ranges, means and standard deviations). Subsequently, these data are interpreted to models of petrogenetic evolution.

Particular emphasis has been placed on geochemical discrimination between individual ash-flow tuffs. This has been done in order to substantiate correlation between disparate outcrops, to identify specific volcanic events within otherwise monotonous sequences and to relate ash-flows to possible parental magmas.

Typically, ash-flow tuffs are heterogeneous mixtures of juvenile vitroclastic material, crystals, and cognate and accidental lithic clasts. Ideally, therefore, any geochemical characterisation of an ash-flow would concentrate on specific phases within the ash-flow (e.g. pumice, blocks or crystals) as is often accomplished in recent volcanic deposits (e.g. Hahn et al., 1979). However, the application of such techniques to the ash-flow tuffs of Snowdonia poses considerable problems as a result of lithification and subsequent low-grade metamorphism. Pumice, for example, is rarely recognised in the field and its extraction for analysis is not a practical proposition. Consequently, an attempt was made, largely on a reconnaissance basis, to characterise individual ash-flow tuffs in terms of their whole-rock geochemistry. This was considered worthwhile following earlier detailed petrographic analysis (Howells et al., 1973; Howells and Leveridge, 1980) of several ash-flow tuffs in north-east Snowdonia which revealed that, with the exception of the relatively crystal and lithic rich basal zone, they are generally fine-grained and homogeneous, with few lithics. Sampling for analysis has avoided basal zones and, where possible, other zones with recognisable lithic enrichment. Although of limited petrogenetic value, the data obtained have proved to be remarkably consistent within the limits of analytical precision. Given the limited scope of the exercise and, with some notable exceptions, the small sample populations for individual ash-flow tuffs, a rigorous statistical approach to discriminance has not been attempted.

Samples (>2 kg) free of surface weathering were collected for geochemical investigation. The samples were jaw-crushed to fragments less than 2 cm and approximately 100 g was grained in a Tema agate disc mill until more than 95 per cent of the sample passed a 200 mesh BSI sieve.

Analyses by X-Ray Fluorescence (XRF) were largely undertaken by Drs P K Harvey and B P Atkins, at the Department of Geology, University of Nottingham, using their published procedures (Harvey and Atkins, 1982). Analyses for REE abundances using inductively coupled plasma source spectrometry were provided by Dr J N Walsh, Department of Geology, New College, London using the procedures outlined by Walsh, Buckley and Barker (1981).

SEDIMENTATION AND BASIN DEVELOPMENT IN SNOWDONIA

Precambrian–Cambrian

The development of the Ordovician basin can be related to deep crustal structures which developed in late Precambrian times (Gibbons and Gayer, 1985) and then influenced sedimentation in Cambrian and Ordovician times. Nowhere in the Welsh Basin is this more apparent than in Snowdonia where an almost complete Cambrian and Ordovician sequence crops out on the south-east side of the Anglesey Horst (British Geological Survey, 1985a,b, 1989). Prominent northeast–south-west faults confined an estimated 2 km thick sequence of lowermost Cambrian acidic ash-flow tuffs (of the Arfon Group) within an extensive graben (Reedman et al., 1984). These tuffs can be correlated across the Menai Straits Fault System as far as the Berw Fault (Figure 1), on Anglesey. This fault plexus influenced subsequent sedimentation, causing major thickness and lithological variations (as in the Arfon Group, British Geological Survey, 1985, 1988).

During early Cambrian times sedimentation was mainly within small restricted fault-bound basins which developed on Precambrian crust (Reedman et al., 1984). Subsequently, this pattern became progressively less complex, with infilling and merging of smaller basins. However, marked variations in lithologies, such as isolated, thick, turbiditic, coarse-grained sandstones in a dominantly silty mudstone sequence (Llanberis Slates Formation), reflect local uplift related to intermittent reactivation of some of the more profound basement fractures. In such an environment correlation of lithological members is likely to be restricted and detailed distant correlations, e.g. with the Lower Cambrian sequence in the Harlech Dome, very difficult.

Widespread tectonism, uplift and arc volcanism in late Tremadoc times, followed by the Arenig marine transgression, separated the Cambrian basinal sedimentation from Ordovician marginal basin development (Kokelaar et al., 1984). In northern Snowdonia, differential uplift and erosion across the major faults of the Menai Straits Fault System occurred prior to the marine transgression (Reedman et al., 1984).

Ordovician

Throughout Ordovician times, about 70 Ma (Figure 8), sedimentation in Snowdonia was dominated by fine-grained, marine, siliclastic deposits. Coarse-grained sediments were associated only with the Arenig marine transgression and locally with the major volcanic episodes. The sedimentary evolution of the basin in central and northern Snowdonia can be divided into four phases:

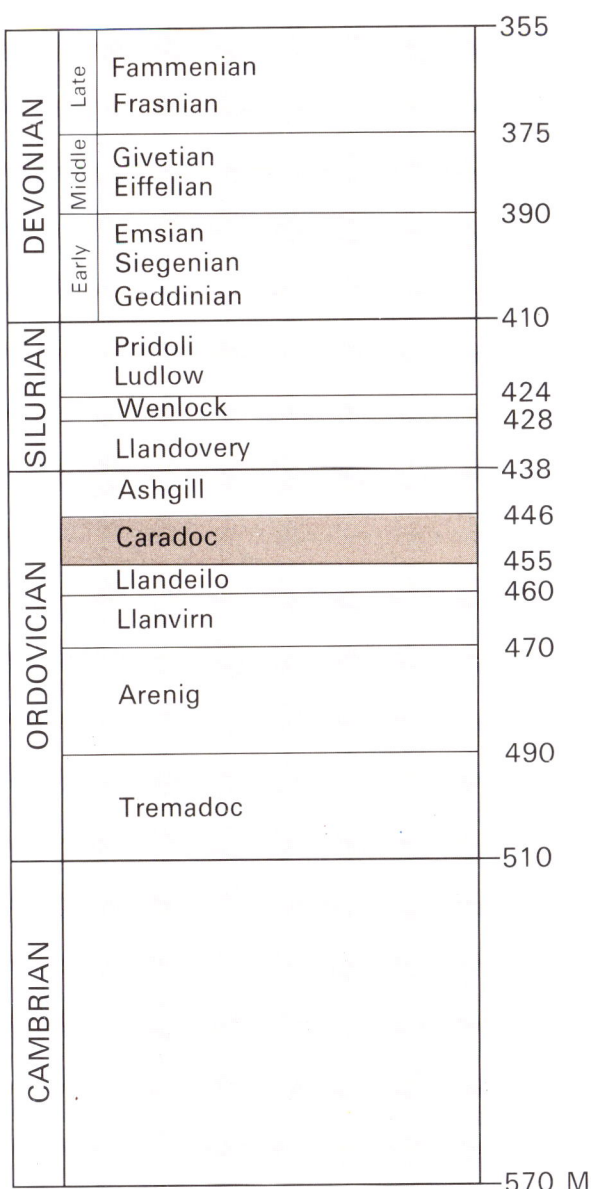

Figure 8   Time-scale, epochs and ages of Ordovician, Silurian and Devonian periods; ages after Snelling, 1987.

Arenig–early Caradoc (Harnagian)

The Arenig transgression was marked by the development of fan-delta complexes. The continued activity of the basement-controlled fractures in the Menai Straits Fault System can be distinguished in facies changes in the basal Arenig sandstones from south-east to north-west in the Bangor District (Reedman et al., 1983; Howells et al., 1985b). The transgression was followed by marine conditions which persisted until the onset of Caradoc volcanism. Mud and silt deposition was interrupted only occasionally by the storm-generated incursion of fine-grained sands and deposition of pisolitic ironstone bands. Throughout this sequence there is little indication of either in-situ volcanic activity or reworking of the deposits of the extensive pre-Caradoc volcanism in southern Snowdonia.

In south central Snowdonia a major olistostromic breccia, involving Upper Cambrian to Lower Caradoc strata has been distinguished (Smith, 1988) and is interpreted as having resulted from uplift on the Harlech Dome fault block (Figure 1). A similar provenance is proposed for the thick sequences of fine-grained sedimentary rocks to the north.

Mid-Caradoc (Soudleyan–Longvillian)

Sedimentation in mid-Caradoc times was intimately associated with the volcanic activity of the two eruptive cycles. The sedimentary sequences are stratigraphically complex, with numerous lateral facies variations related to regional tectonism and the development of individual volcanic centres. Furthermore, the configuration of the basin floor was periodically modified by extensional block and basin tectonism and volcanotectonism.

Sedimentation during the latter part of the 1st Eruptive Cycle (Llewelyn Volcanic Group) was dominated by fluvial and shallow-marine systems at the margin of a tectonically active marine basin (Orton, 1988a,b). The interaction of these influences, and variable rates of sedimentation and subsidence, led to the alternating development of prograding and retrogressive braidplain and fan delta sequences on a south-east-directed palaeoslope.

Subsequent to the 1st Eruptive Cycle, shallow subtidal to deeper offshore environments were established across the area. Prior to the onset of the 2nd Eruptive Cycle, however, there was deltaic progradation from the south-west and the earlier south-east-directed palaeoslope was reversed. The sedimentation during the 2nd Eruptive Cycle was variously fluvial to shallow-marine to deep offshore and locally was influenced by the volcanic activity. However, throughout the cycle there was a general deepening of the sea across the area and a progressive increase of silt and mud deposition.

Late Caradoc (Woolstonian–Onnian) and Ashgill

The fine-grained silt and mudstone sedimentation continued after the cessation of the 2nd Eruptive Cycle. The development of black mudstone accumulation was widespread, as elsewhere in North Wales, and reactivation of one minor volcanic centre had only localised and short-lived influence on the sedimentation. During late Ashgill times, turbiditic incursions of silt and mud became progressively more pronounced.

Silurian

The patterns of marine sedimentation were characterised by pelagic, hemipelagic and turbiditic processes depositing muds, silts and coarse-grained clastics.

BIOSTRATIGRAPHICAL CORRELATION AND PALAEOENVIRONMENTAL INTERPRETATION OF FAUNAS

Biostratigraphical subdivision of the generally sparsely fossiliferous Caradoc rocks in north-west Wales is problematic. While a detailed examination of these problems is not appropriate here, possible relationships between the volcanic event stratigraphy and the biostratigraphy are particularly worth highlighting. Graptolite zonation indicates that the sequence spans parts of the *Diplograptus multidens* and *Dicranograptus clingani* zones (Figure 9), though precision in placing the zonal boundary is prevented due to the sparsity and poor preservation of the fossils.

While the area lies only approximately 100 km from the Caradoc type sections in Shropshire, where its eight stages are defined (Bancroft, 1928, 1933, 1945; Dean, 1958; Hurst, 1979), the diagnostic shelly fossils are generally absent from the sequences in north-west Wales. Facies variations may control these faunal differences, with the generally fine-grained sequences of the lower Caradoc stages in north-west Wales contrasting in particular with the coarser-grained sequences in Shropshire. Other factors, however, such as faunal provincialism and the possible juxtaposition of the two areas by subsequent strike-slip faulting (e.g. Woodcock and Gibbons, 1988) may also be important. Additionally, the almost ubiquitous tectonic deformation of specimens makes specific identification, particularly of statistically defined brachiopod taxa (cf. Williams, 1963), difficult or impossible (see Campbell, 1983). Biostratigraphical subdivision using microfloras has been applied with some success but much remains to be done.

In spite of these problems general conclusions as to the depositional environments of some of the faunas can be drawn. These are facilitated by recent advances in the environmental interpretations of Caradoc shelly faunas (e.g. Pickerill and Brenchley, 1979; Lockley, 1980).

The earliest Caradoc fauna recognised is from low in the Nant Ffrancon Formation (Figure 9) and is of the *Nicollela* assemblage type. An early Caradoc age, possibly Costonian, is indicated by the presence of *Harknesella* cf. *subquadrata* and by the general faunal similarity with that of the Derfel Limestone fauna, described by Whittington and Williams (1955), in the Arenig district to the east.

Most of the Llewelyn Volcanic Group (1st Eruptive Cycle) is unfossiliferous, but poorly preserved brachiopod-dominated faunas occur in strata below the Capel Curig Volcanic Formation and include species of *Rostricellula*, *Dalmanella*, *Howellites*, *Dinorthis* and *Macrocoelia*, which suggest a Soudleyan age.

The strata between the Llewelyn Volcanic Group and the Snowdon Volcanic Group (i.e. between the 1st and 2nd Eruptive cycles) contain the boundary between the Soudleyan and Longvillian stages (Diggens and Romano, 1968; Romano and Diggens, 1969; Wright, 1979) and brachiopod and trilobite dominated faunas are relatively common (Plate 4).

The upper Soudleyan is characterised by a progression through three different faunas. The lowermost of these includes *Dinorthis berwynensis berwynensis* with *Sowerbyella sericea permixta*, followed by *Dinorthis berwynensis angusta* with *Sowerbyella musculosa* and, finally, the topmost fauna is marked by the stratigraphically restricted but locally abundant *Plaesiomys multifida*.

The Soudleyan–Longvillian stage boundary approximately coincides with the horizon of the lower tuff of the Pitts Head Tuff Formation (the first phase of the 2nd Eruptive Cycle), though facies changes associated with this event may in-

Plate 2 The west-facing flank of Tryfan with Llyn Bochlwyd [SH 650 593] in the foreground. The top of the Capel Curig Volcanic Formation is marked by the upper grass limit on the flank of Tryfan and the Pitts Head Tuff Formation forms the feature immediately beyond the lake. (BGS Photograph L1911).

fluence the recognition of the boundary. The Longvillian is typified by the appearance of the brachiopods *Howellites antiquior* and *Macrocoelia expansa*, and by the trilobites *Kloucekia apiculata*, *Broeggerolithus nicholsoni* and *Flexi-calymene planimarginata*.

At higher stratigraphical levels within the Snowdon Volcanic Group (2nd Eruptive Cycle), locally within the reworked facies of the Lower Rhyolitic Tuff Formation, the Lower Crafnant Volcanic Formation and the Bedded Pyroclastic Formation, further upper Longvillian and/or Woolstonian faunal elements appear, including the trilobites *Atractopyge celtica* and *Estoniops alifrons*, and the brachiopods *Dolerothis duftonensis prolixa* and *Nicolella actoniae obesa*.

Although this general stratigraphical picture holds for much of central and northern Snowdonia, there are surprising anomalies between the relationship of the faunas to the effective time planes provided by the primary ash-flow tuff units. Thus, to the south of the Dolwyddelan syncline, supposedly Longvillian elements such as *Broeggerolithus nicholsoni*, *Kloucekia apiculata* and *Howellites antiquior* have been identified at horizons closely associated with the Capel Curig Volcanic Formation, and therefore well below the approximate stratigraphical level of the Pitts Head Tuff, although the tuff itself is absent in this area. Unless there is an, as yet unrecognised stratigraphical break in this sequence, facies control of these faunas might be implicated, with the first appearance of Longvillian faunas being typically associated with the incoming of coarser volcaniclastic sediments. This would cast doubt, therefore, on the value of establishing a boundary between faunas of Soudleyan and Longvillian aspect. Similar suggestions of facies control and diachronism of the faunas are implied by the distribution of the upper Longvillian and/or Woolstonian faunas, which appear within the Lower Rhyolitic Tuff Formation in the south-west and in the Lower Crafnant Volcanic Formation in the north, but do not appear in the vicinity of Snowdon until the onset of the Bedded Pyroclastic Formation.

Plate 3 The north-facing cliffs of Y Lliwedd [SH 623 533] formed of intracaldera ash-flow tuffs of the Lower Rhyolitic Tuff Formation.

Plate 4 Characteristic Soudleyan and Longvillian fossils from sediments within the deposits of the 1st and 2nd Eruptive cycles.

1,2 *Estoniops alifrons* (McCoy) Bedded Pyroclastic Formation, south-east of Llyn Glas [SH 6191 5558]; 1, BGS DJ 4770, head; 2, BGS DJ 4777, pygidium. Both × 1.5.

3,4 *Dolerorthis duftonensis prolixa* Williams, 3, BGS DJ 4179, brachial valve internal, Bedded Pyroclastic Formation, Roman Bridge [SH 7010 5119], × 1.5; 4, BGS DJ 3647, pedicle valve exterior, Bedded Pyroclastic Formation, Y Lliwedd [SH 6343 5301], × 1.5.

5 *Nicolella actoniae obesa* Williams, BGS DJ 3635, pedicle valve external, Bedded Pyroclastic Formation, Y Lliwedd [SH 6343 5301], × 1.5.

6 *Atractopyge celtica* Dean, BGS DJ 3672, Bedded Pyroclastic Formation, Y Lliwedd [SH 6343 5301], × 1.5.

7 *Orthambonites cessata* Williams, BGS DJ 3672, Bedded Pyroclastic Formation, Y Lliwedd [SH 6343 5301], × 2.

8–10 *Dinorthis berwynensis* subsp. nov., BGS DJ 5533, 9, pedicle and 8, 10, brachial valves external, Lower Rhyolitic Tuff Formation, Moel Hebog [SH 5644 4730], × 1.5.

11 *Kloucekia apiculata* (McCoy), BGS DJ 4924, pygidium, breccias at base of Lower Rhyolitic Tuff Formation, north-west of Llyn Edno [SH 6614 5028], × 2.

12 *Flexicalymene planimarginata* (Reed), BGS DJ 5246, Cwm Eigiau Formation, Afon Dylfi, Nantmor [SH 6218 4512], × 2.

13,14 *Plaesiomys multifida* (Salter), Cwm Eigiau Formation; 13, BGS DJ 1428, pedicle valve internal, west of Llyn Idwal [SH 6350 5956], × 1.5; 14, BGS DJ 4438, pedicle valve internal, Carnedd Melyn [SH 6444 4900], × 1.5.

15 *Macrocoelia ?expansa* (Sowerby), BGS DJ 5149, pedicle valve internal, Cwm Eigiau Formation, Clogwyn Aderyn [SH 6384 4678], × 1.5.

16, 17 cf. *Rhipidomena* sp., BGS DJ 4908, 4907, pedicle valve internal, Cwm Eigiau Formation, south-east of Moel Meirch [SH 6618 5018], × 1.5.

18 *Ambonychia* cf. *radiata* Hall, BGS DJ 1955, right valve internal, Capel Curig Volcanic Formation, Pen-yr-ole-wen [SH 6540 6208].

19 *Dinorthis berwynensis berwynensis* (Whittington), BGS DJ 4666, pedicle valve external, Cwm Eigiau Formation, north-west of Cnicht [SH 6384 4678], × 1.5.

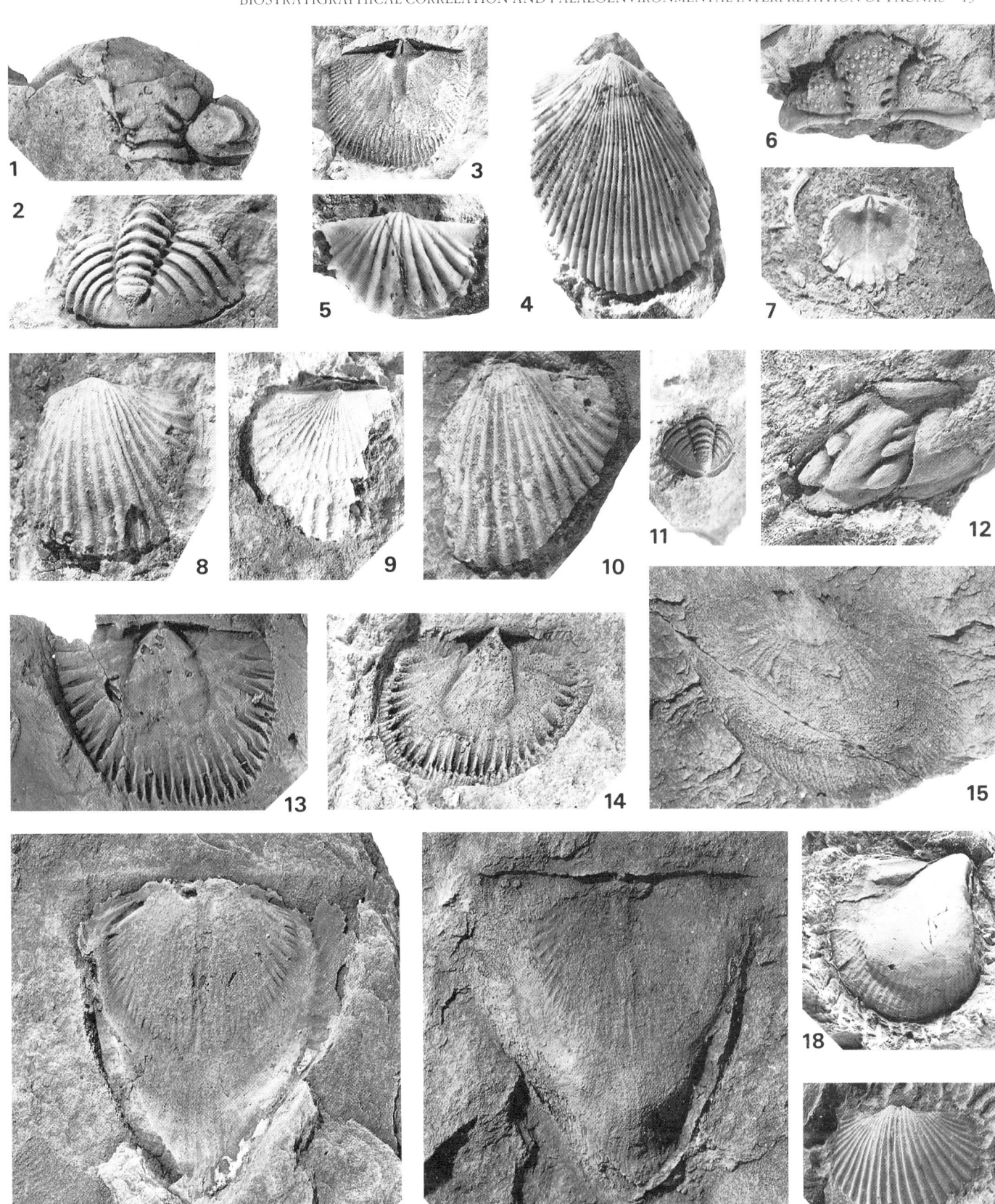

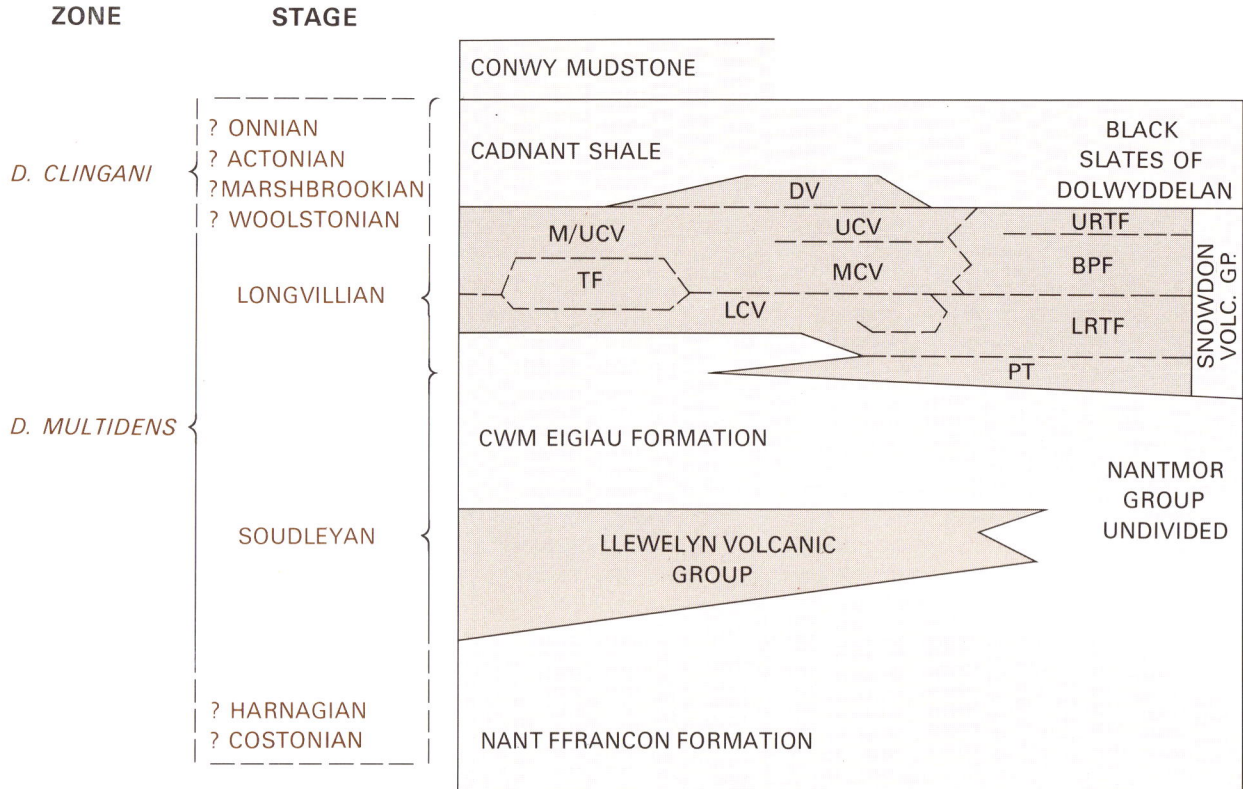

Figure 9   Caradoc volcanic sequence with reference to graptolite zones and shelly stages.

PT: Pitts Head Tuff Formation; LRTF: Lower Rhyolitic Tuff Formation; BPF: Bedded Pyroclastic Formation; URTF: Upper Rhyolitic Tuff Formation; LCV: Lower Crafnant Volcanic Formation; M/UCV: Middle/Upper Crafnant Volcanic Formation; TF: Tal y Fan Volcanic Formation; DV: Dolgarrog Volcanic Formation.

CHAPTER 2

1st Eruptive Cycle (Llewelyn Volcanic Group)

The activity of the 1st Eruptive Cycle lay entirely within northern Snowdonia and is represented by the Llewelyn Volcanic Group (Figure 6). It is underlain by a thick sequence (about 1000 m) of marine strata (Nant Ffrancon Formation) (Howells et al., 1985b), of Arenig to early Caradoc age.

Sedimentation prior to the 1st Eruptive Cycle

The Arenig marine transgression was markedly diachronous across adjacent fault-bounded blocks (Reedman et al., 1983; Beckley, 1987). North-west of the Aber Dinlle Fault (Figures 1 and 5), transgression occurred during middle Arenig (early Whitlandian) times and coarse-grained and channelled basal sandstones in the vicinity of Bangor suggest that temporarily this area remained close to a shoreline in the north-west. This subsequently migrated north-westwards as the basin margin subsided, with Anglesey being transgressed in Fennian times (Beckley, 1987).

Typically, the Arenig sediments display an upward transition from a sublittoral sandstone facies to a silt- and mud-dominated shelf sequence with storm-generated sand layers. The increasing dominance of siltstone and mudstone indicates the development of distal shelf environments. Rates of subsidence, however, varied across major faults, for example south-east of the Aber-Dinlle Fault the basal sandstone facies is locally absent and distal shelf conditions were rapidly established. Temporary shallowing at the end of the Arenig is indicated by the presence of a thin pisolitic ironstone (Howells et al., 1983, 1985b; Trythall et al., 1987).

The thick (>750 m) and rather monotonous Llanvirn to early Caradoc silt- and mud-dominated sequence, the product of distal shelf sedimentation, was punctuated by at least two further episodes of pisolitic ironstone deposition, and consequently shallowing, of probable Llandeilo and early Caradoc age. In the north, towards the top of the sequence, the mudstone component increases, reflecting either basin deepening or a reduction in the influence of the basin margin as a source of coarse detritus. Within these mudstones a thin (<20 cm) metabentonite, with associated *C. peltifer* Zone fauna, appears to represent the earliest evidence of Caradoc volcanism (Roberts and Merriman, 1989). The occurrence of *D. multidens* Zone graptolites and a Soudleyan shelly fauna in these uppermost strata prior to the 1st Eruptive Cycle establish their early Caradoc age.

Locally, the uppermost mudstones display small-scale slumped bedding and contain thick wedges of mudflow breccia with clasts of fine-grained acidic volcanic rocks. The breccias probably developed as a result of seismic activity and the nearby intrusion of exogenous rhyolite domes which marked the onset of volcanism. They were deposited in areas where very thick sequences of extrusive basalt subsequently accumulated within fault-controlled depressions in the sea floor, and are probably related to early movement on those faults.

LLEWELYN VOLCANIC GROUP

The Llewelyn Volcanic Group (Howells et al., 1983; Ball and Merriman, 1989) comprises five volcanic formations interbedded with thick sequences of sedimentary rocks. The earliest volcanic activity developed from at least four, partly contemporaneous centres from which rhyolite lavas and acidic ash-flow tuffs (Conwy Rhyolite Formation), trachyandesite lavas and tuffs (Foel Fras Volcanic Complex), basaltic-andesite lavas (Foel Grach Basalt Formation) and acidic ash-flow tuffs and rhyolites (Braich tu du Volcanic Formation) were erupted (Figure 10). These formations are areally restricted, locally interdigitate and broadly occur from north-west to south-east respectively along the outcrop (Figure 11). Locally, great thicknesses of extrusive volcanic rocks accumulated which lack evidence of subsequent extensive erosion or reworking. The inference is that most of the extrusive rocks were ponded in subsiding, probably fault-bounded sectors of the sea floor. This differential subsidence is clearly indicated by lateral variations in the thickness of strata between successive extensive ash-flow tuffs (Figure 17). Assuming that the tops of the tuffs were approximately horizontal immediately after emplacement, it can be shown, for example, that the basalts of the Foel Grach Basalt Formation infilled

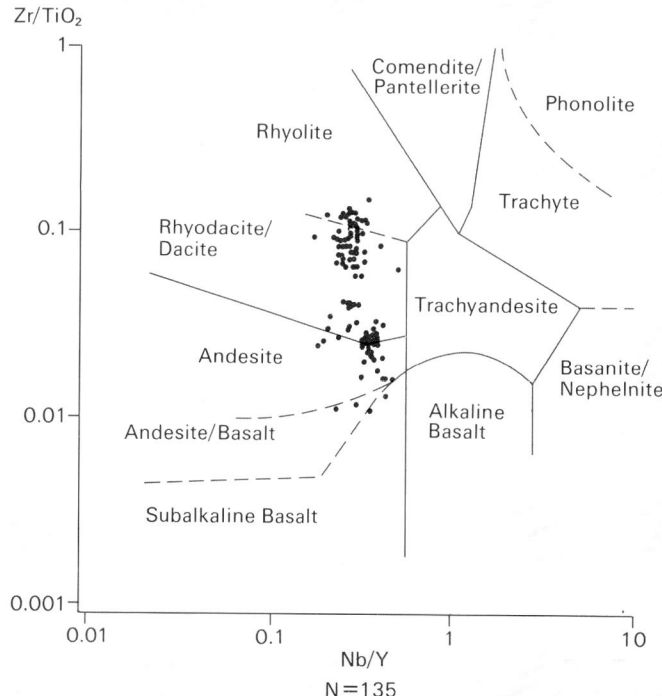

Figure 10 Llewelyn Volcanic Group: analyses of extrusive volcanic rocks plotted on the Zr/TiO_2 vs. Nb/Y discriminant diagram of Winchester and Floyd (1977).

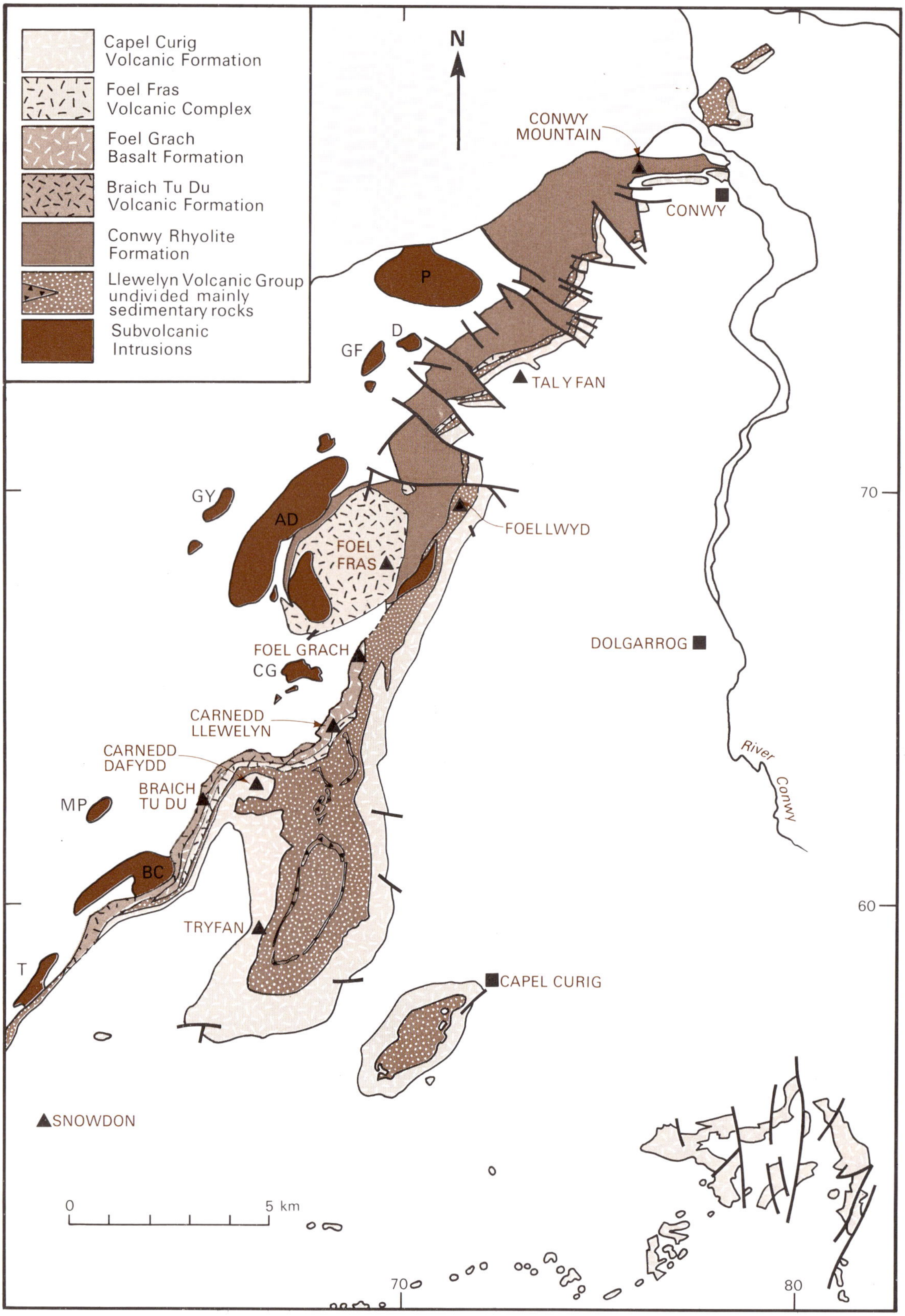

Figure 11 Llewelyn Volcanic Group: distribution of formations and related subvolcanic intrusions.

Intrusions identified as follows: P: Penmaenmawr; D: Dinas; GF: Garreg Fawr; GY: Gyrn; AD: Aber-Drosgl; CG: Carreg y Gath; MP: Mynydd Perfedd; BC: Bwlch y Cywion; T: Talgau.

two small basins or troughs each a few kilometres across (Figure 17). During the interval between the eruption of the ash-flow tuffs of the Braich tu du Volcanic Formation and the Foel Fras Volcanic Complex these troughs subsided by a maximum of 180 and 400 m respectively relative to an intervening stable ridge.

Following the early phase of volcanism, shallow-marine sedimentation predominated. Subsidence was greatest in the south-east and eventually a low, southerly directed palaeoslope was established. Alluvial and fluvial sedimentation in the north passed progressively southwards into beach, inner- and outer-shelf environments. This was the palaeogeographical setting prior to the final and most violent volcanic phase (Capel Curig Volcanic Formation) of the 1st Eruptive Cycle. This volcanism developed at three centres (Howells and Leveridge, 1980) and was dominated by acidic ash-flow eruptions. In northern Snowdonia the tuffs overlie deposits of the early phase of the eruptive cycle, but east and south-east of the Snowdon Massif they are the sole representatives of the cycle. To the west of the massif the products of the 1st Eruptive Cycle wedge out to a feather edge in the enclosing sediments.

Conwy Rhyolite Formation

The most northerly of the early volcanic centres erupted rhyolite lavas and acidic ash-flow tuffs (Conwy Rhyolite Formation, Figure 11), thickest (about 1000 m) about 5 km south-west of Conwy where, close to the eruptive centre, they were intruded by late-stage rhyolite intrusions. Individual rhyolite flow units are better developed to the south-west, extending for some 12 km, and in this direction the flows interdigitate with marine sediments. The rhyolite lavas are characteristically flow-banded (Plate 5) and columnar-jointed in their cores, with autoclastic brecciated surfaces. A possible smaller eruptive centre occurred further south, in the vicinity of Foel Fras (Figure 11).

PETROGRAPHY   Both lavas and intrusive rhyolites are sparsely porphyritic and microporphyritic, with partly crystalline and devitrified glassy groundmass textures. Phenocrysts (<3 mm) of albite are rarely zoned and few show marginal corrosion. Simply twinned alkali feldspar phenocrysts (<2 mm), locally unaltered, are also highly sodic in composition ($Ab_{94}Or_6$); least altered phenocrysts are optically inhomogeneous cryptoperthites. Rare quartz phenocrysts occur as highly corroded relics. The originally least glassy rhyolites comprise a trachytic groundmass of alkali feldspar microlites (Or_{22-61} Ab_{78-89}), rare apatite needles and opaque dust in a siliceous aggregate after a glassy mesostasis. Some banded rhyolites show alternation of bands with trachytic texture and devitrified perlitic textures. Rhyolites with an originally largely glassy groundmass contain subspherulitic intergrowths of quartz and orthoclase (determined by XRD). Chlorite and hematite (±limonite) pseudomorphs of both ferromagnesian minerals and primary opaque oxides rarely exceed 2 per cent of the constituents.

Ash-flow tuffs contain broken and rounded plagioclase (now Ab_{98-100}) phenocrysts with pumice fragments (<4 mm). The latter, replaced by chlorite (ripidolite-brunsvigite), indicate an original glass of intermediate or basic composition. Corroded, skeletal ilmenite grains comprise 1–2 per cent of

Plate 5   Flow-banded rhyolite of the Conwy Rhyolite Formation at Drosgl [SH 706 712].

the constituents. The matrix comprises devitrified, recrystallised cuspate shards in a siliceous aggregate, after vitric dust.

GEOCHEMISTRY   The analysed samples of the rhyolites, representatives of which are presented in Appendix 1, Table 3, mostly plot in the rhyolite field of the total alkali-silica (TAS) diagram of Le Bas et al. (1986) with a few lying on and below the dacite/rhyolite boundary. The large spread of alkali values reflects secondary alteration but generally the suite comprises subalkaline, peraluminous rhyolites. On the Zr/TiO_2 vs. Nb/Y plot of Winchester and Floyd (1977) (Figure 12) the analyses mainly lie within the rhyodacite/dacite field.

Of the HFS elements, considerable variability is seen in Zr (123–474 ppm, average = 246.5 ppm ($\sigma n - 1$ = 83.5), N = 27), P_2O_5 (0.01–0.37 wt%) and TiO_2 (0.1–0.37 wt%), and to a lesser extent in Y (33–80 ppm, average = 56.6 ppm ($\sigma n - 1$ = 12.7)). Variability of Nb is, however, more limited (10–24 ppm, average = 15.2 ppm ($\sigma n - 1$ = 3.5)).

REE-chondrite normalised patterns (Figure 13) of least altered rhyolites (after Ball and Merriman, 1989) show steep patterns for the LREE, with significant LREE/HREE enrichment (La_{Ch}/Yb_{Ch} = 3.38–3.43). Patterns for the MREE and HREE are relatively flat. The profiles are characterised by marked Eu anomalies. LREE variability is similar to that for

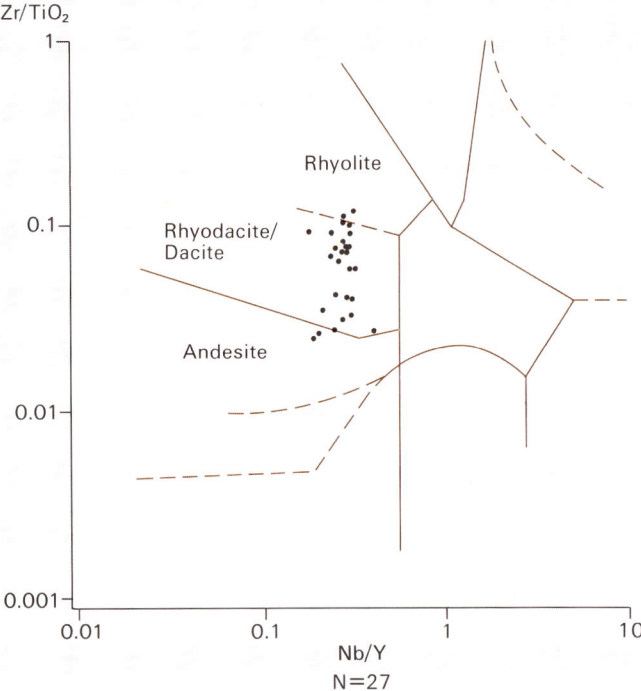

Figure 12 Zr/TiO₂ vs. Nb/Y diagram for the Conwy Rhyolite Formation.

many of the HFS elements (e.g. Ce = 29–105 ppm, N = 27), though some slight enrichment is demonstrably related to hydrothermal alteration (see below).

Certain plots of the relatively immobile trace elements, e.g. Nb vs. Zr (Figure 14), suggest that the Conwy Rhyolite Formation can be subdivided into two geochemical groups which are also geographically distinct. The group relatively enriched in Nb was possibly derived from the subsidiary centre situated near Foel Fras in the south, whereas the other group was derived from the main centre, near Conwy Mountain in the north.

Compared with other rhyolitic rocks of the 1st Eruptive Cycle, the rhyolites of the Conwy Rhyolite Formation have significantly lower HFS element contents than the ash-flow tuffs of the Capel Curig Volcanic Formation, although the 1st Member of the latter has generally lower Zr and TiO₂. Their lower HFS element concentrations imply that they are relatively less evolved, a conclusion consistent with their dacite/rhyodacite composition. The lower rhyolite and ash-flow tuff of the Braich tu du Volcanic Formation are, however, very similar in almost all respects. Some subalkaline rhyolites of the Lleyn Peninsula (Leat and Thorpe, 1986) are also closely similar in many respects (HFS and LIL elements). Most other rhyolitic rocks of Caradoc age differ in having relatively higher HFS element contents (e.g. the Pitts Head Tuff Formation; the mainphase of the Lower Rhyolitic Tuff Formation, etc.). The Yr Arddu tuffs have comparable HFS contents, with the exception of lower P₂O₅, but have significantly higher Th contents.

Two types of alteration, related respectively to hydrothermal activity and low-grade regional metamorphism, can be distinguished (Ball and Merriman, 1989). The hydrothermal

activity is concentrated mainly in the vicinity of the northern eruptive centre and is characterised by intense sericitisation of the rhyolites. Their conversion to aggregates of quartz and white mica resulted in the loss of substantial quantities of Na, Ca and Fe, and minor amounts of Sr and Co, and a slight increase in the LREE (Ball and Merriman, 1989). The effects of low-grade regional metamorphism, involving the partial replacement of phenocrysts and groundmass by white mica and chlorite ± calcite, are most easily discerned in the rhyolites and ash-flow tuffs away from the eruptive centre.

Foel Fras Volcanic Complex

To the south-west, the rhyolite lavas of the Conwy Rhyolite Formation abut against and locally interdigitate with the Foel

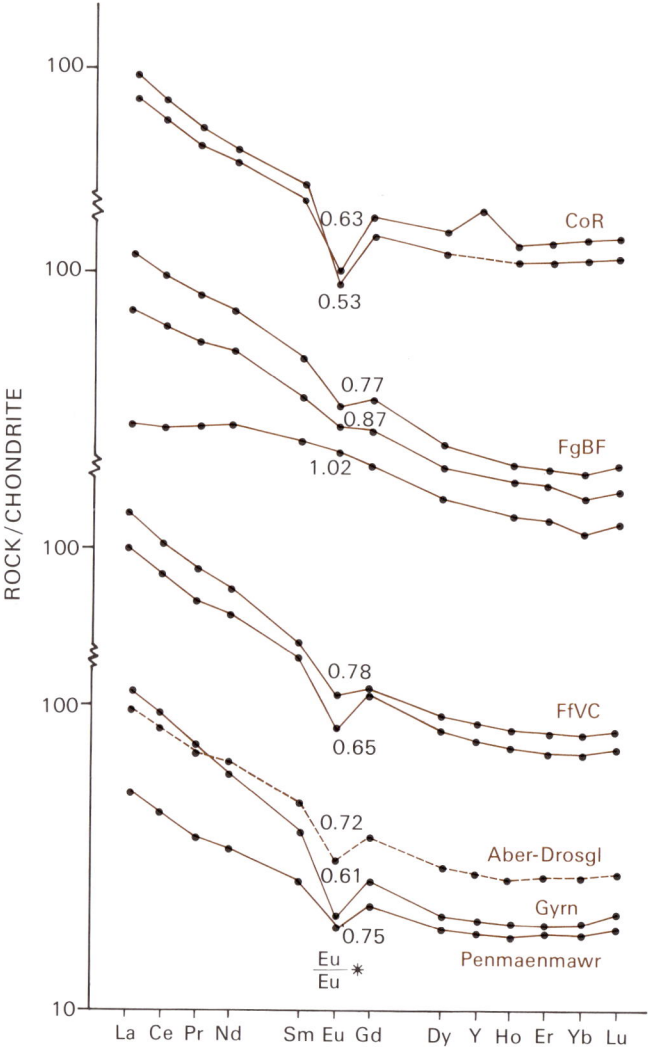

Figure 13 REE-chondrite normalised patterns for the Llewelyn Volcanic Group and selected subvolcanic intrusions (normalising factors after Thompson et al., 1984).

CoR: Conwy Rhyolite Formation; FgBF: Foel Grach Basalt Formation; FfVC: Foel Fras Volcanic Complex.

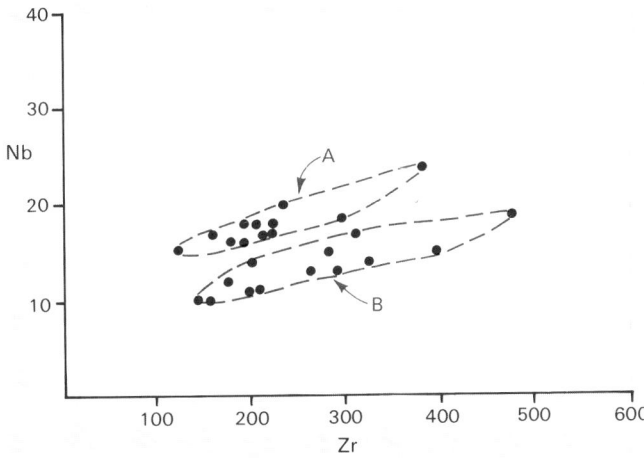

Figure 14 Nb vs. Zr diagram for the Conwy Rhyolite Formation, showing subgroups A, derived from a southern centre near Foel Fras, and B, derived from a northern centre near Conwy Mountain.

Fras Volcanic Complex, comprising lavas, tuffs and intrusions of predominantly trachyandesitic composition (Figure 11) (Howells et al., 1983, Ball and Merriman, 1989; British Geological Survey, 1986). The spatial coincidence and distribution of porphyritic trachyandesite intrusions with the thick accumulation (about 1000 m) of lavas and tuffs of similar composition (British Geological Survey, 1986; Howells et al., 1985b) defines the eruptive centre of the complex. However, the relatively poorly exposed and variable

lithologies about this centre make their inter-relationships difficult to interpret. Also, secondary alteration obscures most of the original textures, apart from feldspar phenocrysts, in both lavas and tuffs, and exposure is relatively poor.

Nevertheless, sporadically developed bedding in the tuffs and weak flow-banding in the lavas and lenses of breccia (Plate 6) display centroclinal dips which define a saucer-shaped structure (Figure 15). Lavas and tuffs, interpreted as a caldera fill, are ponded in the structure. To the north, close to the caldera, a thin trachyandesite lava is intercalated in the rhyolites of the Conwy Rhyolite Formation and a single ash-flow tuff overlying the rhyolites extends for about 5 km. South of the caldera, the complex is also represented by a trachyandesite lava and an ash-flow tuff, which extend for about 8 km, interbedded with marine sandstones and silt-stones. Both lava and tuff are interpreted as outflows from the caldera. The tuff is reworked at its top and locally has been completely removed by erosion. Also southwards, the ash-flow tuff passes laterally from massive, welded tuff to bedded and nonwelded tuff. Further to the south the lava and tuff wedge out into the adjacent sediments.

PETROGRAPHY   The lavas comprise phenocrysts (<4 mm) of plagioclase in a devitrified groundmass densely packed with feldspar microlites, opaque dust and apatite needles. The original glass is devitrified into subspherulitic inter-growths of quartz and alkali feldspar. Former ferromagnesian minerals, up to 5 per cent by volume, are represented by chlorite pseudomorphs (± actinolite ± sphene ± epidote). Plagioclase phenocrysts are commonly zoned, embayed and altered to albite (rarely oligoclase) plus white mica (± epidote ± calcite ± chlorite ± pumpellyite). Most of the

Plate 6 Autobrecciated trachyandesite of the Foel Fras Volcanic Complex cropping out on the south-eastern slopes of Llwytmor [SH 689 691]. (BGS Photograph L2519).

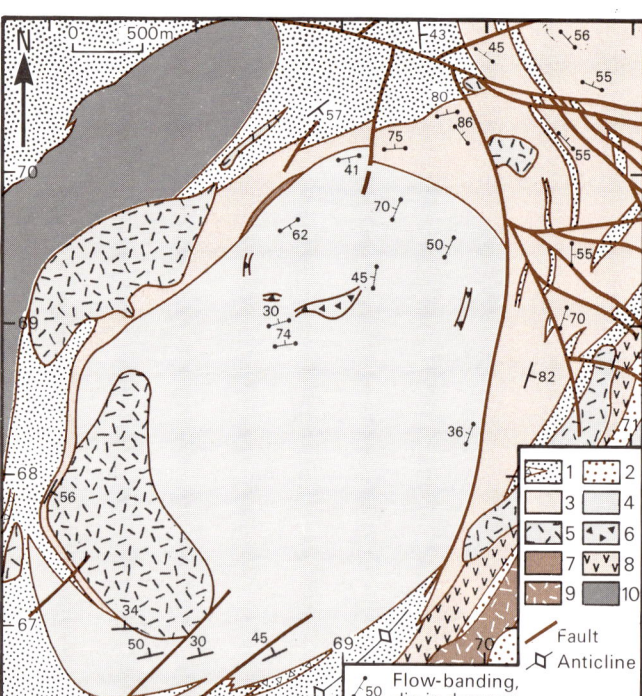

Figure 15 Geological map of the eruptive centre of the Foel Fras Volcanic Complex.

1. Siltstone of the Nant Ffrancon Formation (inset: mudflow breccia); 2. sandstone; 3. rhyolites of the Conwy Rhyolite Formation; 4. trachyandesite (extrusive and intrusive undivided) of the Foel Fras Volcanic Complex; 5. intrusive porphyritic trachyandesite; 6. trachyandesitic ash-flow tuffs of the Foel Fras Volcanic Complex; 7. basalt; 8. acid tuffs; 9. dolerite; 10. microtonalite of the Aber-drosgl intrusion.

lavas contain autoliths (cognate lithic xenoliths) (<5 per cent), up to 6 mm diameter, of a microdioritic intergrowth of plagioclase laths, chlorite pseudomorphs with augite relics and abundant ilmenite and apatite euhedra. Locally, the lavas are intensely autobrecciated (Plate 6) and weakly flow-banded.

The intrusive rocks, also porphyritic, possess a more coarsely crystalline groundmass. Plagioclase phenocrysts (<3 mm) form up to 20–30 per cent of the constituents; they are commonly zoned, and many are embayed. Alteration to albite and white mica (± epidote ± carbonate) is extensive, but relicts (An_{60}) survive in places. Typically, the groundmass comprises plagioclase prisms, euhedral ilmenite (2–3 per cent), chlorite pseudomorphs, apatite needles and a granophyric mesostasis. Relict augite (Wo_{41} Eu_{87} Fs_{22}) survives within a few chlorite ± actinolite pseudomorphs. Other pseudomorphs of fibrous, pale green, actinolitic amphibole together with chlorite probably represent altered hornblende. Ferromagnesian minerals form up to 15 per cent of the constituents and granophyric intergrowths, about 10 per cent, include alkali feldspar (Or_{27} Ab_{71} An_2) which forms overgrowths on plagioclase phenocrysts. Electron microprobe analyses of the granophyre determined a composition of 68–77 per cent SiO_2 and 7–9 per cent total alkalies.

The tuffs consist of broken or rounded plagioclase crystals (<3 mm) and lithic clasts in a recrystallised matrix of albite and quartz + sericite ± calcite ± anastase. Clasts of trachytic-textured lava and microdiorite are fairly common, whereas clasts (<6 mm) of sandstone and silty mudstone are restricted to lithic-rich tuffs. Shards and angular vitric fragments replaced by chlorite with cryptocrystalline silica rims are common. Locally the tuffs show a crude fracture cleavage. Pressure shadows, adjacent to clasts and phenocrysts, are filled with quartz, calcite or chlorite.

The extensive secondary alteration of the Foel Fras Volcanic Complex is consistent with low greenschist facies regional metamorphism (Roberts, 1981). However, the localised occurrence of pumpellyite indicates that the metamorphic grade is close to the prehnite-pumpellyite/greenschist facies boundary, which defines (Liou et al., 1985) an upper limit of 350°C on the temperature of alteration. Early hydrothermal alteration may account for the variable development of sericite but mineralogically such alteration is indistinguishable from that produced by thermal metamorphism.

GEOCHEMISTRY   The Foel Fras Volcanic Complex represents the largest volume of intermediate magma extruded during Caradoc volcanism in north and central Snowdonia. Representative analyses are presented in Appendix 1, Table 3. On a Total Alkalis vs. Silica (TAS) plot the rocks of the complex straddle the andesite, trachyandesite and trachydacite fields. On the Zr/TiO_2 vs. Nb/Y plot (Figure 16) the analyses straddle the boundary between the fields of andesite and rhyodacite/dacite.

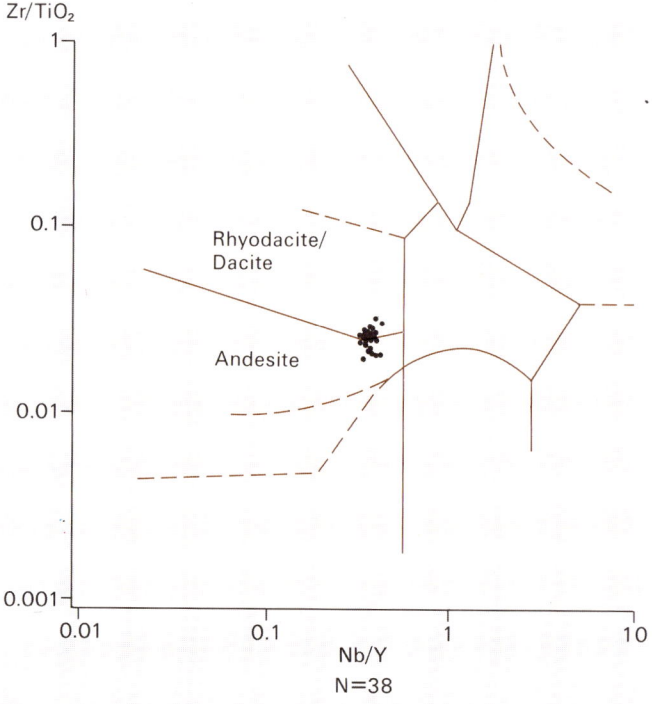

Figure 16 Foel Fras Volcanic Complex, Zr/TiO_2 vs. Nb/Y diagram.

There is very little compositional variation (N = 38), in terms of the immobile elements, within the lavas and ash-flow tuffs, although the latter show slightly greater variation. In particular, variation in Zr (286–419 ppm, average = 361.8 ppm), Nb (17–26 ppm, average = 20.6 ppm) and Y (47–73 ppm, average = 56.6 ppm) is small (sample population standard deviations respectively 31.8 ppm, 2.0 ppm and 4.7 ppm). Similarly, Th (6–14 ppm) and the LREE (e.g. La and Ce) are compositionally restricted. The associated intrusions have very similar concentrations of HFS elements (e.g. Zr = 287–368 ppm, Nb = 17–22 ppm, Y = 50–61 ppm), Th and the LREE, confirming that they are cogenetic. The greater compositional variation of the ash-flow tuffs, compared with the lavas and intrusions, is considered by Ball and Merriman (1989) to reflect the incorporation of epiclastic material and possibly their interaction with sea-water. The only differences between the intrusive rocks and lavas are in the concentrations of Ba, S and Cr, and to a lesser extent V and Sc.

Chondrite-normalised REE patterns of a representative ash-flow tuff and a lava (Figure 13) are similar to those of the Conwy rhyolites, with significant relative LREE-enrichment (La_{Ch}/Yb_{Ch} = 4.0–4.07) and only slight MREE- to HREE-enrichment. Negative Eu anomalies are slightly less than those for the rhyolites of the Conwy Rhyolite Formation.

In terms of the HFS elements, the REE, and their relative ratios (e.g. Nb/Zr), the Foel Fras Volcanic Complex is very similar to the Foel Grach Basalt Formation, though some samples of the latter appear somewhat less evolved. The complex differs only in its relatively lower TiO_2 contents.

Foel Grach Basalt Formation

To the south of the Foel Fras centre, the early phase of the 1st Eruptive Cycle comprised extrusive basalts (Foel Grach Basalt Formation) and rhyolite lavas and ash-flow tuffs (Braich tu du Volcanic Formation). These contrasting eruptions were in part penecontemporaneous. Locally, a breccia at the base of the basaltic sequence abuts against rhyolite lavas and here, it includes rhyolite blocks. The basalts form two separate outcrops (Figure 11); a thick (up to 400 m) restricted accumulation in the Foel Grach–Carnedd Llewelyn basin and, to the south-west, a thinner sequence (up to 180 m) of basalts and basaltic tuffs interbedded with marine sediments in the Foel Meirch–Braich tu du basin (Figure 17). In the thicker accumulation, subdivision of the sequence is difficult owing to relatively poor exposure and individual flows cannot be distinguished. Exposures are mainly of massive, locally pillowed basalt and breccias of pillowed and subangular basaltic blocks in a hyaloclastite matrix.

The laterally restricted but thick basaltic sequences reflect accumulation in small subsidiary troughs in which sub-

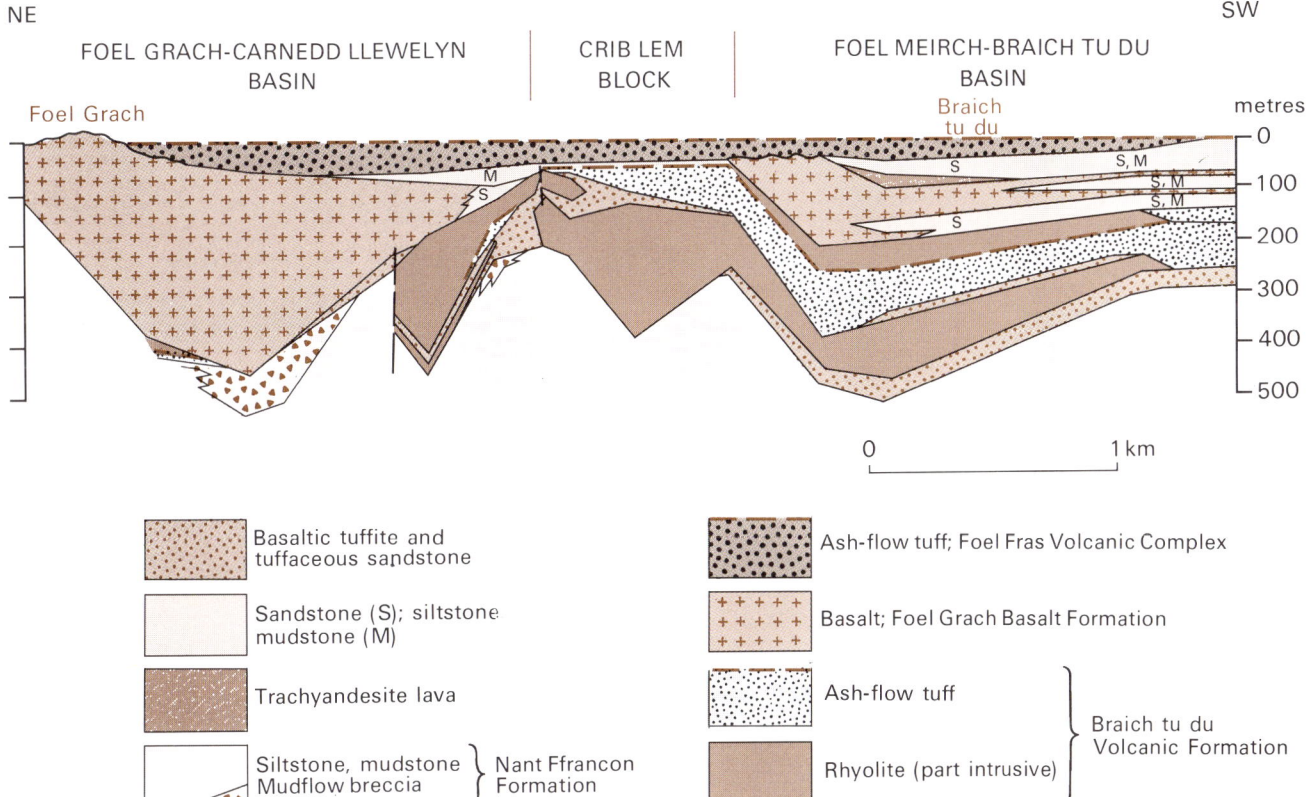

Figure 17 Block and basin topographic control on accumulation of the lower formations of the Llewelyn Volcanic Group at the southern end of its outcrop.

Thicknesses are measured down from the base of the Capel Curig Volcanic Formation (0 m).

sidence, for the most part, kept pace with basalt extrusion. The northerly basalt accumulation of the Foel Grach–Carnedd Llewelyn basin (Figure 17) is overlain by the outflow ash-flow tuff of the Foel Fras Volcanic Complex which locally thins northwards against the basalts, indicating that the basaltic eruptions constructed a low volcanic pile, a few tens of metres high. Marine siltstones directly overlying the basalts on the south side of the trough, however, contain very little volcanic debris and there is no evidence of reworking of a substantial basaltic edifice. The subsidence of this, and other restricted basins of basaltic accumulations in Snowdonia (e.g. Figure 82), may have been in part due to the rapid eruption of magma on to a less dense unlithified sedimentary substrate (Kokelaar et al., 1984), as well as to localised extension and fault block subsidence.

The thinner sequences of basalts in the Foel Meirch–Braich tu du trough to the south (Figure 17) comprise clearly defined flows and basaltic tuffs interbedded with marine sandstones and siltstones. Typically, the flows comprise massive cores, with columnar joints and blocky carapaces. They wedge out into marine sediments to the south. The thickest sequence of basalt is at the north side of the trough, as it is in the Foel Grach–Carnedd Llewelyn trough, and both have the form of half grabens. The basalt effusions were possibly controlled by fissures coincident with the trough margins.

PETROGRAPHY   The grey-green basalts contain a markedly variable frequency of plagioclase phenocrysts (<3.5 mm), up to 25 per cent of which show evidence of magmatic corrosion with embayed outlines and chloritic infillings. Flow alignment of feldspar microlites in the groundmass and phenocrysts is common (Plate 7B), with interstitial chlorite and opaque granules. Autoliths (<5 mm) consisting of microdioritic intergrowths of plagioclase, ilmenite, apatite and chloritic pseudomorphs are present in most samples. Vesicles, variously filled with carbonate, chlorite and white mica, are of frequent occurrence (Plate 7A and C).

Alteration of the lavas is consistent with the low greenschist facies regional metamorphism in the area (Roberts, 1981). Plagioclase is replaced by albite ± carbonate ± epidote ± white mica. Ilmenite is replaced by sphene.

GEOCHEMISTRY   The basalts (Appendix 1, Table 3) were originally designated as such on the interpretation of their mesoscopic and highly altered microscopic characters. The alteration is reflected in the chemistry of the rocks which range through trachybasalt and trachyandesite compositions on the TAS diagram. On the Zr/TiO_2 vs. Nb/Y plot (Figure

18) they range from subalkaline basalt to andesite. Comparison of analyses from the centre (KB 87) and vesicular margins (KB 86) of the same flow shows slight enrichment of K, Na and Ba, depletion of Sr, and higher REE values in the marginal sample. Such mobility possibly reflects low-grade burial metamorphism (Nystrom, 1984) and sea-water/magma interaction (Merriman et al., 1986). Overall enrichment of LIL elements (Sr, K, Rb, Ba, Th) relative to N-MORB (Figure 19) is consistent with andesitic compositions but cau-

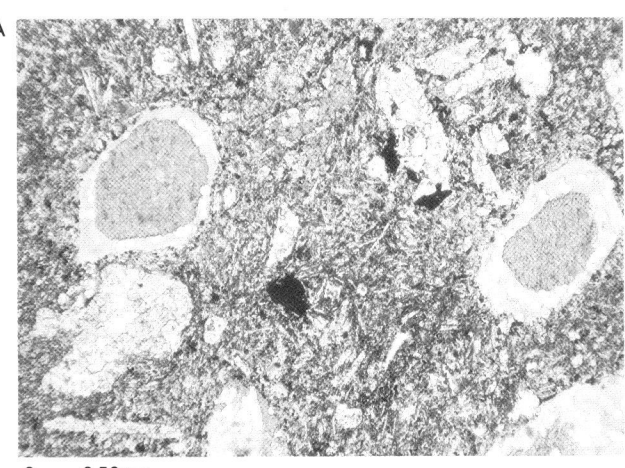

0   0.50mm

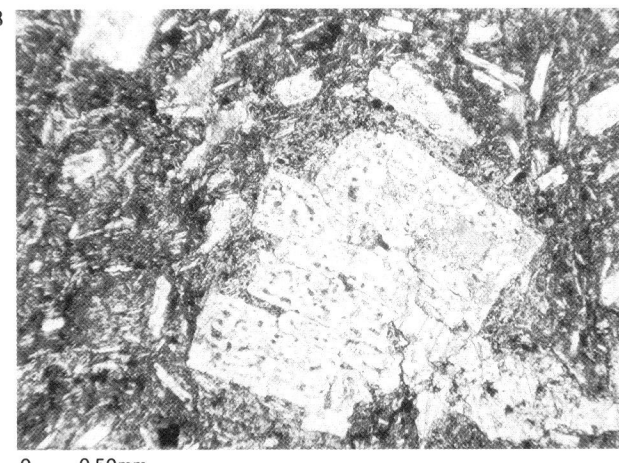

0   0.50mm

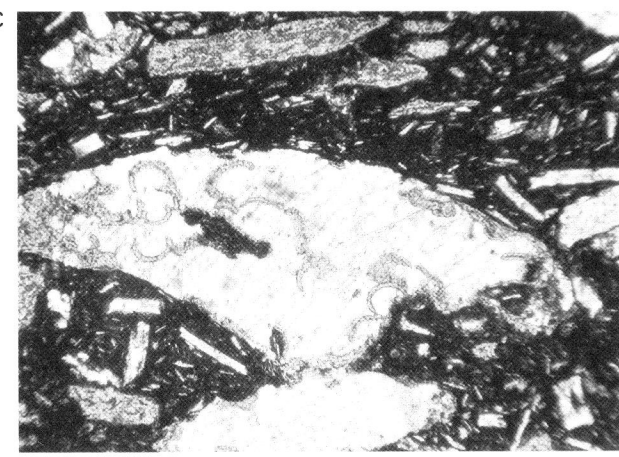

0   0.50mm

Plate 7 Photomicrographs of basalts, Foel Grach Basalt Formation

A   RN 81. Basalt. Carbonated feldspar microphenocrysts and vesicles, with chlorite cores, and rimmed by sericite and carbonate, in a fine-grained matrix of feldspar microlites, chlorite and iron oxide grains. ppl. [SH 6862 6567].
B   RN 225. Basalt. Albitised feldspar phenocryst and flow-orientated feldspar microlites in a matrix of chlorite, carbonate and iron oxide grains. ppl. [SH 6903 6625].
C   RN 225. Basalt. Ferritised groundmass with feldspar microphenocrysts and microlites. The vesicle is infilled with carbonate and vermicular chlorite. ppl. [SH 6903 6625].

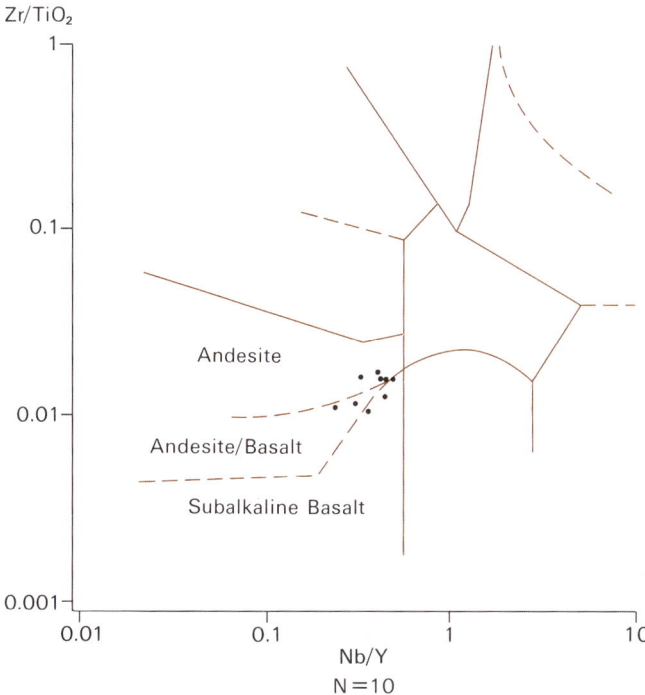

Figure 18   Foel Grach Basalt Formation, Zr/TiO₂ vs. Nb/Y diagram.

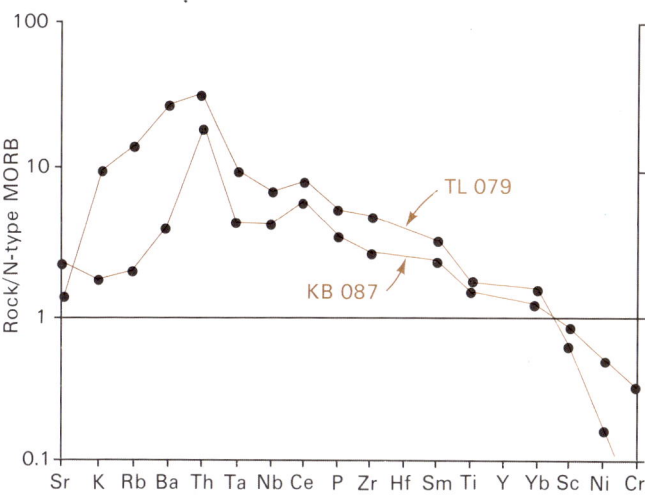

Figure 19   Foel Grach Basalt Formation, multi-element diagram normalised to average N-type MORB (normalising factors after Pearce, 1982a).

tion must be exercised in interpreting the LIL element patterns on MORB normalised multi-element plots, because Rb, Ba, K and Sr are commonly mobile during hydrothermal or low-grade metamorphic alteration (see Chapter 6).

The formation shows significant variation (N = 8) in concentrations of most of the immobile elements, such as the HFS elements (e.g. Zr = 235–514 ppm; Nb = 14–29 ppm; P_2O_5 = 0.39–0.72 wt%), Th and the LREE. The content of TiO₂, though less variable, is relatively high (2.09–2.82 wt%). However, incompatible element ratios are relatively constant, implying cogenetic relationships and probable association in terms of fractional crystallisation. Thus, for example, on a plot of Nb vs. Zr (Figure 20), the analyses approximate to a linear trend. Furthermore, this trend is colinear, as are others based on HFS elements and Th, with that defined by analyses of the Foel Fras Volcanic Complex, whose overall geochemical similarity with these rocks has already been mentioned. The Foel Grach Basalt Formation differs significantly only in its higher TiO₂ content.

Chondrite-normalised REE patterns show considerable variation (Figure 13). The most evolved (REE- and HFS element-enriched) samples have strongly LREE enriched patterns (La_{Ch}/Yb_{Ch} = 4.03–5.19) with slight Eu anomalies. The least evolved sample shows a flatter pattern with only slight LREE-enrichment (La_{Ch}/Yb_{Ch} = 2.29) and no Eu anomaly. These variations are also consistent with fractionation involving plagioclase.

On the Zr-Ti-Y diagram of Pearce and Cann (1973) the data indicate a calcalkaline character and the three samples for which appropriate analyses are available plot in the Volcanic Arc Basalt Field on the Hf-Th-Ta diagram of Wood (1980).

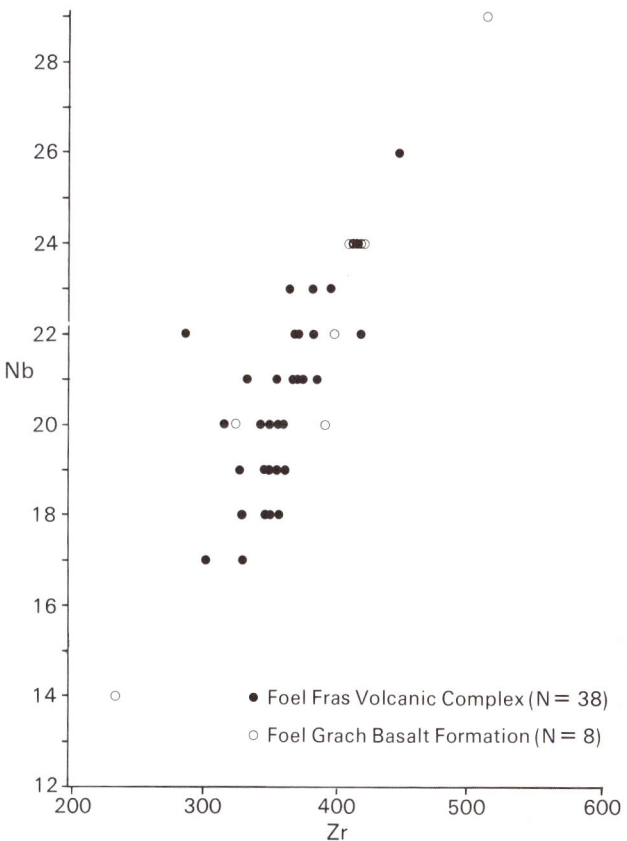

Figure 20   Foel Fras Volcanic Complex and Foel Grach Basalt Formation, Nb vs. Zr diagram.

Braich tu du Volcanic Formation

A sequence of rhyolites with an acidic ash-flow tuff (Braich tu du Volcanic Formation) is the most southerly expression of the early phase of the 1st Eruptive Cycle. The closely associated basaltic tuff and basalt at the type section, previously considered (Howells et al., 1983) to be a component of the formation, are now interpreted as an interdigitation of the Foel Grach Basalt Formation. The formation crops out (Figure 11) to the south of the main basalts of the Foel Grach–Carnedd Llewelyn basin and underlies those of the Foel Meirch–Braich tu du basin (Figure 17). The thickest sequence of rhyolites occurs between the two basins and constituted part of a relatively stable block or ridge, the Crib Lem block, at the time the two basins subsided. The eruptive centre, or centres, of the rhyolites and ash-flows were probably located within the vicinity of the Crib Lem block. To the south, in the type section at Braich tu du, the upward sequence of rhyolite lava (60 m thick), acidic ash-flow tuff (90 m thick) and rhyolite lava (45 m thick) is underlain by, interbedded with, and overlain by water-lain basaltic tuffs and tuffites. All are overlain by the outflow trachyandesite lava of the Foel Fras Volcanic Complex.

The rhyolite lavas typically show even flow-banding in the lower parts of the flows (Plate 8), contorted flow-banding in the cores, and autoclastic brecciated tops. Polygonal jointing is locally well developed. Basal parts of the upper flow contain small angular clasts of perlitic-fractured rhyolite enclosed in the fine-grained, devitrified and recrystallised matrix in which banding is still discernible. The clasts probably resulted from fragmentation caused by rapid quenching of the flow surface in contact with water.

The central ash-flow tuff (Plate 9) contains lithic and cognate chloritic clasts at its base, with the chloritic clasts above forming distinctive flattened fiamme up to 10 cm long. The flow is weakly welded at its base but is strongly welded above 6–7 m with quartzose recrystallization accentuating the foliation (Plate 10). The core of the flow is rheomorphosed with contorted and locally brecciated foliation, and the top is pervasively autobrecciated. The ash-flow tuff contains small, euhedral, prismatic, albite phenocrysts which impart a strong magmatic character to the fabric.

The lateral variations in thickness and distribution of the rhyolites and ash-flow tuff suggest that the emplacement was affected by topographic variation, possibly controlled by active fault lines, especially in the early stages. A local boss-like intrusion of feldspar porphyry in the vicinity of the Crib Lem block has been regarded as a late-stage intrusion at a possible eruptive centre (Howells et al., 1985b) .

PETROGRAPHY   Texturally, the rhyolites are characterised by a relatively high proportion (up to 25 per cent) of phenocrysts (<2 mm) of both sodic plagioclase and cryptoperthitic alkali feldspar. The devitrified groundmass comprises subspherulitic intergrowths of quartz and feldspar, and flow-banding is commonly accentuated by thin ribs of secondary quartz mosaics. Sphene, after accessory ilmenite, apatite and rare zircon together form 1 per cent. Secondary white mica (± chlorite) is commonly concentrated along flow bands. Plagioclase phenocrysts are patchily replaced by calcite.

GEOCHEMISTRY   The analyses (Appendix 1, Table 3) plot in the rhyolite field on the TAS diagram and mainly in the rhyodacite/dacite field on the Zr/TiO_2 vs. Nb/Y diagram (Figure 21).

In terms of the HFS elements, Th, the LREE (La and Ce), and their various ratios, the lower rhyolite and ash-flow tuff are very similar, although the ash-flow tuff is slightly

Plate 8 Even flow-banding in rhyolitic lava of the Braich tu du Volcanic Formation [SH 6480 6238]. (BGS Photograph L1934).

depleted in Zr and Nb. They both differ markedly from a single analysis (KB 180, Appendix 1, Table 3) of the upper rhyolite which is considerably enriched in Zr, Nb, Y and Th and depleted in TiO_2.

The lower rhyolite and ash-flow tuff are very similar, in terms of the immobile elements, to the Conwy Rhyolite Formation. Comparison of the tuff with other ash-flow tuffs of the 1st Eruptive Cycle is considered below.

Gwern Gof Tuff

The Gwern Gof tuff crops out in the core of the Tryfan Anticline (Figure 22) within a thick sequence of sedimentary rocks (see below). It comprises a non- to weakly welded acidic ash-flow tuff, 30–45 m thick, overlain by cross-laminated, well-bedded, coarse- to fine-grained, tuffaceous sandstone, locally up to 20 m thick. The latter formed by the local reworking of the underlying ash-flow tuff. The full extent and provenance of the primary ash-flow is unknown. However, its possible correlatives (Penamnen Tuffs and Clogwyn Gottal Tuff; Appendix 1, Table 4) (British Geological Survey, 1989) are widely distributed in eastern central Snowdonia.

Plate 9   The lower rhyolite (R) and ash-flow tuff (T) members of the Braich tu Du Volcanic Formation separated by a thin sequence of marine siltstone, sandstone and basic tuffite (S) on the north-east slopes of Nant Ffrancon [SH 648 621]. The columnar joints in the welded tuff (T) are perpendicular to its base which dips steeply to the right.

Interpretation of the facies of the underlying sediments (Orton, 1988a,b) indicates that the tuff was emplaced in shallow-marine conditions on a westerly facing palaeoslope. The base of the tuff is even and undisturbed where it rests on siltstones and mudstones but is locally irregular where it overlies sandstones. The basal 2 m of the tuff is commonly well cleaved. Siliceous nodules are abundant throughout the tuff.

The ash-flow tuff comprises shards and crystals (up to 5 per cent) of albite and quartz in a matrix of microcrystalline quartz and feldspar. The shards display a weak normal grading and are predominantly replaced by a quartzofeldspathic aggregate which is slightly coarser than that of the matrix. A few coarser shards and pumice fragments are replaced by chlorite and carbonate, and shards in the basal zone are commonly replaced by a micaceous aggregate. Garnet crystals are sparsely scattered throughout.

The ash-flow tuff (Appendix 1, Table 4) plots in the rhyodacite/ dacite field of the Zr/TiO_2 vs. Nb/Y diagram (Figure 24). Their geochemical discriminance with respect to other ash-flow tuffs of the 1st Eruptive Cycle is considered below.

The overlying tuffaceous sandstones are characterised by an abundance of albite crystals (<45 per cent) and lesser amounts (about 10 per cent) of quartz crystals and coarse (up

RHYOLITE

Blocky, autobrecciated top

Flow folding and local ramping

Even foliation, accentuated by silicification

0 Metres 10

Silicified foliated tuff with small-scale flow folding

Silicified, foliated tuff

Fiamme-rich tuff

Bedded acid tuff
Basaltic tuffite

Plate 10   Section through the ash-flow tuff member of the Braich tu Du Volcanic Formation with details of its texture. Locality as Plate 9.

to 1.8 m) and fine shard fragments, all derived from the underlying tuff.

Sedimentation within the 1st Eruptive Cycle (Llewelyn Volcanic Group)

The volcanic formations of the early phase of the 1st Eruptive Cycle contain local thin sedimentary intercalations. However, a thick sedimentary sequence overlies the volcanic formations of the early phase of activity, envelops the Gwern Gof Tuff, and underlies the Capel Curig Volcanic Formation which represents the final climactic phase of volcanism (Figure 6).

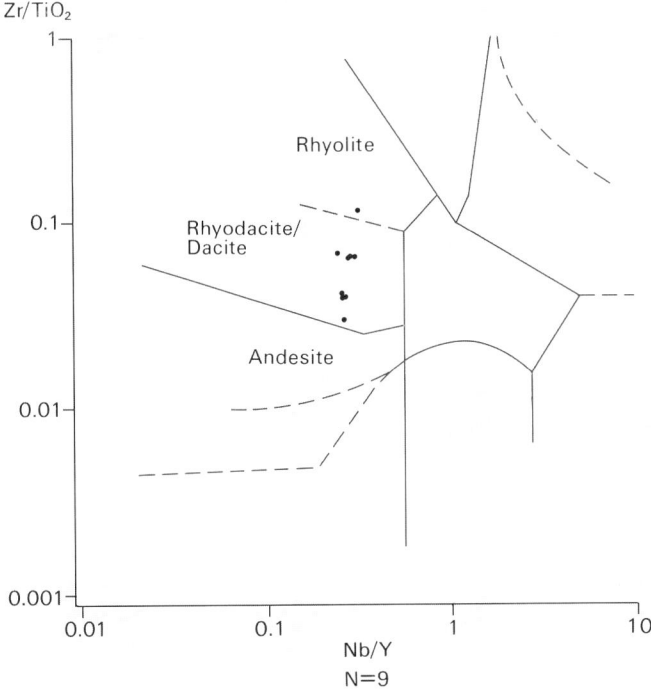

Figure 21 Braich tu du Volcanic Formation. Zr/TiO_2 vs. Nb/Y diagram.

This sedimentary sequence varies in thickness from less than 50 m to over 500 m and, in contrast to the rather monotonous succession of mudstones and siltstones underlying the 1st Eruptive Cycle volcanic rocks, its composition reflects the periodic input of large volumes of coarse clastic sediment. Facies variations resulted from the complex interplay of alluvial, fluvial and shallow-marine sedimentary environments, with provenance and palaeoslope directions varying through time. Volcanic activity was relatively low.

In the north, where the sequence is thinnest (Figure 11), shallow-marine siltstones, which overlie the early phase volcanic rocks, are superseded by up to 40 m of poorly sorted, heterolithic, pebble conglomerates. These are interpreted as alluvial, fan-base, sheet-flood deposits (Howells and Leveridge, 1980). Locally, in the top few metres, the beds are reddened by haematite. Southwards, the conglomerates thin and pass up into sandstones and laminated siltstones of fan-delta and lacustrine origin respectively (Howells and Leveridge, 1980). These in turn are overlain by fluvial sandstones deposited in a high-energy regime indicative of braided stream deposits. The north to south facies and thickness variations, and palaeocurrent indicators in this northern part of the outcrop are consistent with deposition on a gentle, southerly dipping palaeoslope, coarse debris being supplied in response to intermittent uplift in the north and west.

The outcrops further to the south, in the Tryfan anticline (Figure 22), are linked to those of the northern crop by the Gwern Gof Tuff (Plate 11). In the north, the tuff lies a few metres above the trachyandesitic outflow tuff of the Foel Fras Volcanic Complex and about 200 m below the base of the Capel Curig Volcanic Formation. However, in the core of the Tryfan anticline, it is underlain by at least 150 m of sedimen-

tary rocks and, in the eastern limb of the anticline, it is overlain by almost 400 m of sedimentary rocks.

Sedimentary rocks below the Gwern Gof Tuff in the core of the Tryfan anticline (Plate 11) comprise shallow-marine siltstone and sandstone, locally containing a brachiopod fauna dominated by *Rostricellula* assemblages. In the lowest exposures, two beds of acid tuff and tuffite, up to 10 and 25 m thick respectively, are tentatively correlated with the Braich tu du Formation, the southernmost expression of acidic volcanism during the early phase of activity in the 1st Eruptive Cycle. The association of these tuffs and tuffites with fossiliferous siltstones suggests emplacement in a shallow-marine environment, the finer units by fall out and the coarser units by gravity flows of volcaniclastic debris. Above, two coarsening-upward cycles of siltstone to sandstone (Figure 22; Plate 13A and B), each representing an episode of delta progradation separated by delta abandonment sequences, have been identified (Orton, 1988a,b). Magnetite-rich sandstones (Plate 12) associated with delta abandonment facies indicate significant nearshore wave energy. These sandstones, together with similar magnetite-rich sandstones higher in the sequence, are considered (Evans and Greenwood, 1988) to be the cause of the aeromagnetic anomaly over Snowdonia. Lateral facies variations and palaeocurrent indicators imply sediment provenance and progradation of the deltas from the east, the effusive volcanics of the early phase of the 1st Eruptive Cycle forming a relatively stable, subaqueous ridge to the north-west on which a thin contemporaneous sequence of muds and silts was deposited.

The Gwern Gof Tuff is generally overlain by siltstones with some thin sandstone beds and acidic ash-fall tuff beds. The occurrence of *Rostricellula* within the siltstones above the Gwern Gof Tuff indicates the persistence of the shallow-marine conditions. Above is a succession, up to 250 m thick (Plate 14), dominated by medium- and coarse-grained sandstones, with grey siltstones forming intercalations between major sandstone units and thin partings between individual sandstone beds. Orton (1988a,b) interprets the sequence as two successive, south and south-easterly prograding, braidplain delta systems fed by the alluvial fans that developed in the north (Figure 23). Transgressive shoreface facies form the uppermost part of both cycles in the eastern limb of the Tryfan anticline. Locally, magnetite-rich sandstones are common with magnetite distributed both along wave-swash lamination and as disseminations.

Further to the south-east, the uppermost part of the sedimentary interval is exposed below the Capel Curig Volcanic Formation in the core of the Capel Curig anticline (Figure 23). Here, interbedded laminated mudstone and fine-grained siltstone, regarded as marine shelf deposits (Francis and Howells, 1973) are overlain by a prograding braidplain coastline sequence (Orton, 1988a,b). At the south-west closure of the anticline, this sequence, comprising about 30 m of medium- to coarse-grained sandstones with trough-, planar- and swaley-cross-stratification, reflects the interaction between fluvial and marine processes at the southernmost termination of the braidplain. Large wave ripples (Plate 13C) and amalgamated sets of hummocky cross-stratification (Plate 13D) have been described by Fritz and Axelrod (1987) and Orton (1988a,b) and indicate high energy, shallow water conditions on a marine tidal flat.

Plate 11 View of Ddeugwm Ridge with a well-exposed sequence of sandstones and siltstones (S) underlying the Gwern Gof Tuff (GGT) in the core of the Tryfan Anticline on the east side of Cwm Tryfan [SH 675 595].

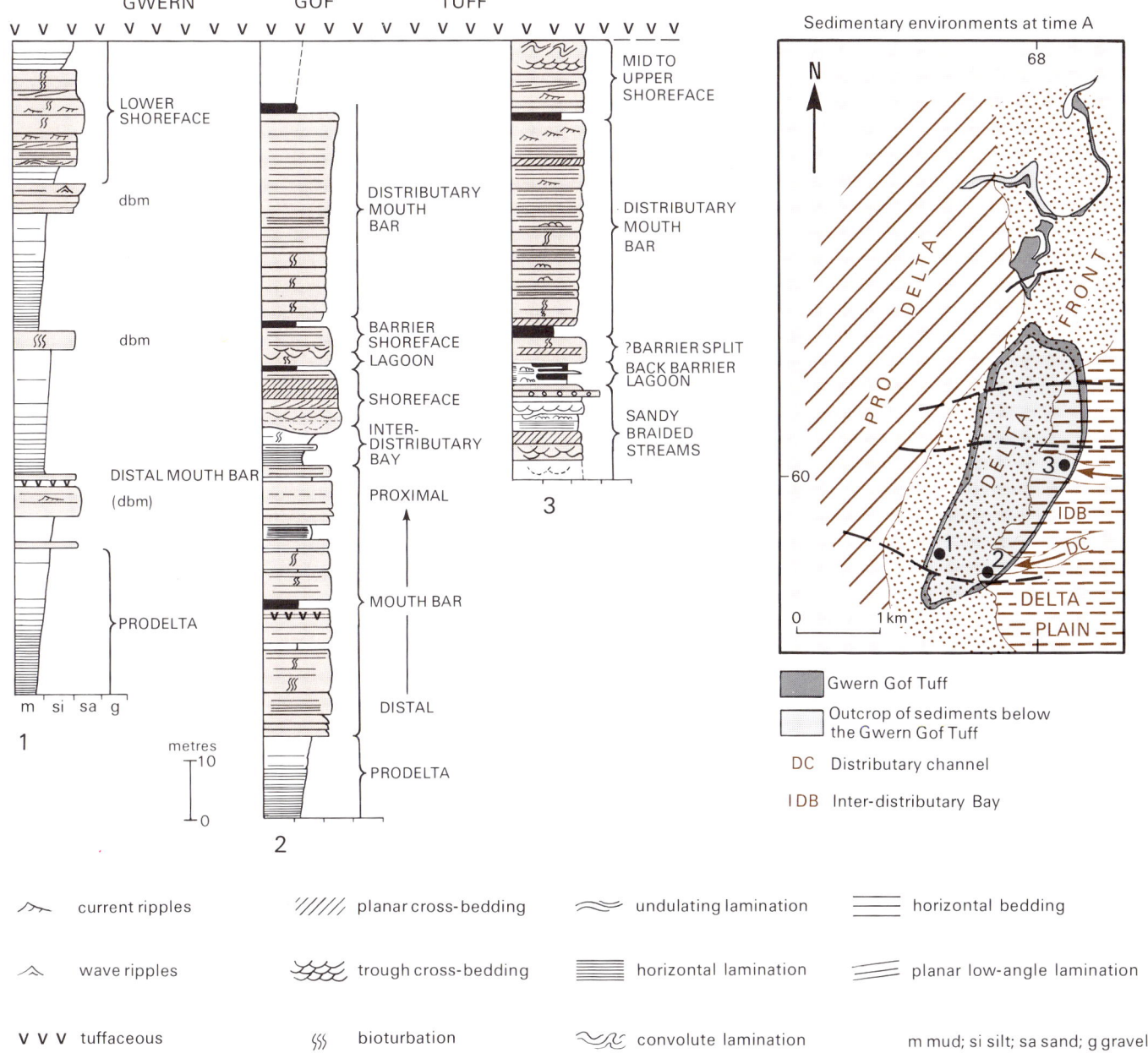

Figure 22   Sedimentological and environmental interpretation of sequences below the Gwern Gof Tuff in the core of the Tryfan anticline (after Orton, 1988a).

Concentrations of magnetite along internal laminae are common.

In summary, sedimentation in the interval between the early and late phases of volcanic activity of the 1st Eruptive Cycle indicates that, initially, marine conditions persisted over the main outcrop of the early phase volcanic rocks while, to the south-west, fan delta systems prograded westwards and the Gwern Gof Tuff was emplaced on the westerly dipping palaeoslope. Later, uplift in the north and west led to the development of subaerial alluvial fan sequences in the north passing southwards into braidplain fan deltas with shallow-marine tidal flats in the south and south-east. Palaeoslopes during this period were predominantly to the south and south-east.

Capel Curig Volcanic Formation

The climactic volcanism of the 1st Eruptive Cycle was dominated by voluminous acidic ash-flow eruptions represented by the Capel Curig Volcanic Formation which crops out extensively in northern and eastern Snowdonia (Figure 25) (Howells et al., 1979; Howells and Leveridge, 1980). The tuffs were erupted from three centres (Figure 26), two in a subaerial environment in the north and a third in the south, which was to a large extent subaqueous. The ash-flows which were erupted subaerially transgressed southwards across a shoreline into a subaqueous environment. The tuffs at the extreme north-eastern limit of the outcrop were also emplaced subaqueously. In the subaerial environment a continuous se-

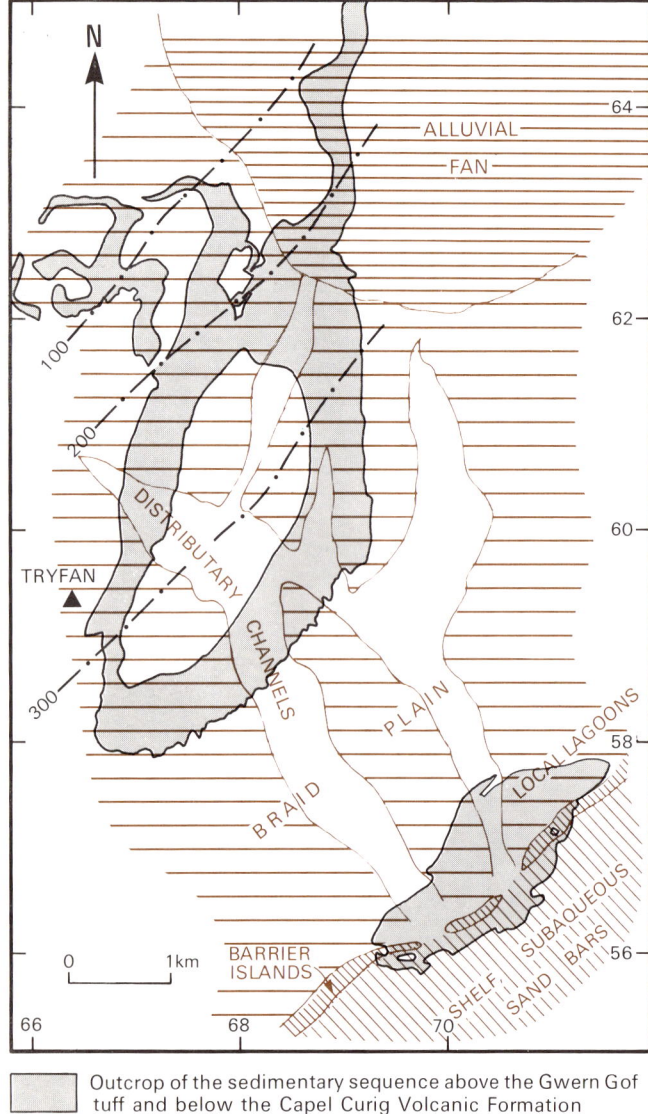

Outcrop of the sedimentary sequence above the Gwern Gof tuff and below the Capel Curig Volcanic Formation

Thickness of sequence in metres

Figure 23 Environmental interpretation of the sequence above the Gwern Gof Tuff and below the Capel Curig Volcanic Formation (after Orton, 1988a).

quence of tuffs accumulated, whereas in the subaqueous environment marine sedimentary rocks are intercalated within the tuffs.

Four volcanic members are distinguished, of which the lowest two, the 1st and 2nd members, are the most voluminous (about 25 km³) and extensive. These lithologically similar members are single, primary, welded ash-flow tuffs, up to 150 m and 130 m thick respectively (Figure 25). Their fabrics vary from eutaxitic to parataxitic and in the southern outcrop they are locally rheomorphosed.

In the subaerial environment, towards the north, the bases of both the 1st and 2nd members are even and nonwelded. Above, the planar welding foliation is accentuated by lenses of recrystallised quartz, and the matrix is overprinted by a

mosaic of recrystallised platy feldspar crowded with fine inclusions. In the subaqueous environment, in the south and in the extreme north of the outcrop, chloritic fiamme are common. Here, both members are welded to their basal contacts, which are extremely irregular, with flames of sediment intruding, up to 40 m, the body of the tuff (Figure 27). Large load casts of tuff protrude into the underlying sediment and are locally detached (Figure 27). These irregular, transgressive bases are interpreted (Francis and Howells, 1973) as resulting from the rapid emplacement of the hot ash-flow tuff on unlithified, water-saturated sediments, a process probably facilitated by fluidisation of the sediment at the contact (Kokelaar, 1984). The greater abundance of siliceous nodules (Plate 15D) in the tuffs in the subaqueous setting probably also reflects interaction of the tuff and sea-water. Many are centred on lithophysae, possibly produced by the upward migration of water vapour and volatiles entrapped at the interface between tuff and wet sediment. Such migration caused rheomorphic disruption of the welding fabrics, both on the microscopic and mesoscopic scale (Howells and Leveridge, 1980).

The most distal facies of these ash-flow tuffs in the subaqueous environment, in the southernmost outcrops (Figure 25), is represented by isolated pods of welded ash-flow tuffs (Howells et al., 1978; Campbell, 1983; Howells et al., 1985a) (Plate 15C). These pods are similar to those detached from transgressive lobes at the bases of the continuous ash-flow tuff sheets in the subaqueous environment immediately to the north. They are interpreted (Howells et al., 1985a) as having been detached, at a change of slope, from the front of the ash-flow and then to have travelled into deeper parts of the basin. Their transport was probably facilitated by film boiling (Leidenfrost Effect) at the tuff/water interface. In some instances the pods, on coming to rest, retained sufficient heat to fluidise, and collapse into, the wet sediment substrate (Figure 27).

Branney (1986) reinterpreted the pods as having resulted from the drainage of the parent ash-flow sheet during its waning stages of transport and emplacement, thus requiring no change of slope. Cas and Wright (1987) speculate, on the basis of published descriptions, that the parent ash-flow passed over the surface of the sea and that the pods represent tuff which eventually settled down through the water column.

The 3rd Member, 0–170 m thick, comprises a restricted accumulation of primary welded ash-flow tuffs (Figures 26B). Their distribution indicates that they were erupted from a centre in the subaerial environment and transgressed only a limited distance into the sea. The lowest tuff, up to 50 m thick, is the most extensive, with the tuffs above forming well-defined planar-bedded flow units up to 4 m thick. The tuffs are characterised by the distinctive platy feldspar mosaic recrystallisation recognised in the subaerial expression of the lower two members.

The 4th Member, 0–200 m thick, is similarly restricted in distribution (Figure 26B). It comprises slumped tuffs, block and ash-flow tuffs, debris flows which are locally reworked, accretionary lapilli tuffs (Plates 15B, 15E and 16A) and a few thin primary ash-flow tuffs. They occur in the subaqueous environment to the south of the shoreline established at the commencement of their eruption. However, the extensive occurrence of accretionary lapilli, especially towards the top of the member, indicates that at least part of the volcanism was subaerial. The distribution of the accretionary lapilli, and the thickness and facies variations of the slumped and reworked tuffs, define the position of the eruptive centre (Figure 26B) which is marked by a late-stage intrusive rhyolite. Although the centre was initiated subsequent to the emplacement of the 1st Member it only reached its peak of activity following the eruption of the 3rd Member.

Within the subaqueous environment, sediment intercalations occur above both the 1st and 2nd members. These sediments thicken to the south (Figure 23) and, in both instances, the primary ash-flow tuffs show only limited reworking which is restricted to the topmost 1 m of the tuffs. However, the sediments above comprise mudstones, siltstones and locally cross-bedded sandstones, with no indication of volcaniclastic debris derived from the underlying tuffs. They contain marine shelly faunas, including *Dinorthis*, which favoured shallow, high-energy regimes (cf. Pickerill and Brenchley, 1979), dalmanellids, which are typical of quieter depositional regimes and, further south-east, faunas typical of a deeper water, outer-shelf environment.

The distribution of tuffs of this final phase of the 1st Eruptive Cycle, and the associated sediments, indicate that they were profoundly influenced by subsidence south of a line, possibly a hinge zone above a deeper fault plane, which lay close to the shoreline (Howells and Leveridge, 1980). It is proposed that similar subsidence caused the marked thickness variation and facies changes of the strata subjacent to the acidic ash-flow tuffs and continued to exert a similar influence on sedimentation following the 1st Eruptive Cycle.

SUBAQUEOUS WELDING   The model of the evolution of the Capel Curig Volcanic Formation implies that ash flows can transgress from a subaerial to subaqueous environment and retain sufficient heat to weld on emplacement (Figure 27). In the subaqueous environment, the presence of a welding fabric in the tuffs immediately adjacent to wet sediment suggests that in this situation welding was facilitated by water vapour. The locally developed micaceous replacement of shards at the contacts may also be a result of reaction with water, with hydrolisation alteration being accompanied by desilicification and the resultant silica migration enhancing nodule development. If welding takes place at a lower temperature when water vapour is present, this temperature difference may be a further factor in controlling the recrystallisation features described.

Theoretical work by Sparks et al. (1980) suggests that welding might take place very readily in the presence of water. They argue that the viscosity of a flow decreases in the presence of steam in pore spaces, under pressure equivalent to hydrostatic pressure, and the cooling, due to the incorporation of steam, is limited. In addition, the presence of water vapour increases diffusivity which facilitates the welding process.

PETROGRAPHY   The primary ash-flow tuffs comprise shards, crystals of feldspar and quartz, fiamme and lithic clasts in a fine-grained matrix of quartz, feldspar, sericite and chlorite probably after vitric dust.

Shards are predominantly of slender cuspate form with fewer squat and angular fragments. They are entirely

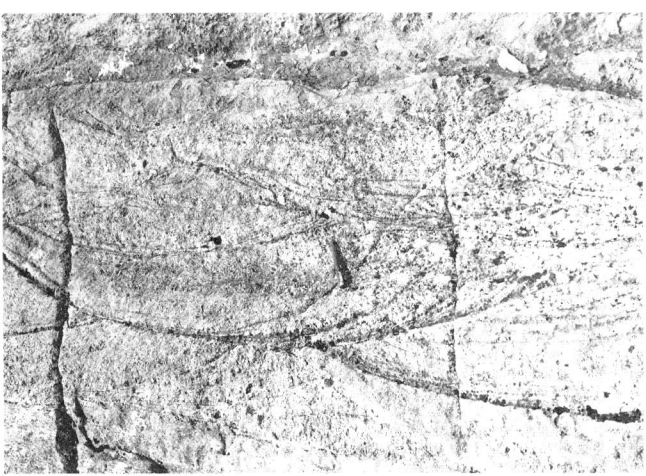

Plate 12   Magnetite-rich layers in trough cross-bedded sandstones above the Gwern Gof Tuff on the western limb of the Tryfan Anticline in Cwm Tryfan [SH 670 600].

devitrified, pseudomorphed by quartzofeldspathic mosaics, with their peripheries commonly accentuated by sericite shreds. In places, particularly near the base of flows in the southern outcrops, the shards are replaced by a fine micaceous aggregate.

Crystals of feldspar and quartz are common. The feldspars

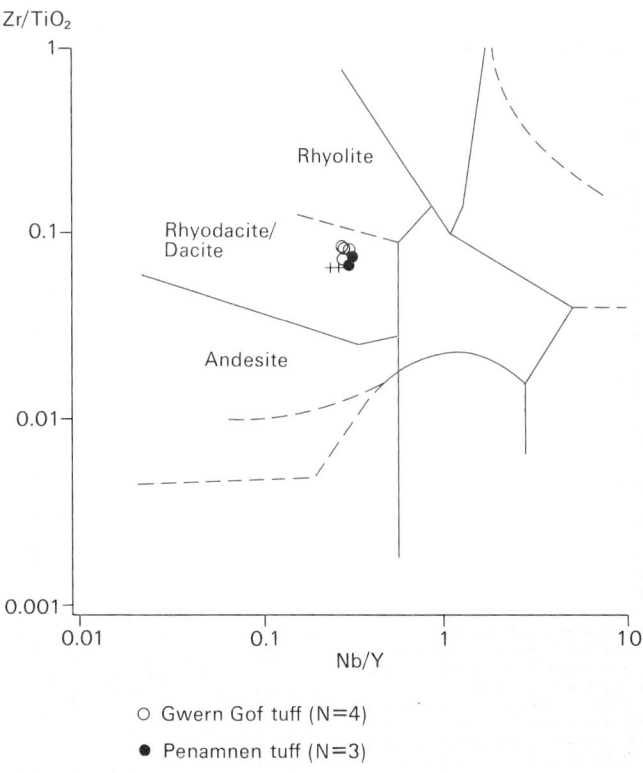

○ Gwern Gof tuff (N=4)

● Penamnen tuff (N=3)

+ Clogwyn gottal tuff (N=2)

Figure 24   Gwern Gof, Penamnen and Clogwyn Gottal tuffs, Zr/TiO_2 vs. Nb/Y diagram.

Plate 13   Sedimentary structures in sediments below and within the Capel Curig Volcanic Formation.

A   Trough cross-stratification and horizontal lamination at the base of a distributary channel in a delta-plain sequence (Orton, 1988) beneath the Gwern Gof Tuff, cropping out on Ddeugwm ridge [SH 677 594] on the eastern limb of the Tryfan Anticline. (Photograph by G Orton).

B   Thin (5 – 10 cm), normally graded beds of medium- to fine-grained sandstone, interpreted (Orton, 1988) as stream effluent flood deposits of a proximal mouth bar in a fan-delta sequence, cropping out below the Gwern Gof Tuff on Ddeugwm ridge [SH 677 594], eastern limb of the Tryfan Anticline.

C   Large wave ripples with sinuous to straight crestlines and varying asymmetry of ripple profiles in coarse-grained sandstone cropping out below the lowest tuff member of the Capel Curig Volcanic Formation at Cwm Clorad Isaf [692 562]. (BGS Photograph A14590).

D   Amalgamated sets of hummocky cross-stratification transitional to scour-and-fill-type structures in sandstones, interpreted as upper shoreface deposits (Orton, 1988), below the lowest tuff member of the Capel Curig Volcanic Formation at Cwm Clorad Isaf [692 562] in the Capel Curig Anticline. (Photograph by G Orton).

E   Normally graded beds of fine-grained sandstone to siltstone, interpreted (G Orton, 1988) as inner shelf deposits reflecting waning storm events, cropping out between the first and second tuff members of the Capel Curig Volcanic Formation, north-west limb of the Capel Curig Anticline. (Photograph by G Orton).

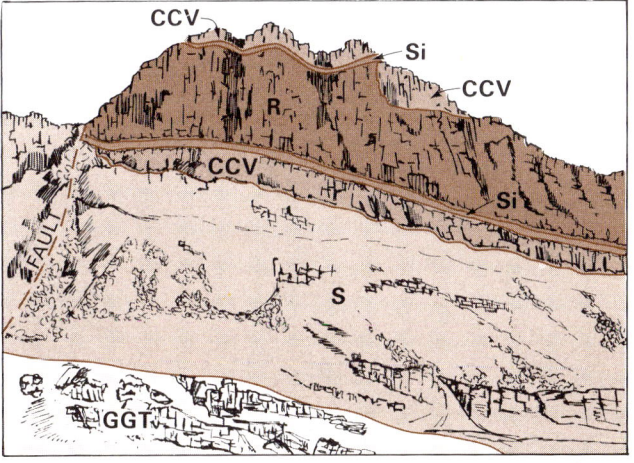

Plate 14   The east-facing flank of Tryfan [SH 665 595] with ash-flow tuffs of the Capel Curig Volcanic Formation (CCV), rhyolite (R) and thin siltstone intercalations (Si) forming the upper crags and overlying a thick sequence of sediments above the Gwern Gof Tuff (GGT).

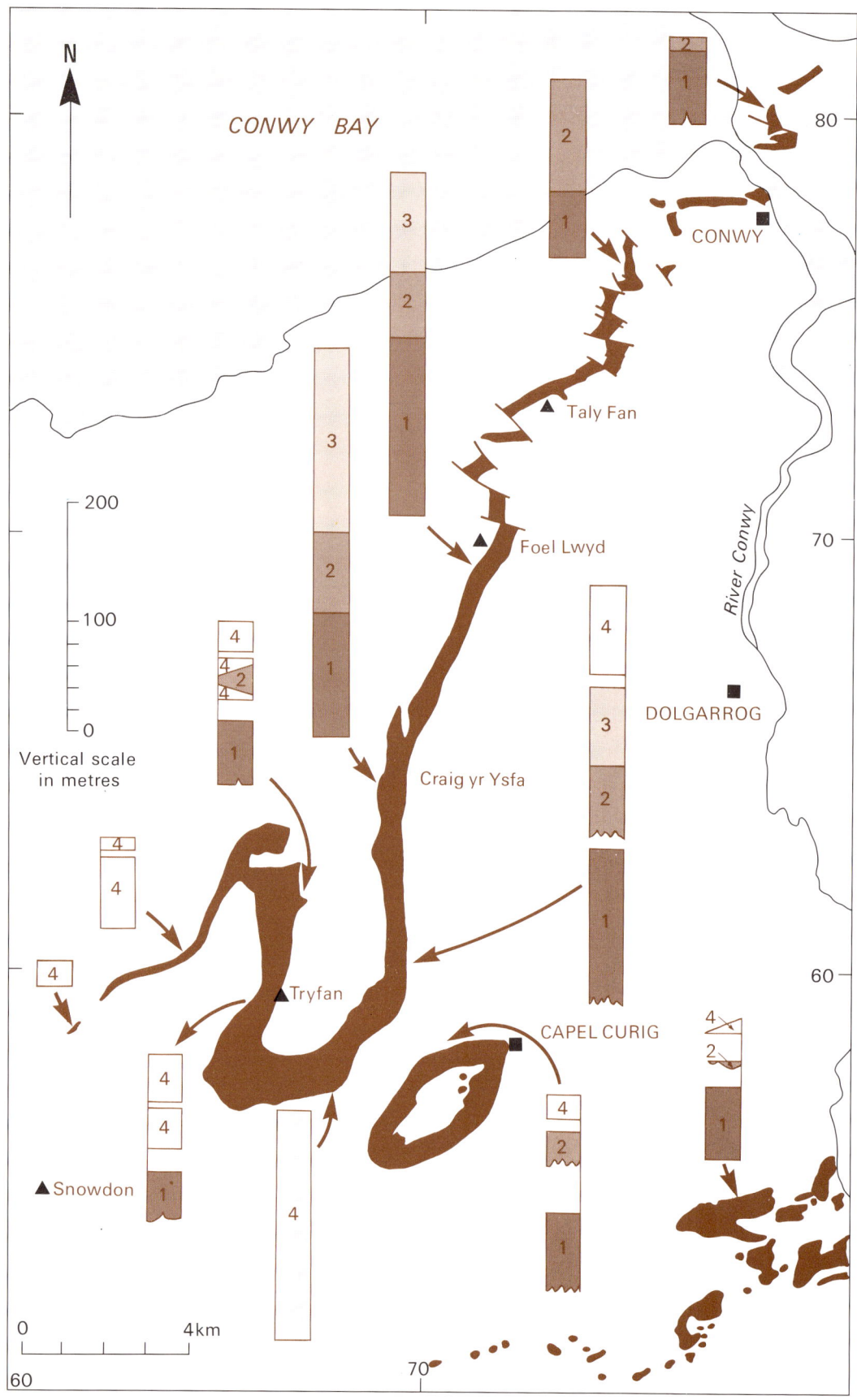

Figure 25 Outcrop and measured sections of the Capel Curig Volcanic Formation. Numbers on sections refer to 1st, 2nd, 3rd and 4th Members.

Figure 26 A. Interpretation of the depositional environments of the 1st and 2nd members of the Capel Curig Volcanic Formation, showing flow directions and the distribution of isolated pods of the 1st Member.
B. and C. Distribution of the 3rd and 4th members of the Capel Curig Volcanic Formation (after Howells and Leveridge, 1980).

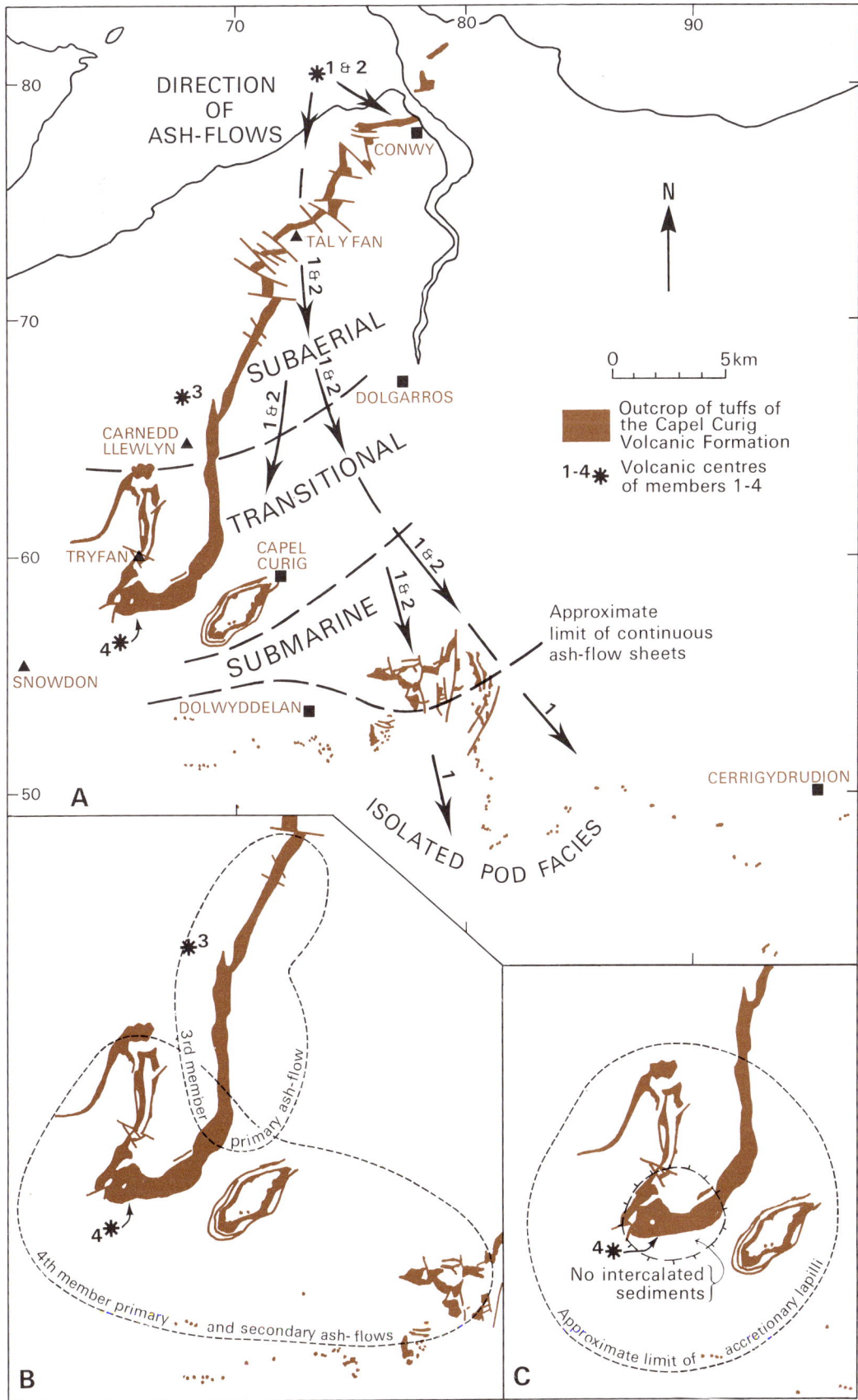

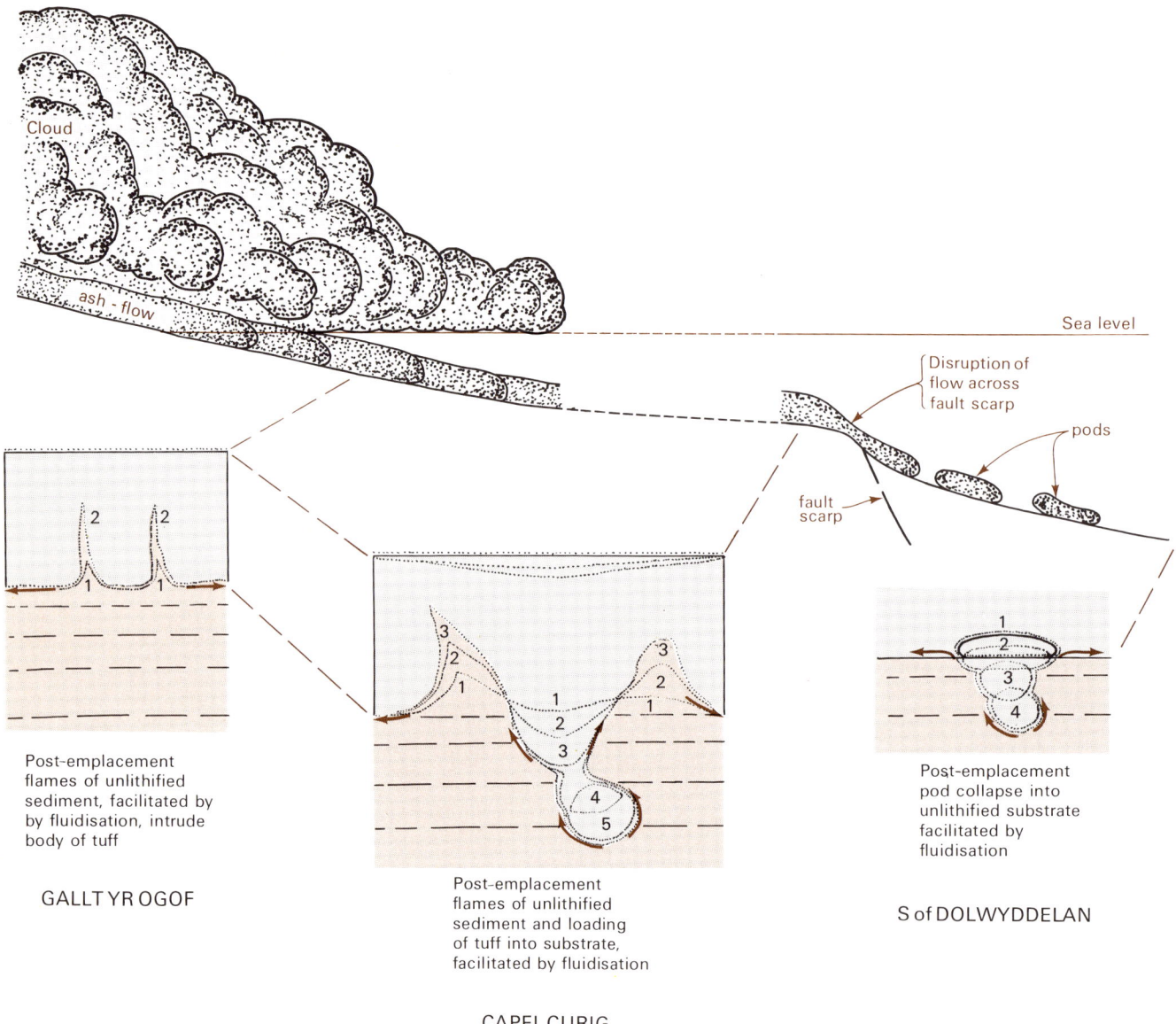

Figure 27   Model for the emplacement of the 1st Member, primary ash-flow tuff, Capel Curig Volcanic Formation.

are most abundant and dominantly of albite-oligoclase composition. They occur as rounded euhedral, resorbed and fragmented crystals, up to 4.3 mm in length. Some crystals are embayed on their fractured inner surfaces, indicating original resorbed cores. The feldspars are generally sericitised to some degree and locally the alteration is complete. Elsewhere, the feldspars are turbid with numerous small inclusions of iron oxide and bubbles.

Subhedral, resorbed and fragmented quartz crystals, up to 1.5 m in diameter, are less common. In the lowest two members, garnet occurs sporadically in small, euhedral and rounded crystals up to 0.8 mm in diameter, invariably embayed and sieved with chlorite, and with iron ore accentuating the rims and fractures. The garnets are pink, slightly pleochroic, and are consistently associated with coarse quartz

segregation in the intensely welded parts of the flows, suggesting a secondary origin.

Sphene, skeletal ilmenite, magnetite, epidote and apatite are common accessories.

Exotic lithic clasts are a minor constituent concentrated mainly in the basal zone. They include mudstone, siltstone, tuff and both acid and basic intrusive rocks. Higher in the flows, clasts are mainly of pumice although they are commonly obscured by recrystallisation. In the subaqueous environment chloritic fiamme are common.

GEOCHEMISTRY   Analytical data for the Capel Curig Volcanic Formation primary acidic ash-flow tuffs are shown in Appendix 1, Table 4. On a plot of Zr/TiO$_2$ vs. Nb/Y (Figure 28) the analyses mainly straddle the rhyolite-rhyodacite/

A

0   5cm

B

0   5cm

C

D

E

Plate 15   The Capel Curig Volcanic Formation

A   Basal welded tuff of the 2nd Member emplaced beyond a shoreline, Capel Curig Anticline [7135 5809]. The tuff is in irregular contact with mudstone at the bottom left of the photograph and incorporates dark patches of mud.

B   Accretionary lapilli tuffs at the top of the 4th Member, Tryfan [6621 5950].

C   Prominent outcrop of welded tuff at Carreg Alltrêm [SH 740 507] forming part of an isolated pod of the 1st Member.

D   Siliceous nodules at the top of the 1st Member at Tal-y-Llyn Ogwen [SH 688 611]. (BGS Photograph L2380).

E   Accretionary lapilli tuffs of the 4th Member at Gallt yr Ogof [SH 686 582]. (BGS Photograph L2389).

A
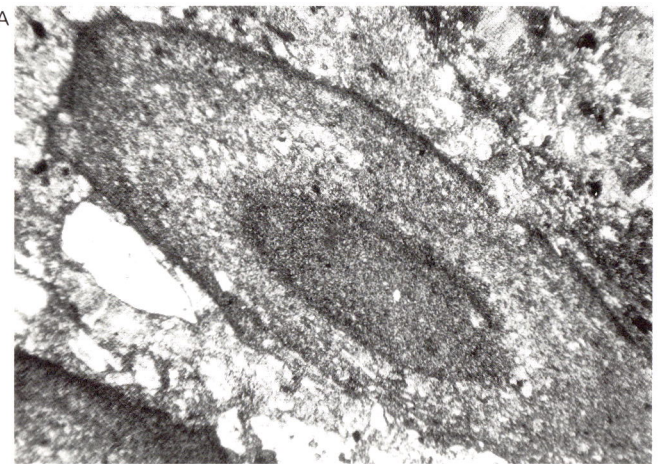

B
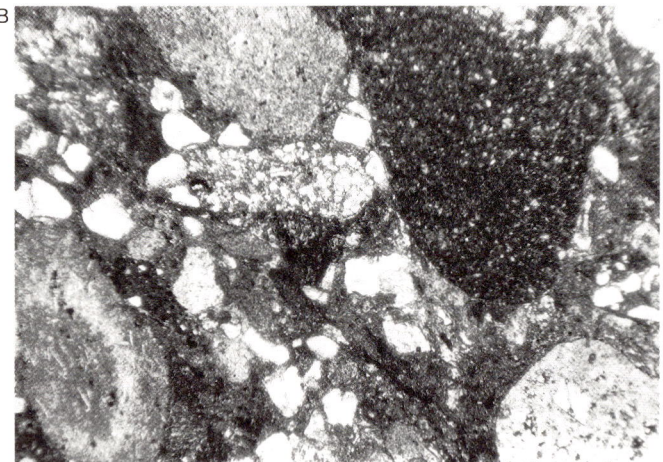

C
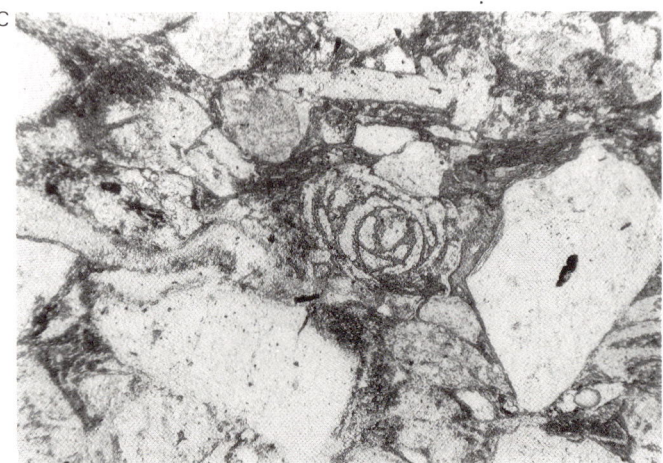

Plate 16 Photomicrographs, reworked tuffs and volcaniclastic sandstones

A BP 150. Accretionary lapillus in reworked tuffs of the 4th Member, Capel Curig Volcanic Formation, Tryfan, ppl. [SH 6592 5895].
B RN 1105. Volcaniclastic sandstone. Well-rounded clasts of microlitic basalt, acid tuff and siltstone in finer-grained quartz, feldspar, sericite matrix. Sub-Pitts Head Tuff Formation, Moel Hebog, ppl. [SH 5677 4721].
C KB 1211. Volcaniclastic sandstone overlying exogenous rhyolite dome (A1 composition). Clasts of recrystallised rhyolite, perlitic rhyolitic glass and subhedral feldspar crystals occur in silty matrix. Cwm Eigiau Formation, ppl. [SH 6608 5211].

dacite field boundary and the tuffs include the most evolved rocks of the 1st Eruptive Cycle.

The four ash-flow tuff members of the formation are sufficiently distinct compositionally in terms of their immobile trace elements to enable their discrimination. This is most clearly demonstrated on a plot of Nb vs. Zr (Figure 29), with considerable resolution of the sample populations of all four members, although there are slight overlaps between the field for the 2nd Member and the fields of the other three members. Additionally, the geochemistry supports the correlation of the isolated pods of welded tuff with the 1st Member.

The 1st Member is markedly different from the 2nd, 3rd and 4th members (Appendix 1, Table 4) in its relatively lower concentrations of TiO_2, Ga, and Zr, its commonly lower concentrations of Nb, Y, La and Ce, and to a lesser extent Th. It particularly differs from the 3rd and 4th members in its higher Nb/Zr, Th/Zr and Ce/Zr ratios and further differs from the 4th Member in having higher ratios of Zr, Nb and Y to TiO_2. Resolution with the 2nd Member is the most problematic though this member has, in addition to generally lower Nb/Zr ratios, slightly lower Th/Zr ratios. The lower concentrations of Nb, Zr and Th indicate that the 1st Member represents a less evolved magma than the 2nd, 3rd and 4th members.

The 2nd Member differs significantly from the 4th Member (Appendix 1, Table 4) in its generally lower Zr and Y contents, and higher Th content. Also it has higher

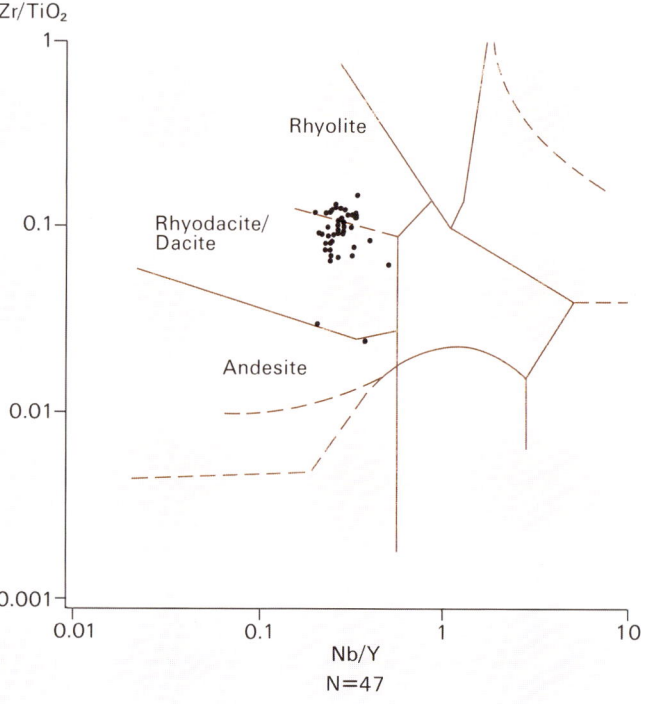

Figure 28 Capel Curig Volcanic Formation, Zr/TiO_2 vs. Nb/Y diagram.

TiO$_2$/Zr, Y/Zr and Ga/Zr ratios and a lower Y/TiO$_2$ ratio. Discriminance with respect to the 3rd Member is poorer, with the 2nd Member differing only in its generally lower Zr, Nb and TiO$_2$ contents and higher TiO$_2$/Zr ratio. Strong separation of the two fields is seen, however, on plots of Nb/Y, Zr/Th, Zr/Y, Zr/TiO$_2$, Zr/La and Zr/V.

The 3rd Member differs from the 4th Member (Appendix 1, Table 4) in having higher Y and Nb, slightly higher Th and Ga, and lower V. Ratios of Nb/Zr, Ga/Zr, Y/TiO$_2$ and, to a lesser extent, Y/Zr and Nb/TiO$_2$ are also higher.

Analyses of two samples of rhyolite (TL 90, 91) (Appendix 1, Table 4), intruded close to the eruptive centre of the 4th Member, contain similar concentrations of Nb, Zr, Th, La and Ce to those in the 2nd, 3rd and 4th members and the Bwlch y Cywion microgranite (p.00).

Geochemical comparison of the four members of the Capel Curig Volcanic Formation with other ash-flow tuffs of the 1st Eruptive Cycle (Appendix 1, Tables 3 and 4) suggests that compositionally, both the Gwern Gof Tuff and the ash-flow tuff of the Braich tu du Volcanic Formation are similar in many respects to the 1st Member (Figure 29). Consequently, both can be readily resolved from the 3rd and 4th members and to a lesser extent from the 2nd Member, but not with respect to the 1st Member, given the small sample sizes available. For example, the Gwern Gof Tuff has lower TiO$_2$, Th, Zr, and Ce. It differs from the 1st Member in having a higher Ga content and significantly higher P$_2$O$_5$ content

(X = 0.257 wt%, N = 4 compared with 0.90 wt%, σn-1 = 0.029, N = 46 for the Capel Curig Volcanic Formation) in which it also differs from the Braich tu du ash-flow tuff. It defines distinct fields relative to the 2nd, 3rd and 4th members, and to the Braich tu du ash-flow tuff on all plots with respect to Zr, and on plots of Th/Y and Th/TiO$_2$. Further separation compared with the 3rd and 4th members of the Braich tu du ash-flow tuff is seen on plots of Nb/Y, Nb/TiO$_2$, Nb/Th, Y/TiO$_2$ and TiO$_2$/V.

Intrusions related to the 1st Eruptive Cycle

A number of subvolcanic intrusions are spatially associated with the extrusive rocks of the 1st Eruptive Cycle (Figure 11). These are mainly of intermediate, andesite-dacite composition, with fewer rhyolites, and many display geochemical characteristics which suggest that they are comagmatic with the overlying extrusive rocks.

The intrusions can be broadly divided into three sets from north to south. A boss-like intrusion (Penmaenmawr) and two smaller plugs (Carreg fawr and Dinas) occur in the north (Figure 11). Petrographically, they are sparsely porphyritic microdiorites, comprising andesine laths (An$_{36-40}$), bastite pseudomorphs after enstatite, subhedral augite and hornblende with interstitial, partly granophyric intergrowths of quartz and alkali feldspar, and accessory biotite, ilmenite and apatite. Scattered autoliths of very fine-grained, pyroxene-rich microdiorite possibly represent a chilled facies. Secondary alteration is extensive, plagioclase being replaced by albite + prehnite, white mica ± pumpellyite ± calcite ± epidote, orthopyroxene by chlorite and both augite and hornblende by fibrous amphibole.

Three intrusions (Aber-Drosgl, Carreg y Gath, Gyrn) are clustered close to the small trachyandesite caldera of the Foel Fras Volcanic Complex (Figure 11). The largest intrusion (Aber-Drosgl) is dominantly a porphyritic microtonalite, comprising plagioclase prisms (<3 mm) contained in a groundmass of anhedral quartz and alkali feldspar (Or$_{68-74}$). The groundmass is locally granophyric. Augite, hornblende, biotite, ilmenite and apatite occur as accessories and are commonly concentrated in scattered autoliths of coarse plagioclase euhedra. Hydrothermal alteration is locally recognised, with the complete replacement of plagioclase by white mica ± carbonate and ferromagnesian minerals by chlorite. Elsewhere, alteration is more typically that of low greenschist facies metamorphism, with plagioclase replaced by phengite ± epidote ± stilpnomelane ± carbonate; augite and hornblende partially or completely replaced by actinolite ± chlorite; ilmenite by sphene and biotite by chlorite ± stilpnomelane.

In the smaller intrusions (Gyrn and Carreg y Gath) the textures are commonly granophyric, with phenocrysts (<5 mm) and microphenocrysts of both sodic plagioclase and microperthite. Euhedral hornblendes (up to 0.4 mm) form up to 4 per cent and most are unaltered.

The largest of the southerly group of intrusions, Bwlch y Cywion (Figure 11), consists of microgranite, with later marginal intrusion of rhyolite. The microgranitic texture comprises sodic plagioclase euhedra (<1.5 mm), subhedral microperthite and anhedral interstitial quartz. Primary biotite (<1 per cent) is locally replaced by chlorite and other

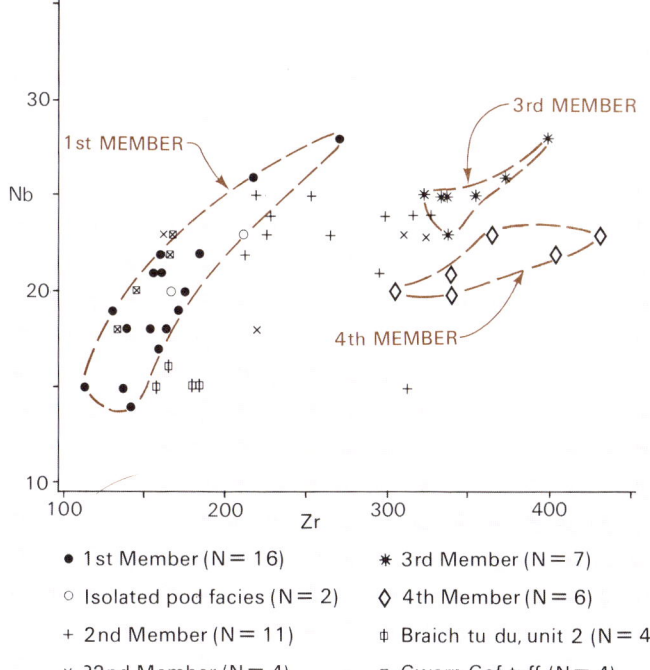

Figure 29 Capel Curig Volcanic Formation, Nb vs. Zr diagram, showing the correlation of the isolated pod facies with the 1st Member, the extent of discrimination of the 4 members and comparison with unit 2 of the Braich tu du Volcanic Formation and the Gwern Gof Tuff.

accessory minerals include apatite and ilmenite, with rare zircon and sphene. Both the smaller intrusions, Mynydd Perfedd and Talgau, are locally intensely altered, with white mica (± calcite) after plagioclase phenocrysts in a fine-grained intergrowth of white mica, chlorite, calcite and sphene, after original glass.

GEOCHEMISTRY   Selected analyses of the principal intrusions associated with the 1st Eruptive Cycle are presented in Appendix 1, Table 5. On a plot of Zr/TiO_2 vs. Nb/Y (Figure 30), Penmaenmawr and its associated intrusions Carreg Fawr and Dinas, together with Aber-Drosgl, plot in the andesite field. All other intrusions fall within the dacite/rhyodacite field, with Carreg y Gath near the andesite boundary and Bwlch y Cywion close to the rhyolite boundary.

Penmaenmawr, Dinas and Carreg Fawr are significantly depleted in Zr and P_2O_5 (Appendix 1, Table 5) relative to the other major intrusions. Its chondrite-normalised REE profile (Figure 13) is moderately LREE-enriched (La_{Ch}/Yb_{Ch} = 3.06), with a significant Eu anomaly and comparatively flat HREE pattern. Aber-Drosgl, by comparison, is relatively more enriched in total REE but its chondrite-normalised profile is otherwise similar. Aber-Drosgl is notable for its higher TiO_2 and P_2O_5 content than other intrusions (Appendix 1, Table 5) associated with the 1st Eruptive Cycle. The chondrite-normalised pattern for the Gyrn granophyre is very strongly LREE-enriched (La_{Ch}/Yb_{Ch} = 5.96), with a greater Eu anomaly than for the Penmaenmawr and Aber-Drosgl intrusions. Its overall REE content is not, however, very different from the latter two. Gyrn has broadly similar HFS and LREE contents and relative ratios to Carreg y Gath, Mynydd Perfedd and Talgau, although there are greater variations in their respective LIL element concentrations (particularly Th). Bwlch y Cywion differs from all of the other intrusions in its low TiO_2, very low P_2O_5 and high Th contents (Appendix 1, Table 5).

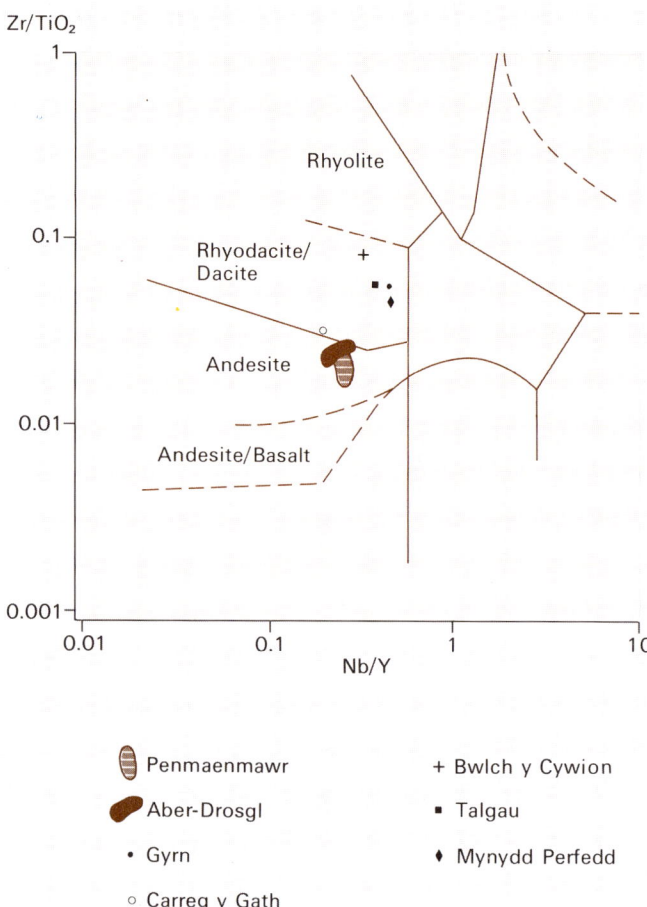

Figure 30   Intrusions associated with the Llewelyn Volcanic Group, Zr/TiO_2 vs. Nb/Y diagram (after Ball and Merriman, 1989).

PETROGENESIS OF THE 1ST ERUPTIVE CYCLE: LLEWELYN VOLCANIC GROUP AND RELATED INTRUSIONS

The Llewelyn Volcanic Group comprises extrusive volcanic rocks displaying a wide range of compositions when plotted on the Zr/TiO_2 vs. Nb/Y diagram (Figure 10) of Winchester and Floyd (1977). The Foel Grach Basalt Formation includes compositions ranging from subalkaline basalt to andesite (Figure 18), the Foel Fras Volcanic Complex from andesite to dacite (Figure 16), the Conwy Rhyolite Formation and the Braich tu du Formation predominantly dacite and rhyodacite (Figures 12 and 21) and the Capel Curig Volcanic Formation dacite/rhyodacite to rhyolite (Figure 28). Amongst the associated intrusions (Figure 30), Aber-Drosgl plots in the andesite field, Penmaenmawr straddles the boundary between the andesite and dacite/rhyodacite fields and the remainder plot in the dacite/rhyodacite field. On an AFM diagram (Figure 31) both extrusive and intrusive rocks plot as a broad and relatively continuous band, predominantly within the calcalkaline field but in part transitional to tholeiite, though this plot must be interpreted with caution because of the possible effects of element mobility, par-

ticularly of Na and K, during post-emplacement alteration (see Chapter 6).

The continuity of compositional variation suggests the possibility that the various intrusive and extrusive products were originally comagmatic and related by fractional crystallisation. Variation diagrams (SiO_2 index) of the same rocks (Ball and Merriman, 1989) clearly display curvilinear relationships, with negative correlations for the incompatible elements (e.g. Fe, Mg, Mn, P, Ti and, to a lesser extent, Al and Ga) and positive correlations for the incompatible elements (e.g. K, Rb and Th) with the exceptions of Zr and Nb whose distribution patterns are flat. These relationships are consistent with compatible element depletion and incompatible element enrichment as a result of fractionation of a common parent magma.

REE profiles for all of the rock units (except for the Braich tu du and the Capel Curig Volcanic formations, for which no REE data are available) are generally very similar (Figure 13), with pronounced LREE enrichment and gentle slopes towards the less enriched HREE, interrupted only by pronounced negative Eu anomalies. The Conwy Rhyolite Formation and the Gyrn granophyre have the most marked Eu anomalies, indicating that their genesis probably involved the greatest

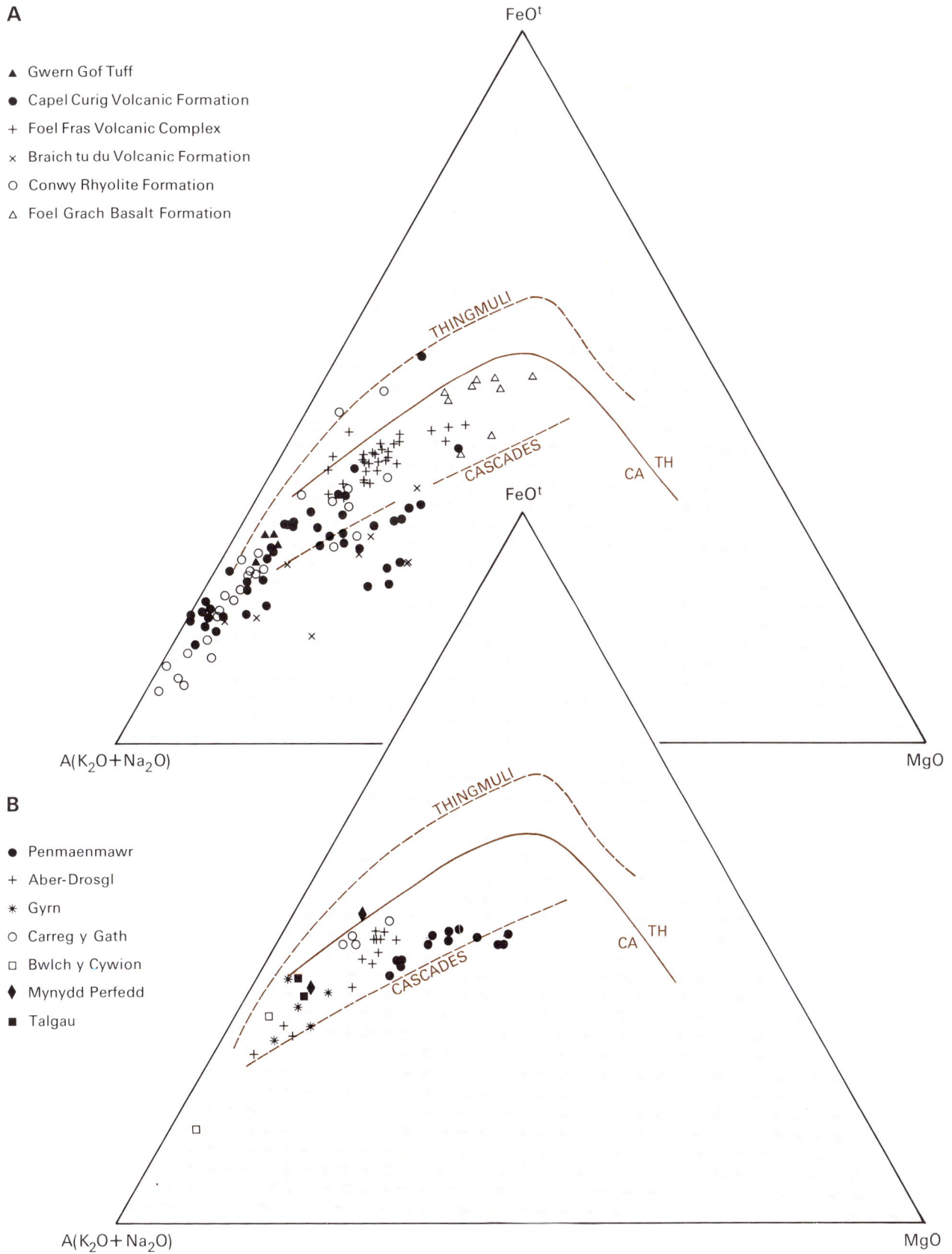

A

▲ Gwern Gof Tuff

● Capel Curig Volcanic Formation

+ Foel Fras Volcanic Complex

✕ Braich tu du Volcanic Formation

○ Conwy Rhyolite Formation

△ Foel Grach Basalt Formation

B

● Penmaenmawr

+ Aber-Drosgl

✳ Gyrn

○ Carreg y Gath

□ Bwlch y Cywion

◆ Mynydd Perfedd

■ Talgau

Figure 31 Llewelyn Volcanic Group and related intrusions, AFM diagram.

CA: Calcalkaline; TH: Tholeiitic.
A   Extrusive volcanic units.
B   Intrusions spatially related to the Llewelyn Volcanic Group.

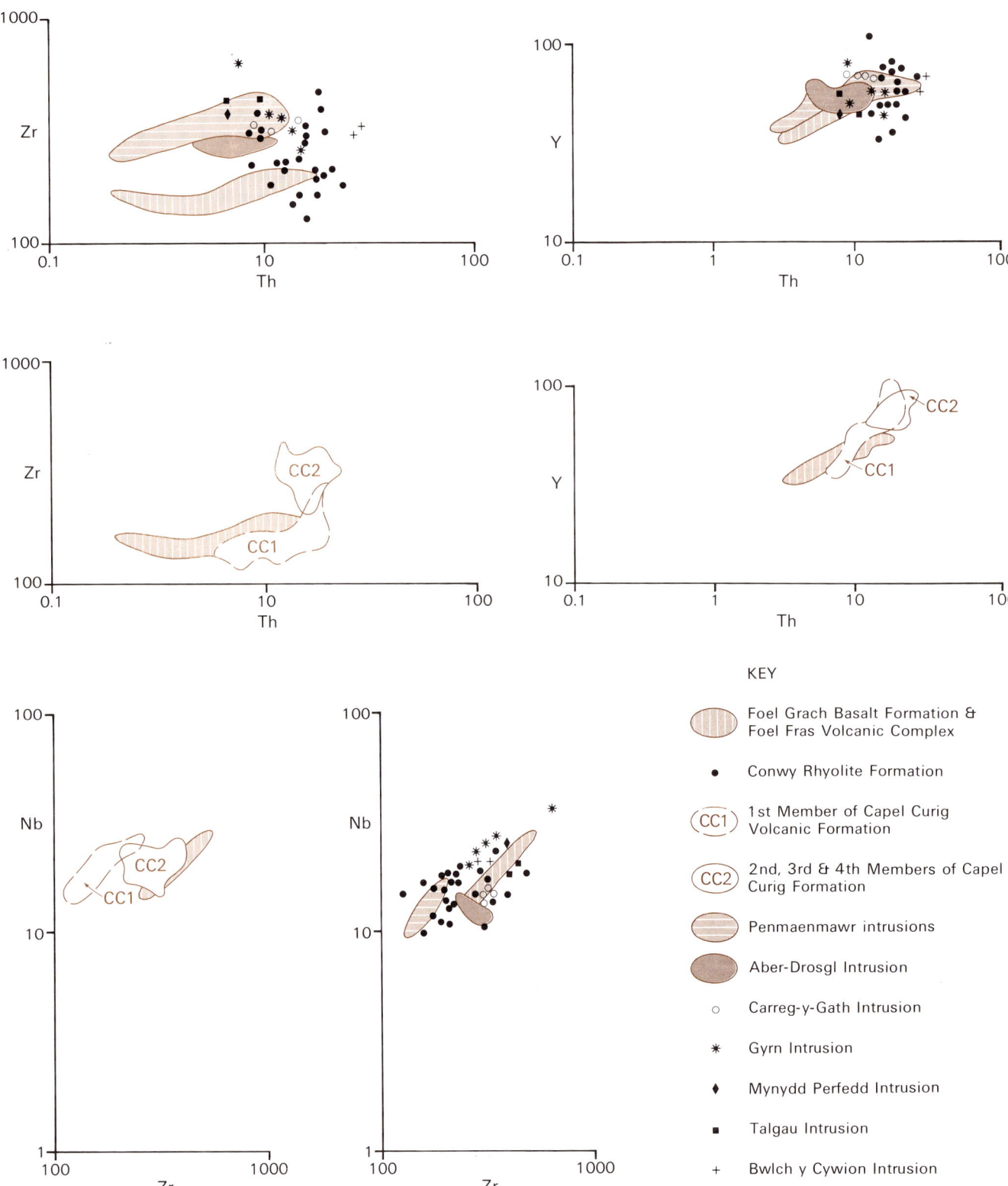

Figure 32   Llewelyn Volcanic Group, various element log-log plots (modified after Ball and Merriman, 1989).

relative degree of plagioclase fractionation. The highest HREE concentrations also occur in the Conwy Rhyolite Formation, again consistent with it representing the most evolved rocks of the group. The relative decline in LREE + MREE/HREE in the more evolved rocks can be interpreted in terms of apatite fractionation (Ball and Merriman. 1989).

Many volcanic rock series can be shown to be related by crystal fractionation as the dominant first order process, following the closed system Rayleigh fractionation law (Cl/Co = $F^{(D-1)}$). Such series, when plotted on a logarithmic variation diagram versus a highly incompatible element, should define straight line variation trends whose slope approximates to $(1 - D)$ where D is the bulk partition coefficient (Allegre et al., 1977). The most highly incompatible elements within subalkaline magma series are usually Rb, U and Th. The relative mobility of Rb and U makes them unreliable indices for the rocks under consideration and Th is therefore more useful. In addition, Zr can be used as a fractionation index, though it is not truly incompatible in basic–intermediate series related by significant clinopyroxene fractionation (i.e. D_{Zr} about 0.7) and hence the slopes of the variation trends are unreliable. Furthermore, Zr is not incompatible within silicic compositions which crystallise zircon (e.g. the Aber-Drosgl intrusion and the Conwy Rhyolite Formation).

A log-log plot of Zr vs. Th (Figure 32a) resolves most of the data, with the exception of that for the Conwy Rhyolite Formation, into two distinct groups with differing ranges of zirconium content. The group characterised by higher zirconium values comprises the Foel Grach Basalt Formation and Foel Fras Volcanic Complex, which together define a straight-line trend, the upper three members of the Capel Curig Volcanic Formation (Figure 32c) and all the intrusions with the exception of the Penmaenmawr microdiorite. The low zirconium group comprises the Penmaenmawr microdiorite, the data points for which also define a straight-line trend (Figure 32a), and the lowest member of the Capel Curig Volcanic Formation (Figure 32c). The Conwy Rhyolite Formation has a range of zirconium values which overlaps both the above groups, this compositional heterogeneity perhaps being explained by the eruption of the formation from two centres (Ball and Merriman, 1989), the most southerly of which erupted magma relatively depleted in zirconium. The same two groups of rocks are similarly resolved by plotting Nb vs. Th but not by Y vs. Th (Figure 32b) though, in the latter case, straight-line correlation of the data, excepting that for the Conwy Rhyolite Formation, is clearly displayed.

The case for closed system Rayleigh fractionation is further enhanced by considering log-log plots of Nb vs. Zr (Figure 32e,f). These again indicate that the Foel Grach Basalt Formation and the Foel Fras Volcanic Complex clearly define a straight-line trend. Most of the intrusions lie close to this trend except for the Penmaenmawr intrusion, and, in this instance, also the Gyrn granophyre, which defines a closely sub-parallel trend (Figure 32f), together with the 1st Member of the Capel Curig Volcanic Formation (Figure 32e). The Conwy Rhyolite Formation again straddles the two trends (Figure 32f) and the 2nd, 3rd and 4th members of the Capel Curig Volcanic Formation, though somewhat scattered, are more closely related to the Foel Fras–Foel Grach trend (Figure 32e). Their compositional variation may reflect increasing compatibility of Zr in these rocks (Ball and Merriman, 1989). The two trends, relatively Zr-enriched (Foel Fras–Foel Grach etc.) and Zr-depleted (Penmaenmawr etc.) can be repeated on other diagrams (e.g. TiO_2 vs. Zr).

The straight-line relationships described above are consistent with fractionation and removal of an assemblage of plagioclase, K-feldspar, biotite, amphibole and Ti-magnetite (Pearce, 1982b) to produce residual rhyolite magmas. Plagioclase, augite, ilmenite and apatite are all observed phases concentrated in autoliths within the more basic and intermediate rocks of the Llewelyn Volcanic Group (Ball and Merriman, 1989), and biotite and hornblende may be represented by chlorite pseudomorphs. Theoretical calculations (Ball and Merriman, 1989) confirm that a two-stage fractionation model can generate 'trachyandesite' from a Foel Grach Basalt Formation parent (by removal of 23 per cent plagioclase, 17 per cent olivine and some ilmenite) and that rhyolite similar to Conwy Rhyolite compositions can then be generated by removal of plagioclase, augite, ilmenite and apatite. The latter process is supported petrographically by the observation of a granophyric groundmass in samples of the Foel Fras Volcanic Complex. Petrochemical comparison of the Conwy Rhyolite Formation with experimental data (Ball and Merriman, 1989) suggests that they evolved in a comparatively shallow crustal regime (P_{H_2O} < 500 bars).

The case for the significant involvement of fractional crystallization in generating the acidic magmas is further strengthened by examination of the geochemical relationships within the 2nd Eruptive Cycle (Snowdon Volcanic Group), where it is concluded that there were at least two distinct basaltic parental compositions (also present as extrusive phases) involved in the basalt–icelandite–subalkaline rhyolite–peralkaline rhyolite magma suite. The recognition of two discrete fractionation trends within the Llewelyn Volcanic Group may therefore be a manifestation of two similar basaltic parents during the earliest phase of Caradoc volcanism in North Wales.

CHAPTER 3

Sedimentation between the 1st and 2nd Eruptive cycles (Cwm Eigiau Formation)

The interval between the 1st and 2nd Eruptive cycles was predominantly a phase of shallow and offshore marine siliciclastic sedimentation which is represented in the Cwm Eigiau Formation (Figure 9). The influence of volcanism during this phase was limited, although deposits of both minor acidic and basaltic volcanism are represented. The main features were the changes in palaeoslope directions and sedimentary facies, related to the shift in loci of the volcanic centres from the 1st Eruptive Cycle to those of the 2nd Eruptive Cycle.

Much of the detailed interpretation in this account derives from the work of Dr Geoff Orton. His research (Orton, 1988a,b) concentrated on the sedimentology of the sequences deposited during the 1st Eruptive Cycle and in the interval between the 1st and 2nd Eruptive cycles.

Precise definition of the interval between the two eruptive cycles is problematic over parts of central and northern Snowdonia (Figures 6 and 9). This is largely because there is only limited areal overlap of the deposits of the final phase of the 1st Eruptive Cycle (the Capel Curig Volcanic Formation) and the initial phase (Pitts Head Tuff Formation) of the 2nd Eruptive Cycle.

CENTRAL SNOWDONIA

In central Snowdonia (Figure 6), the sedimentary interval between the 1st and 2nd Eruptive cycles is most easily described by reference to the base of the 2nd Eruptive Cycle, marked by the lower Pitts Head Tuff (Figure 6).

Over much of the area, sedimentary environments were predominantly marine although in the south-west, about Moel Hebog (Figure 39), fluvial and deltaic conditions prevailed and from here the palaeoslope, prior to the emplacement of the Pitts Head Tuff, generally dipped to the north-east. This contrasts with the south-easterly and southerly dipping palaeoslope which prevailed earlier to the north, prior to, during and immediately after the emplacement of the Capel Curig Volcanic Formation. There was a general marine regression before the emplacement of the lower Pitts Head Tuff and coarse siliciclastic sediments were introduced from the west and north-west. However, the acme of fluvio-deltaic progradation may have been reached at some stage significantly before, rather than immediately prior to lower Pitts Head Tuff emplacement. The source areas included an important basaltic component (Orton, 1988a,b) in addition to significant acidic tuffs and tuffites.

The facies belts are discussed in the context of three separate areas with respect to the outcrop of the Pitts Head Tuff Formation (Reedman et al., 1987a; Orton, 1988a,b).

1 In the south, near Moel Hebog, the sediments below the Pitts Head Tuff Formation (Figure 33) comprise a prodelta sequence dominated by mudstone and siltstone, overlain by distributary mouth bar sediments and an alluvial plain and fan sequence immediately subjacent to the lower Pitts Head Tuff (Figure 39A).

The alluvial plain sediments form tabular sheets dominated by trough cross-bedded, medium- to coarse-grained, pebbly sandstones typical of braided stream deposits of mid to distal alluvial fans. A proximal-fan facies is also developed. Immediately subjacent to the lower Pitts Head Tuff, debris flows and a greater variation of grain size imply the formation of smaller, steeper, or short-headed fans (cf. Flores, 1975) than had prevailed previously.

2 In the central part of the outcrop (Figure 39A), sand-dominated sequences underlying the lower Pitts Head Tuff are generally thinner than their equivalents in the south. The north-easterly progradation of alluvial fans is inferred, with their deposits passing initially into alluvial plain and distal alluvial fan deposits and then into delta-front deposits. The alluvial fans comprise poorly sorted, massive, coarse-grained, locally conglomeratic sandstones deposited by debris flows and sheet floods. The distal alluvial fan deposits are dominated by stacked sets of tangentially based, tabular cross-bedded, medium-grained sandstone. The delta-front deposits comprise siltstones which sharply coarsen upwards into medium-grained sandstones containing abundant wave-swashed, reworked concentrates of detrital magnetite and il-menite (cf. Reid and Frostick, 1985). These sandstones represent the progradation of a wave-influenced beach.

3 In the northern part of the outcrop (Figure 39A), the substrate of the Pitts Head Tuff is dominated by shallow-marine sedimentary rocks. In the vicinity of Llanberis Pass, a clast-supported conglomerate, up to 7 m thick, comprising inversely graded or non-graded beds, 1–2 m thick, with rhyolite clasts up to 30 cm, represents a short-headed fan-delta dominated by debris flows derived from the west and north-west. Finer sediments along strike to the north and south probably represent shallow-marine conditions between the salients. Elsewhere, the sequence generally comprises a coarsening-upwards profile of sandstones, commonly containing brachiopods in the bases of scours. Amalgamated sets of hummocky cross-stratification and low-angle cross-bedding are common. At the top of the sequence sandstones with large, straight-crested, symmetrical wave-ripples pass into swaley cross-stratified sandstones. The sequence is interpreted as a prograding, storm-dominated shoreline, with lower shoreface sandstones having been deposited immediately prior to the eruption and emplacement of the lower Pitts Head Tuff. Shelf-ridge sandstones with interbar siltstones and mudstones occur with periodic fossiliferous, massive or cross-bedded storm sandstones, 5–20 cm thick. The sandstones contain relatively common brachiopods typical of the *Dinorthis–Macrocoelia* association of Pickerill and Brenchley

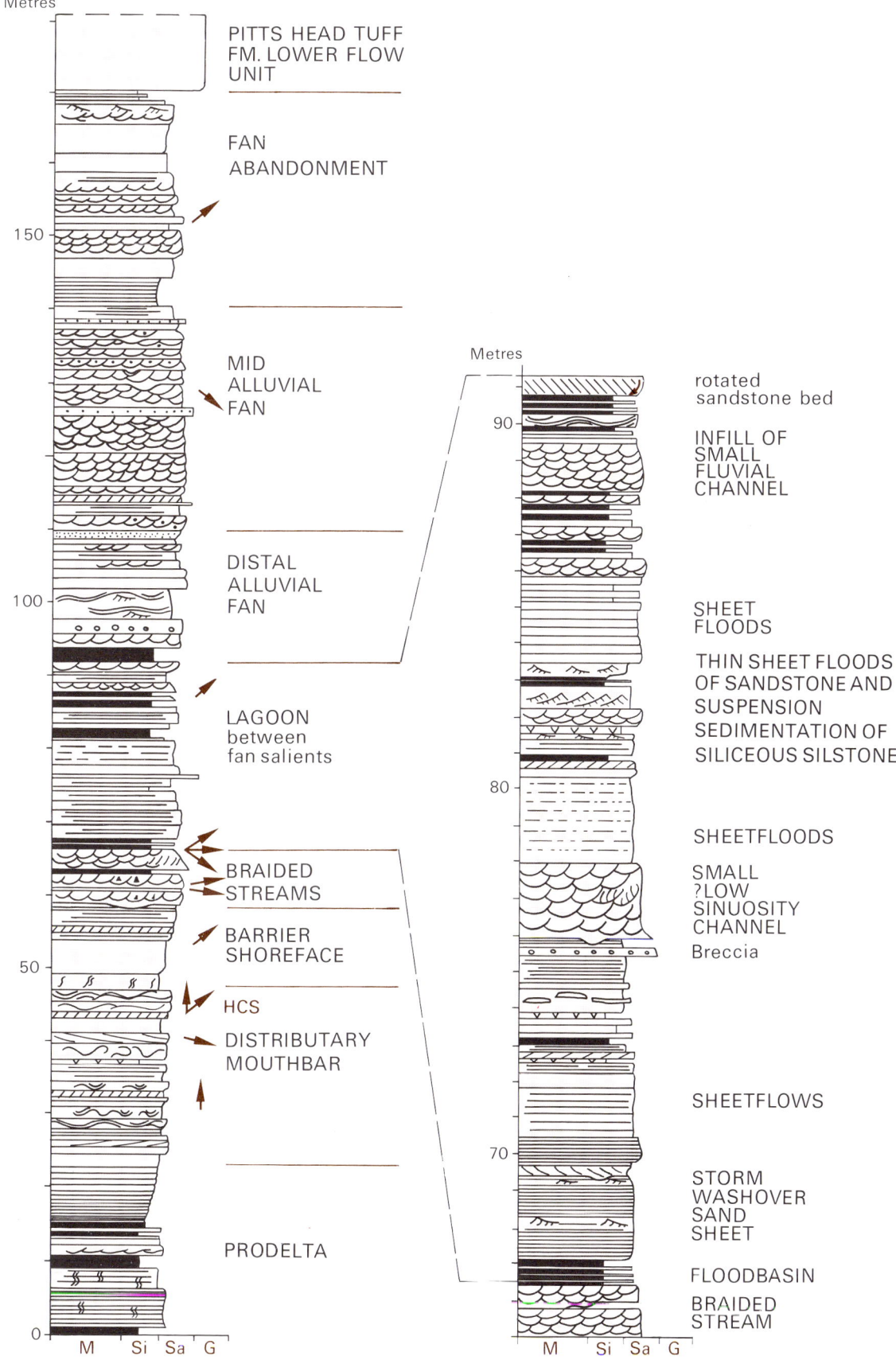

Figure 33 Graphic log and environmental interpretation of the sequence below the Pitts Head Tuff Formation, at Moel Hebog (base to top [SH 5705 4680 to 5683 4690]) (after Orton, 1988b).

Ornament as for Figure 22; arrows indicate compass-rose directions of currents; HCS: Hummocky cross-stratification.

(1979), implying water depths possibly as shallow as 25 m or less.

NORTHERN AND EASTERN SNOWDONIA

In northern and eastern Snowdonia, the last phase of the 1st Eruptive Cycle (Capel Curig Volcanic Formation) is well-defined (Howells et al., 1978; Howells and Leveridge, 1980). However, the Pitts Head tuffs did not reach this area and as a result the base of the Lower Crafnant Volcanic Formation (Figure 6) has been used to define the local base of the 2nd Eruptive Cycle. With the absence of the Pitts Head Tuff Formation, correlation of the sediments between the two eruptive cycles is made with reference to the Soudleyan–Longvillian boundary which approximately coincides with the stratigraphic level of the Pitts Head Tuff Formation.

The interval between the two eruptive cycles is represented by a variable sequence, up to 975 m thick, of marine mudstones, siltstones and sandstones with locally subordinate debris flow deposits, tuffs and tuffites. It is only in the north of the area, between Foel Lwyd and Tal y Fan (Figure 11), that nonmarine sandstones and associated conglomerates occur low in the sequence. Hence, the extensively developed subaerial surface, prior to and subsequent to the final episode of ash-flow tuff emplacement (Capel Curig Volcanic Formation) of the 1st Eruptive Cycle in the north, was inundated by a marine transgression. The sequence thickens from north to south and in this direction there is a general facies transition from dominantly shallow-marine, coarse- and fine-grained sandstones to offshore siltstones with laterally impersistent sandstones, indicating temporary persistence of the north to south palaeoslope.

In the north, near Foel Lwyd (Figure 11), up to 600 m of cross-bedded sandstones and conglomerates are overlain by about 150 m of siltstones, muddy siltstones and thin, cross-bedded sandstones with, at the top, dark grey mudstones underlying the lowermost ash-flow tuff of the 2nd Eruptive Cycle. The lower sandstones and conglomerates are interpreted as being fluviodeltaic, with facies variations and current direction data suggesting derivation from a landmass to the north-west (Howells and Leveridge, 1980). The overall fining-upward character of the sequence reflects a general transgression which affected much of central and northern Snowdonia following the onset, in the extreme south-west, of the 2nd Eruptive Cycle.

Further to the south, between Llyn Cowlyd and Capel Curig, the proportion of sandstone in the sequence diminishes and it is largely restricted to the middle part of the interval between the two eruptive cycles (Figure 34). The sequence can be subdivided into three units. A lower unit, up to 600 m thick, consists of mudstones and siltstones with few laterally impersistent sandstones. It is succeeded by a middle unit, up to 250 m thick, of shallow-marine sandstones with subordinate acid tuffs, tuffites, basic tuffs and siltstones. The sandstones are typically fine grained, well bedded, with well-developed, low-angle, hummocky and swaley cross-stratification, and are locally richly fossiliferous; the Soudleyan–Longvillian stage boundary has been determined close to the top of the sequence (Multiplicata Sandstone; Diggens and Romano, 1968). The sequence represents fluctuating inner and outer shelf regimes, characterised by fair-weather silt and mud deposition, into which transgressive storm-generated sheet sands were frequently introduced. There were, in addi-

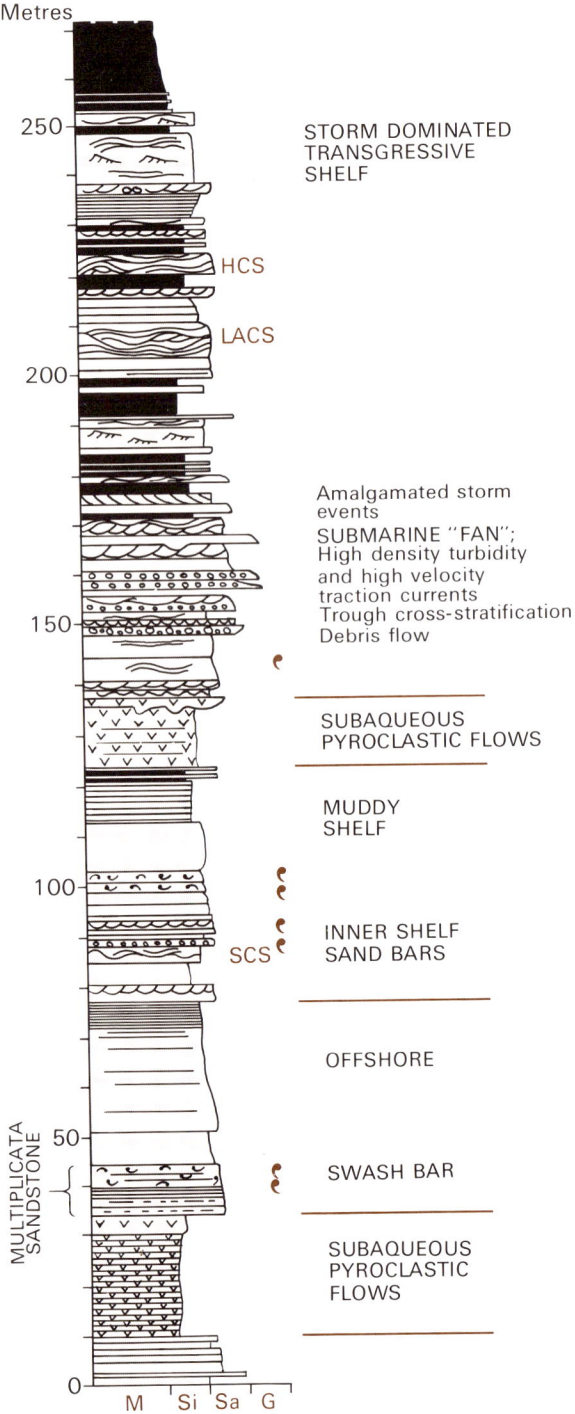

Figure 34 Graphic log and environmental interpretation of part of the sequence in the upper part of the Cwm Eigiau Formation, Llyn Cowlyd [SH 713 615] (after Orton, 1988a).

Ornament and abbreviation as for Figures 22 and 33; in addition, LACS: Low-angle cross-stratification; SCS: Swaley cross-stratification.

tion, phases of inner shelf sand bar construction and periods during which deposition was dominated by small-scale, volcaniclastic mass flows reflecting the secondary redistribution of pyroclastic debris. The upper unit, some 75 m thick, comprises mudstones, siltstones and fine-grained sandstones, reflecting a deeper offshore environment, and few thin acid tuffites.

Significant, but laterally restricted basaltic tuff accumulations, up to 200 m thick, occur within the middle unit (Howells et al., 1978, 1981a). The tuffs are composed entirely of fine- to coarse-grained, poorly sorted and rarely graded basaltic fragments. Locally, parallel-bedding is picked out by zones of thin alternations of coarse- and fine-grained layers and block concentrations. Elsewhere, isolated blocks are randomly scattered throughout the massive tuffs. The basaltic debris is completely altered to aggregates of chlorite, carbonate and iron-oxide, although albitised feldspar phenocrysts, basalt fragments and basaltic scoria can be distinguished. In the southernmost outcrop, Curig Hill (Figure 35), the beds display centroclinal dips and have been interpreted (Howells et al., 1978) as reflecting the upper level of a funnel-shaped volcanic neck beneath a tuff cone. Zones of slumped bedding, planes of discordance and penecontemporaneous faults are common. The overlying uniformly bedded tuffs, which wedge out laterally into the enclosing sandstones, represent reworking of the pyroclastic debris during the destruction of the tuff cone. The limited contamination of the associated sediments with basaltic debris suggests that the eruptions were small and that the debris was mainly confined to depressions above the vent.

The acid tuffs, interbedded with siltstones and sandstones of the middle unit, are characteristically fine grained, and are composed of devitrified, recrystallised fine shards and dust. Locally, the tuffs, with admixture of epiclastic debris, grade into tuffites. The tuffs represent distal ash fall-out which settled into the marine environment and was locally reworked. Locally, repeated thin, flaggy-bedded tuffs and tuffites suggest possible accumulation resulting from sloughing of unstable tuff, temporarily stored (cf. Carey and Sigurdsson, 1984) in shallower parts of the basin. Disruption of primary pyroclastic deposits and secondary emplacement by transport as debris flows is more clearly determined when coarser grade debris is involved. Some fine-grained tuffs include clasts of tuff and siltstone, in places concentrated in the basal layers. Elsewhere, beds composed throughout of angular clasts and rafts of sediments and tuffs in a matrix of mixed coarse to fine epiclastic and pyroclastic debris indicate deposition from high density debris flows.

To the east and south-east of the Snowdon massif, the tripartite division of the interval between the two eruptive cycles is not maintained. Furthermore, there is a contrast in the sequences to the north and south of the axis of the Dolwyddelan syncline (Locality 13, Figure 6). To the north, the entire succession, 700 m thick, comprises alternating thick siltstones and sandstones. Many of the sandstones are fossiliferous and Longvillian faunas occur seemingly very low in the sequence (Romano and Diggens, 1969), close to the stratigraphic level of the Capel Curig Volcanic Formation. Individual sandstone units are up to 70 m thick. The sandstones are variably fine to medium grained, and massive to weakly bedded and cross-bedded. The faunal content is dominated by dalmanellid and sowerbyellid brachiopods with subordinate trilobites. Comparison with the *Howellites* 'community' of Pickerill and Brenchley (1979) and with their *Dinorthis* and *Macrocoelia* 'subcommunities' can be made,

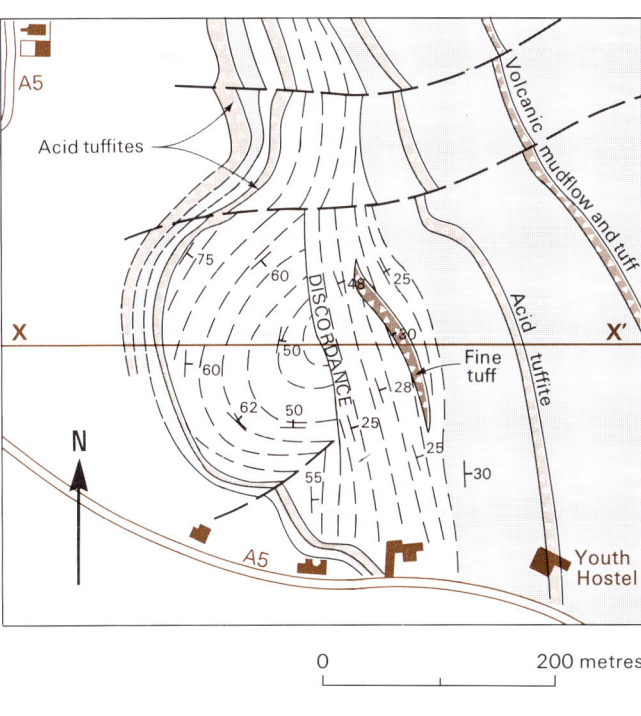

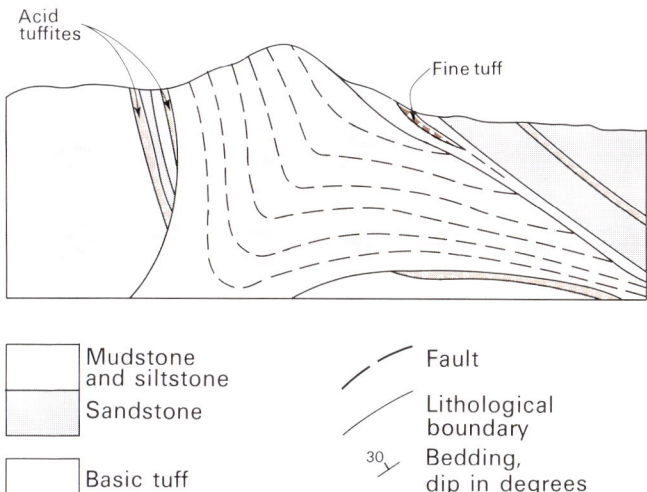

Mudstone and siltstone		Fault
Sandstone		Lithological boundary
Basic tuff		Bedding, dip in degrees

Figure 35 Sketch map and W-E cross section of basaltic vent, Curig Hill (modified from Howells et al., 1978). Youth Hostel [SH 7260 5791].

implying shallow water depths (?less than 30 m) for much of the sequence. However, little of the material is unequivocally in situ and estimates of water depth must be treated with considerable caution. Acidic tuffites up to 15 m thick are relatively common in the topmost 250 m of the sequence.

Further to the east, the sequence can be subdivided into a lower siltstone-dominated sequence, up to 220 m thick, of Soudleyan age, and an upper more sandstone-rich sequence (up to 800 m thick) of Longvillian age. The former has been interpreted in terms of deeper shelf environments than the latter (Pickerill and Brenchley, 1979; Campbell, 1983) with varying degrees of storm-influenced sedimentation an important factor.

CHAPTER 4

2nd Eruptive Cycle (Snowdon Volcanic Group)

The activity of the 2nd Eruptive Cycle is expressed in acidic ash-flow tuffs, intrusive and extrusive rhyolites, basalts, hyaloclastites and basic tuffs, which are variably associated with marine sediments and together comprise the Snowdon Volcanic Group (Figures 6, 9). It can be traced for some 45 km, from north-east to south-west Snowdonia (Figure 36). The sequence reflects a complex development of eruptive centres where the activity shifted temporally and spatially, but which was contained mainly within a shallow- to offshore-marine environment. Three main centres of activity have been defined, the Llwyd Mawr Centre, the Snowdon Centre and the Crafnant Centre (Figure 36). Major caldera collapse was a feature in the development of each of these centres.

Deep-seated, north-east–south-west and north-north-east–south-south-west-trending fractures were influential in determining the loci of the eruptive centres. They influenced the form of subsequent volcanotectonic subsidence about the centres and defined a broad north-east–south-west belt of maximum extension. This belt confined voluminous dolerite sills (Figure 87) which were mainly intruded late in the 2nd Eruptive Cycle (Campbell et al., 1988). The structural belt has been termed the 'Snowdon Graben' (Kokelaar, 1988) or 'failed rift' (Campbell et al., 1988).

The Snowdon Volcanic Group was defined (see Howells et al., 1983 for details) to modify the Snowdon Volcanic 'Suite' (Williams, 1927) and the broadly equivalent Crafnant Volcanic 'Series' (Davies, 1936) in order to meet modern requirements of stratigraphical nomenclature. The group is amended here to comprise in central and south-western Snowdonia, in upward succession, the Pitts Head Tuff Formation, the Lower Rhyolitic Tuff Formation (including the Yr Arddu Tuffs), the Bedded Pyroclastic Formation and the Upper Rhyolitic Tuff Formation (Figure 6). The Pitts Head Tuff Formation (Reedman et al., 1987a) was previously considered (Howells et al., 1983) to be a member of the Cwm Eigiau Formation which lies below the Snowdon Volcanic Group. In north-eastern Snowdonia the group is redefined to include the Lower, Middle and Upper Crafnant Volcanic Formations, the Tal y Fan Volcanic Formation and the Dolgarrog Volcanic Formation which were previously excluded (Figure 6).

Geochemically, a plot of Zr/TiO_2 vs. Nb/Y (Figure 37) for the entire Snowdon Volcanic Group indicates that three main magmatic compositions are represented; subalkaline basalt, rhyolite and a transitional rhyolite-comendite/pantellerite. A near continuum of compositions exists between them although intermediate compositions are comparatively rare.

The penecontemporaneity of the volcanism at different centres throughout the cycle precludes description in an ordered sequential fashion. As a result, the cycle is described in terms of its constituent formations, which reflect the activity at different centres, their sedimentary context and their relationships to tectonic and volcanotectonic structures.

LLWYD MAWR CENTRE: THE PITTS HEAD TUFF FORMATION

The 2nd Eruptive Cycle was initiated by acidic eruptive activity at the Llwyd Mawr Centre in south-western Snowdonia where there is no indication of the activity of the 1st Eruptive Cycle. Its deposits (Pitts Head Tuff Formation) have two distinct expressions, an intracaldera facies and an outflow facies.

Intracaldera facies

The eruptive centre of the tuffs is interpreted (Roberts, 1969; Reedman et al., 1987a) to have lain close to a restricted accumulation of welded acidic ash-flow tuff, up to 700 m thick, in the vicinity of Llwyd Mawr (Figure 38). The tuff forms an elongate synclinal outlier, partly concordant and part discordant with much older (Llanvirn) siltstones and silty mudstones (Nantmor Group; Figure 9). Within the tuff, Roberts and Siddans (1971) interpreted, from compactional variations in component lithic clasts and pumice, two distinct eruptive pulses. However, the few geochemical analyses of samples in serial sections do not suggest any related compositional change. The tuff is intruded by large, high-level rhyolite domes and is considered to have been entrapped in a volcano-tectonic depression, a graben or a caldera, the uppermost part of which has not been preserved. The absence of early Caradoc strata below the intracaldera tuffs, compared with the thick development (about 1 km) below the proximal outflow tuffs, implies local uplift and erosion in the vicinity of the centre prior to caldera collapse. Evidence from the sediments subjacent to the proximal outflow tuffs suggests that the caldera developed in a subaerial setting.

Outflow facies

In the adjacent Moel Hebog syncline to the east (Figure 38), two acidic ash-flow tuffs are lithologically identical to the thick intracaldera tuffs and are interpreted as outflows from the centre. Here, evidence of their substrate shows that both tuffs (up to 90 m thick) were emplaced subaerially, but further to the north-east, the lower flow transgressed a shoreline and continued for at least 10 km in a marine environment (Figure 39) (Reedman et al., 1987a). In both environments the flow retained sufficient heat to weld on emplacement.

In the subaerial environment, up to 1 m of bedded, non-welded vitroclastic tuffs is locally present below the lower outflow tuff (Plate 18). The latter is characterised by a non-welded base, passing rapidly up into welded tuff with chloritic fiamme, a prominent zone of siliceous nodules, a concordant, parataxitic, welding fabric accentuated by siliceous segregation in the main body of the flow unit (Plate 24A) and an eroded top (Figure 40). The growth of the large, 10–40 cm, siliceous nodules (Plates 18 and 24B) in a zone 3 to 5 m above the base resulted from the entrapment of volatiles below the zone of most intense welding. The rela-

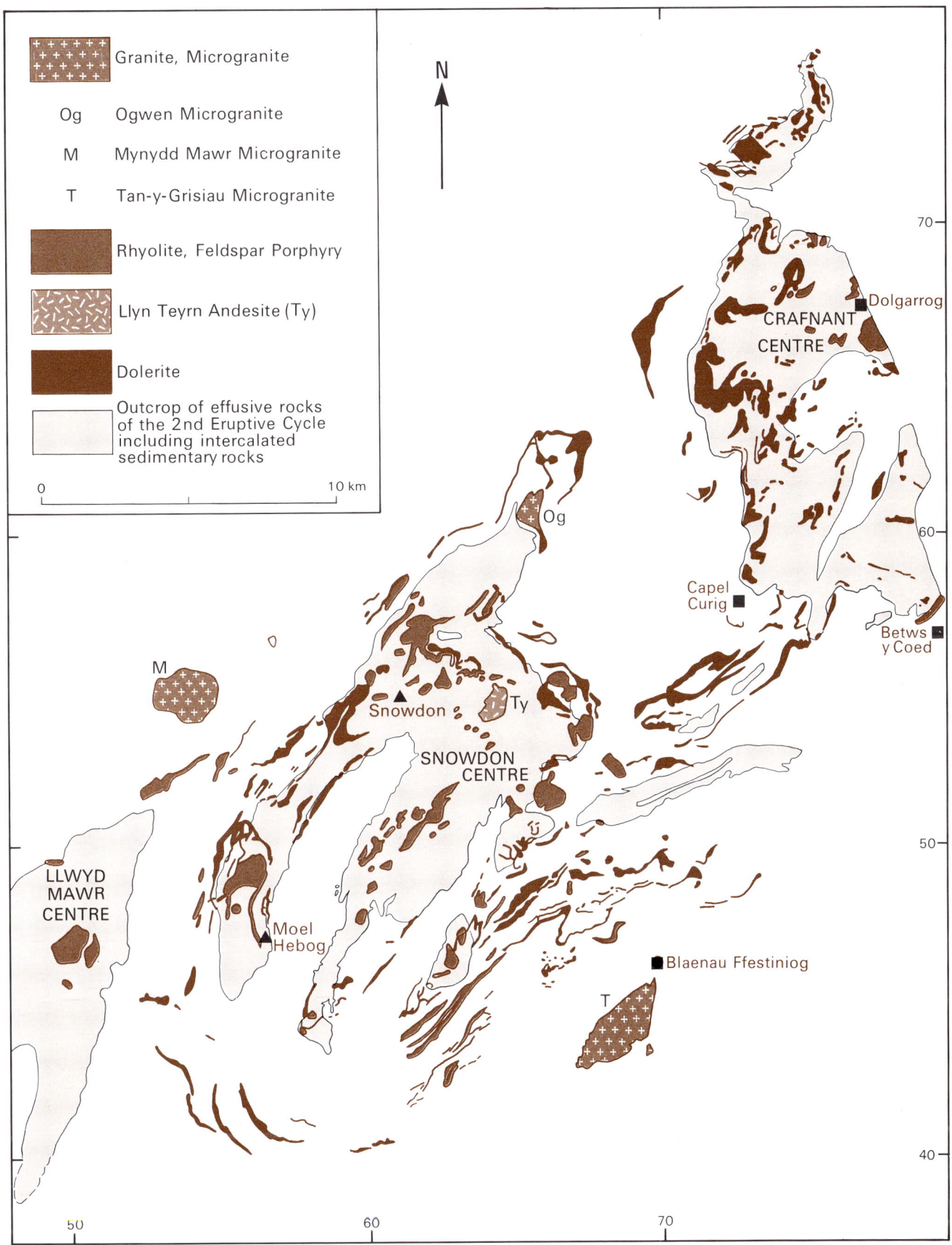

Figure 36   Outcrop of the Snowdon Volcanic Group (2nd Eruptive Cycle) with main eruptive centres and the distribution of associated intrusions.

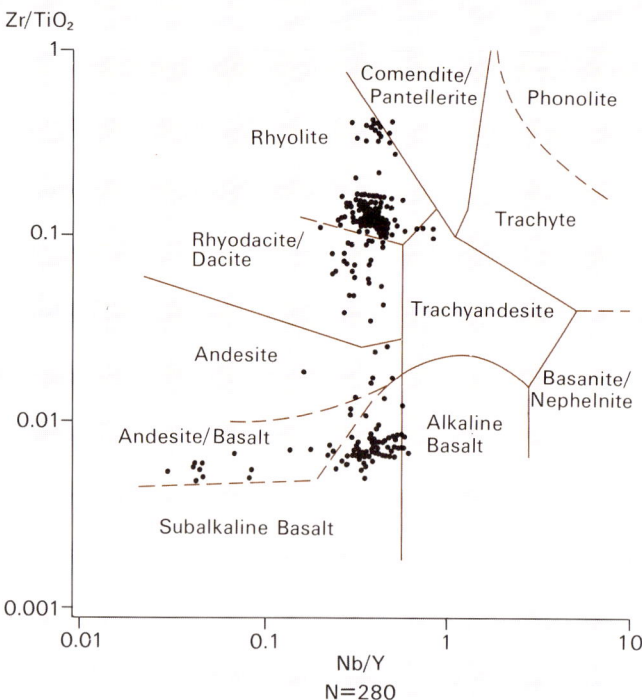

Figure 37 Snowdon Volcanic Group: analyses of extrusive volcanic rocks plotted on the Zr/TiO_2 vs. Nb/Y diagram of Winchester and Floyd (1977).

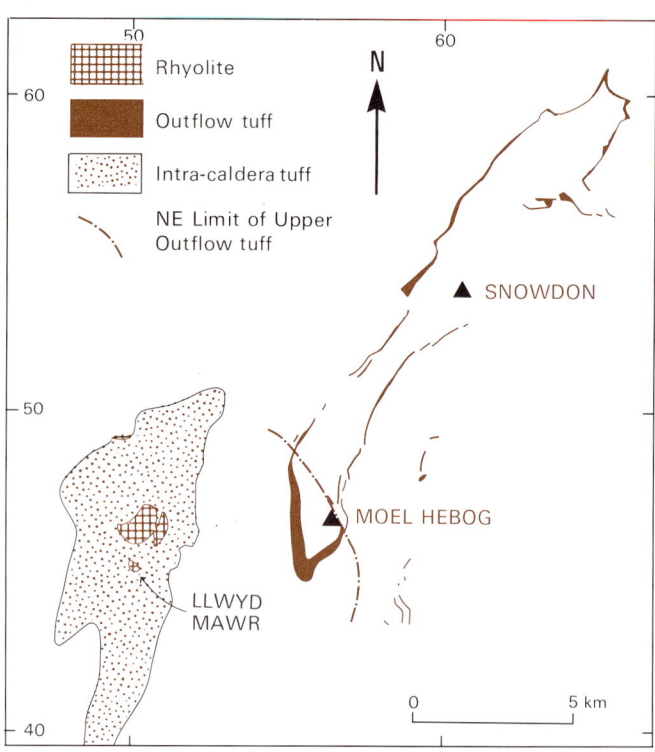

Figure 38 Outcrop of the Pitts Head Tuff Formation showing the relationship of the Llwyd Mawr intracaldera facies, with associated rhyolite intrusions, to the outcrop of the outflow ash-flow tuffs.

tionship of some of these nodules to the welding foliation indicates that their growth occurred during the late stages of compactional welding but before it was completed. Locally they are prolate in the welding foliation as a result of laminar shear during rheomorphic flow.

In the submarine environment, the tuff comprises a feldspar crystal-rich, nonwelded or weakly welded basal zone, an eutaxitically welded central zone with distinctive chloritic fiamme, rheomorphic folds and irregular zones of siliceous nodules, and a locally reworked, nonwelded, vitric dust-rich tuff top (Figure 40). The ductile deformation of the welding foliation (Figure 40, Plate 18B) and the formation of crude diapir-like structures associated with intense nodule development have been ascribed (Wright and Coward, 1977) to the upward streaming of volatiles, mainly water vapour, from the base of the tuff. On this interpretation, the nodules would represent original gas cavities (lithophysae) which were later infilled with silica.

The lower tuff is restricted to the west side of Snowdonia. To the east, it passes laterally into reworked tuffs interbedded with marine sandstones and, yet further east, the passage of the flow was restricted. This restriction is considered to have been caused by the earliest expression of north-east–south-west faults (Beddgelert Fault Zone) (Rast, 1969; Beavon, 1963; Howells et al., 1986) (Figure 49), which controlled the sites of the later Lower Rhyolitic Tuff Formation volcanism.

In the south-west, near Beddgelert and Moel Ddu (Figure 38) and on the eastern side of Moel Hebog (Plate 19), the tuff occurs as large autochthonous rafts within the post-lower Pitts Head Tuff sequence (Reedman et al., 1987a). The tuff in most of these rafts is either partially or completely brecciated,

and some display welding foliations oriented at a high angle to their lower and upper contacts, thus indicating rotation of the rafts during mass movement. The mass gravity movement and brecciation of the tuff was related to contemporary tectonism in the eastern area (Reedman et al., 1987a) and occurred while the tuff was still plastic. In the body of the flow, the transition from ductile to brittle deformation is represented by conjugate sets of kink bands in the foliation, and later extensive brittle fracture resulted in pervasively brecciated tuff (Figure 40).

Subsequent to the emplacement of the lower tuff, the north-east-directed palaeoslope, extending from a subaerial to a submarine environment, was maintained. However, evidence of storm-generated, outer shelf facies sediments overlying the lower tuff in the north, at its distal end, indicates rapid sea-floor subsidence following its emplacement. In the south-west, a coarsening-upwards sequence, up to 40 m thick, of medium- and coarse-grained sandstone and boulder conglomerate overlies the lower tuff (Reedman et al., 1987a). This sequence is interpreted as a proximal alluvial fan, derived from the south-west and thinning rapidly to the north-east (Figure 39B). It is overlain by the upper outflow tuff.

The upper outflow tuff, up to 50 m thick and restricted to a relatively small area about Moel Hebog, in the south-west, is lithologically similar to the lower tuff. Its transport to the north-east was limited, presumably by topography, in the area of the subsequent development of the Lower Rhyolitic Tuff Formation caldera.

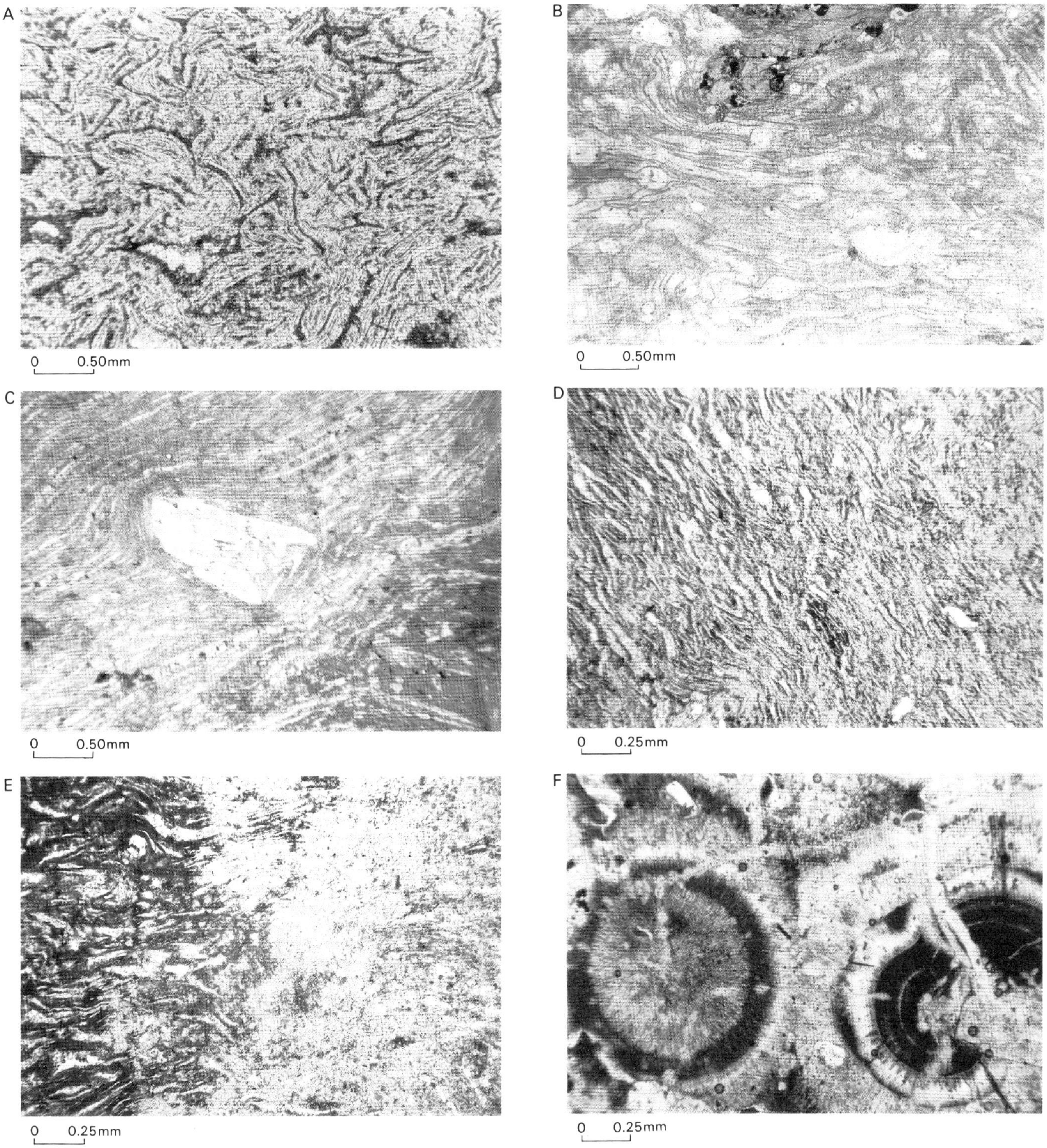

Plate 17 Photomicrographs, welded, acidic, ash-flow tuffs

A   KB 666. Eutaxitic texture; devitrified shards, recrystallised to a fine-grained quartzose mosaic, and form accentuated by chlorite and sericite flakes, upper flow-unit, Pitts Head Tuff Formation, Cwmystradlyn, ppl. [SH 5582 4571].

B   MH 668. Parataxitic texture; devitrified shards, recrystallised fine-grained aggregate of sericite, quartz and feldspar. Pitts Head Tuff Formation, west of Llyn Idwal, ppl. [SH 6411 5969].

C   RN 872. Parataxitic texture; fine-grained, quartz segregations accentuating foliation which is deflected about a sericitised, albite feldspar phenocryst. Lower Rhyolitic Tuff Formation, Cwm Tregalan, ppl. [SH 6127 5360].

D   MH 1275. Parataxitic texture; compacted fabric of devitrified shards, recrystallised, quartzofeldspar, accentuated by sericite and chlorite flakes. Lower Rhyolitic Tuff Formation. Moel y Dyniewyd, ppl. [SH 6225 4834].

E   MH 1275. Parataxitic texture overprinted by a 'cloud' of fine quartzose recrystallisation. Lower Rhyolitic Tuff Formation. Moel y Dyniewyd, ppl. [SH 6225 4834].

F   MH 1183. Spherulites; fine, fibrous, recrystallised aggregates of quartz, sericite and chlorite, overgrowing a small albite feldspar phenocryst at bottom right. Yr Arddu tuffs, ppl. [SH 6246 4611].

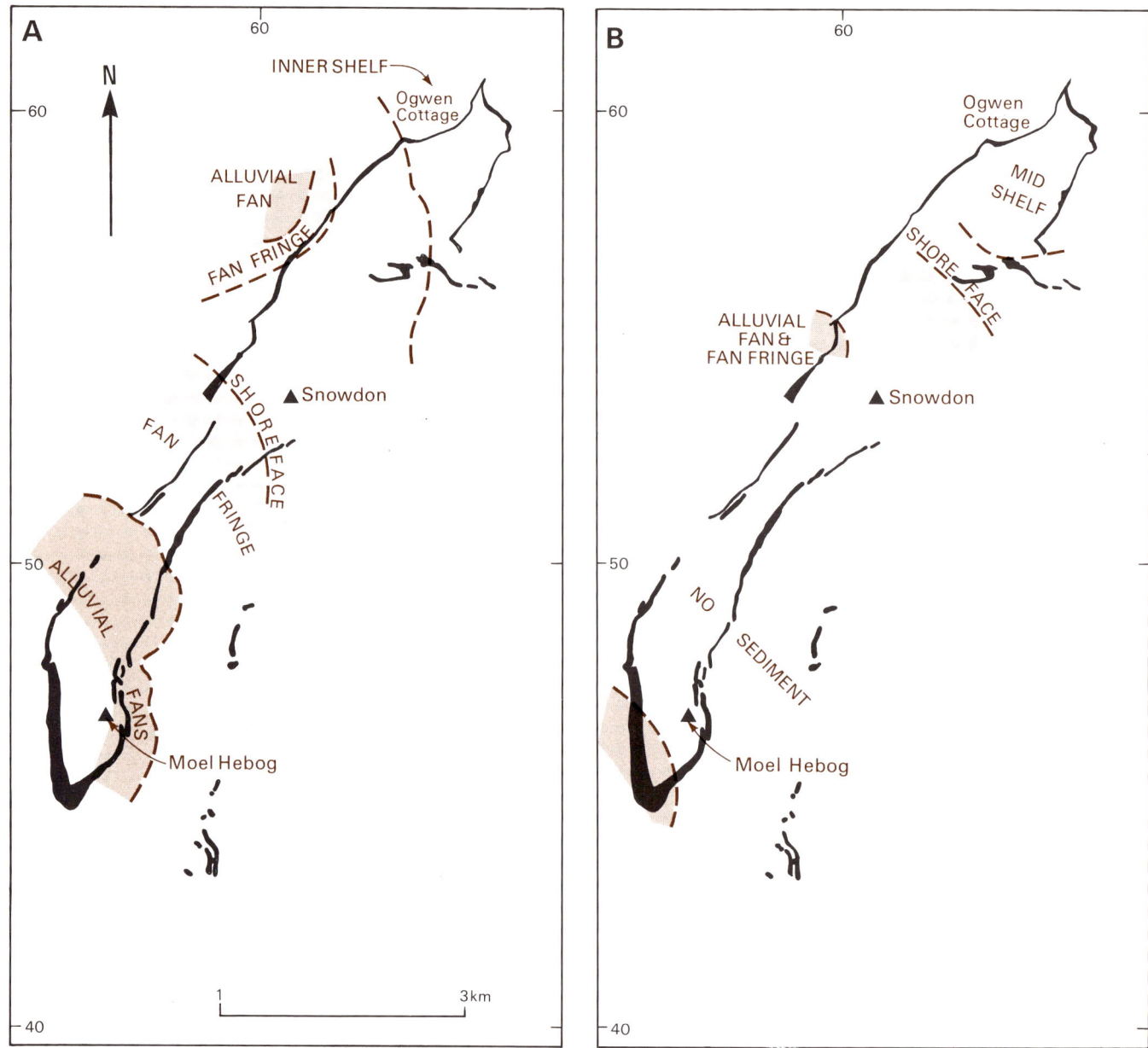

Figure 39 Environmental interpretation of (A) sediments below and (B) sediments above the lower Pitts Head tuff (after Reedman et al., 1987a).

Plate 18   The lower outflow tuff of the Pitts Head Tuff Formation, Moel Hebog.

The ash-flow tuff overlies coarse-grained sandstone (S) and bedded tuffs (Be). The base of the ash-flow tuff comprises nonwelded tuff (T_1) and is overlain by columnar-jointed welded tuff (T_2). A prominent zone of siliceous nodules (N) is overlain by densely welded tuff (T_3) with a conspicuous, silicified, welding foliation [SH 5684 4694].

SW   NE

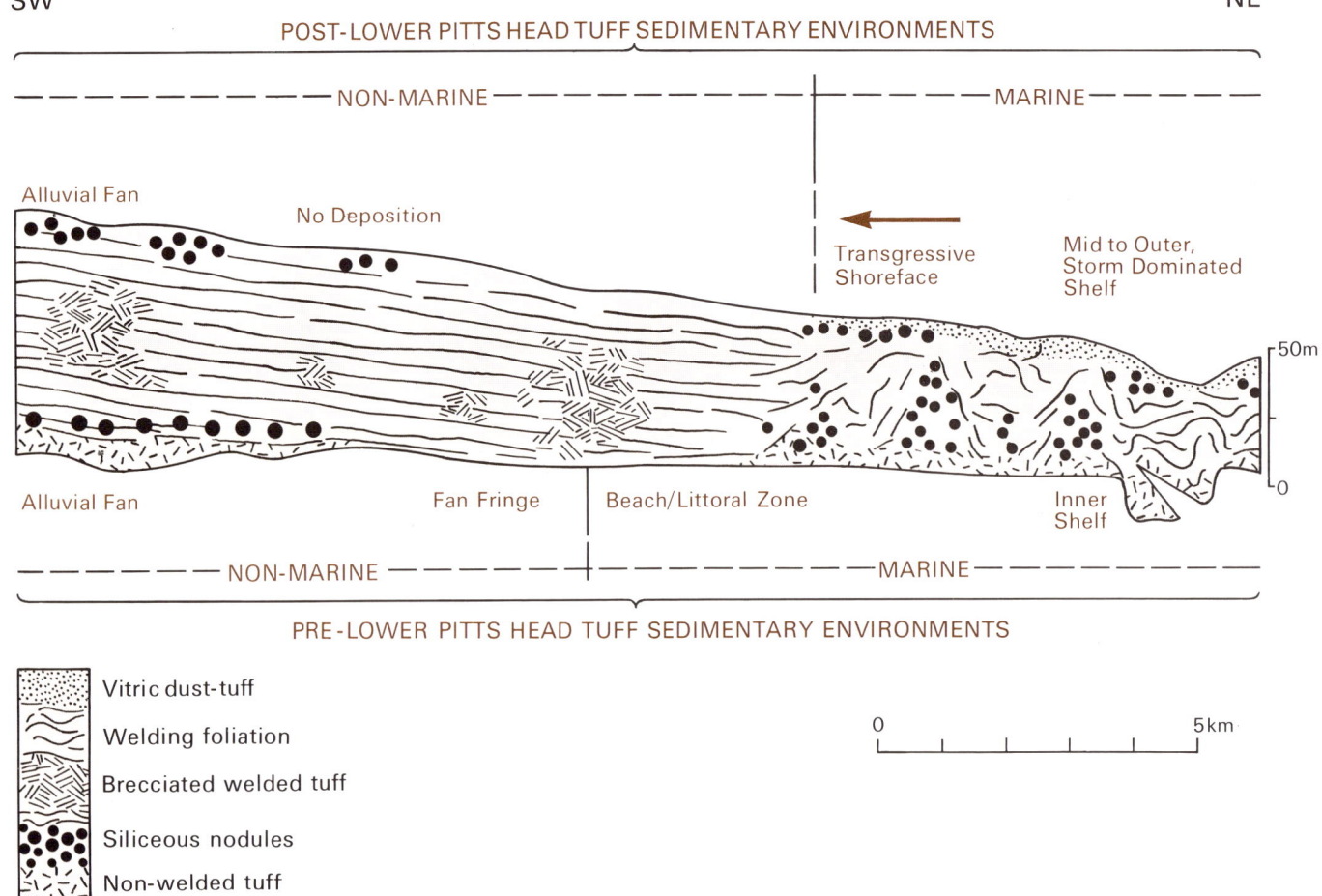

Figure 40   Model of lithological characters of the lower Pitts Head outflow tuff related to environments interpreted from sediments pre- and post-emplacement (modified after Reedman et al., 1987a).

The current disposition of the proximal outflow tuffs, relative to the intracaldera tuffs of Llwyd Mawr, has been affected by up to 2 km of post-volcanic dextral movement on the Cwm Pennant Fault Zone (Figure 111; Smith, 1988); the intracaldera tuffs were originally situated further south.

PETROGRAPHY   The devitrified welded tuffs comprise a fine aggregate of quartz, feldspar, sericite and chlorite in varying proportions (Plate 17B). In places, impersistent ribs of coarser recrystallised quartz accentuate the welding fabric. Shards are locally well defined although generally the original fabric is overprinted by recrystallisation.

Phenocrysts are predominantly of euhedral, subhedral and fragmented albite-oligoclase feldspar (< 2 mm) with very few quartz. They are particularly concentrated in the basal zones of the tuff, whilst above they are isolated and rarely exceed 5 per cent of the volume of the rock.

Lithic fragments are mainly of siltstone or fine sandstone, some with indented margins indicating incorporation into the flow when only partly lithified. Such clasts occur mainly in the basal zone in the northern outcrop. Perlitic fractures in the matrix are common in the central part of the flow in the

north, but on the southern flanks of Snowdon they are restricted to the lowest few metres, indicating welding to its base.

Siliceous nodules, as small as 0.4 mm, comprise a quartz mosaic and the welding foliation can be distinguished both passing into and deflected around the nodules, indicating that they developed during compactional welding.

Nonwelded vitroclastic textures are present only in the bedded tuffs below the lower tuff at Moel Hebog, in the basal crystal-rich tuff near Moel Hebog and in the north of the outcrop.

GEOCHEMISTRY   On the Zr/TiO_2 vs. Nb/Y diagram (Figure 41) the analyses (Appendix 1, Table 6) of the intracaldera and outflow Pitts Head tuffs plot mainly in the rhyolite field. The analysis of a rhyolite intrusion within the intracaldera tuffs sampled from the north-west of the outcrop plots in the comendite/pantellerite field. On this diagram, the fields of the intracaldera and lower outflow tuffs are closely coincident; the upper outflow tuff analyses form a cluster at the edge of the main concentration.

In terms of the relatively immobile trace elements and their

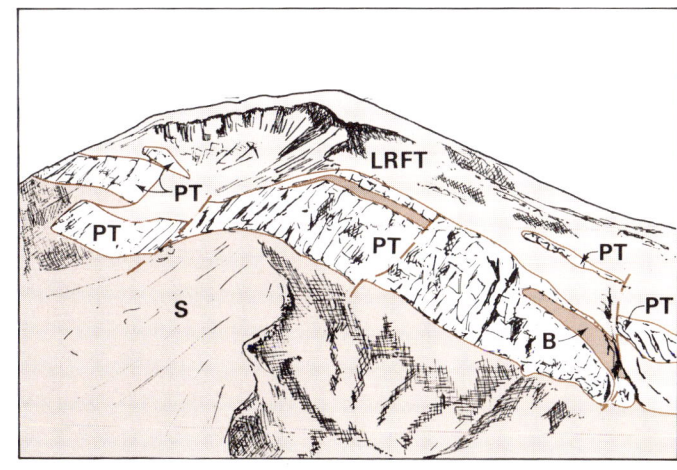

Plate 19 Moel Hebog [SH 565 469], viewed from the north-east, showing primary and reworked tuffs of the Lower Rhyolitic Tuff Formation (LRTF) overlying and enclosing disrupted rafts and blocks of the Pitts Head Tuff Formation (PT) near the southern margin of the Lower Rhyolitic Tuff Formation caldera. The Pitts Head Tuff Formation is underlain by sediments (S) and intruded by basalt (B).

respective ratios, the lower portion of the intracaldera tuffs (all data are from the lowest 200 m) is closely comparable to the outflow tuffs. The upper outflow tuff differs significantly from the intracaldera and lower outflow tuffs in its relatively lower TiO_2 content (Figure 42) and, consequently, higher ratios with respect to TiO_2, most notably Nb/TiO_2 and Y/TiO_2 (Appendix 1, Table 6). It is also enriched in Nb compared with the intracaldera and lower outflow tuffs, for example with respect to Zr (Figure 43, Appendix 1, Table 6). The lower flow unit differs significantly from the intracaldera tuffs only in its higher Ba and to a lesser extent Sr contents, these differences possibly reflecting intracaldera hydrothermal alteration.

No significant differences have been recognised between the subaerial and subaqueous facies of the lower flow unit except with respect to relatively mobile elements, e.g. the lower K_2O content (respective means; subaerial: 7.08 ($n = 8$, $\sigma n - 1 = 0.76$) and subaqueous: 4.39 ($N = 9$, $\sigma n - 1 = 1.98$)) and the slightly lower FeO content of the subaqueous facies. However, these differences may reflect metasomatism associated with the evolution of the Lower Rhyolitic Tuff Formation caldera (Chapter 6) rather than sea water interaction. The correlation of the isolated rafts of Pitts Head tuffs to the south-east of the main outcrop with the lower outflow tuff is substantiated particularly by the HFS element/TiO_2 ratios.

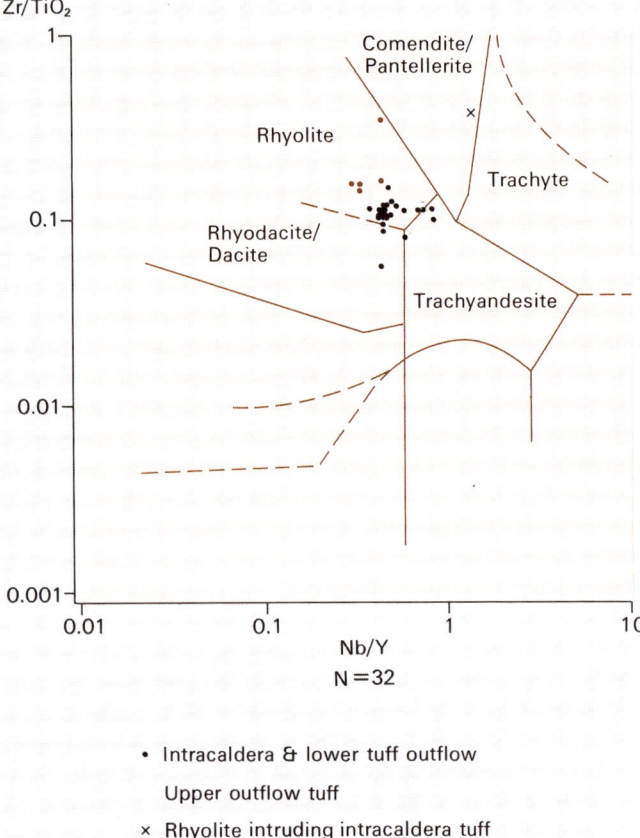

Figure 41 Pitts Head Tuff Formation, Zr/TiO_2 vs. Nb/Y diagram.

In serial section (Figure 44), the lowest 120 m of the intracaldera tuffs show little variation in terms of the immobile trace elements. The break (about 67 m above the base) between a lower and upper flow unit, as indicated by Roberts and Siddans (1971) on compactional evidence, does not reflect a compositional change equivalent to the lower and upper outflow tuffs. The serial geochemistry of the lower outflow tuff is more variable than the intracaldera tuffs with respect to height (Figure 44). There is a weak tendency for a decrease in incompatible element concentration with height above base in the two most distal sections of the lower outflow tuff, but the sampling interval is too crude to draw any firmer conclusions.

Discriminance with respect to the other ash-flow tuffs of the 2nd Eruptive Cycle are considered below.

Strata between the Pitts Head Tuff Formation and the Lower Rhyolitic Tuff Formation

About the south-west side of the Snowdon massif, the Pitts Head Tuff Formation is unconformably overlain by either non-pillowed basalt or by ash-flow tuff of the Lower Rhyolitic Tuff Formation. Here, a virtually flat subaerial surface persisted after the emplacement of the Pitts Head Tuff Formation. To the north-east, where the upper outflow tuff is missing, the interval between the lower outflow tuff of the Pitts Head Tuff Formation and the Lower Rhyolitic Tuff Forma-

tion is represented by a sequence of marine sediments which gradually thickens northwards. Eventually, north of Llanberis Pass, up to about 200 m of coarse- to medium-grained, cross-bedded sandstones with intercalations of siltstone, fine-grained acid dust tuffs, basalts and hyaloclastites intervene. Hence the north-east-directed palaeoslope, which had developed prior to the emplacement of the Pitts Head ash-flow tuffs, was maintained, with a north-west–south-east shoreline in the vicinity of Snowdon.

In the north-east, the nonwelded, dust-rich top of the lower outflow tuff is locally reworked and is sharply overlain by cross-laminated, coarse-grained and locally pebbly sandstones (Plate 20). The sandstones contain a marine shelly fauna, of the *Dinorthis–Macrocoelia* community (cf. Pickerill and Brenchley, 1979). Most distinctively, however, the shells are disarticulated and dispersed along low-angle, foreset laminae which indicates at least some redistribution. The sandstones contain very little material from the dust-rich top of the underlying ash-flow tuff. The lower outflow tuff approximately coincides with the Soudleyan–Longvillian stage boundary, although it is questionable as to what extent this boundary is here identified on the basis of facies-controlled faunas and therefore diachronous on a regional scale.

Orton (1987) interpreted the fining-upward sequence immediately above the lower outflow tuff in Llanberis Pass as representing a transgressive, non-barred, wave-influenced shoreline (Figure 49). Fine-grained sandstones subjacent to the Lower Rhyolitic Tuff Formation are parallel-laminated and bioturbated, with minor swaley cross-stratification and wave ripples. Orton interpreted them as typical of the lower shoreface to offshore transition zone at possible water depths of 9–18 m. Similar facies are also recognised further north. Yet further north-east, close to the limit of the outcrop (Figure 39), hummocky cross-stratified sandstones, turbidites and siltstones (Figure 45) indicate greater water depths.

Fine, acid, vitric dust tuffs are a distinctive component of the sequence in the north of the outcrop. They occur in thin flaggy beds with cross-lamination, some normal grading and locally scoured tops. Typically, they are interbedded with, and grade into thin siltstones. The bedforms indicate emplacement as tuff turbidites from fine pyroclastic debris temporarily stored in shallower parts of the basin.

LATE-STAGE BASALTIC VOLCANISM (SUB-LOWER RHYOLITIC TUFF FORMATION BASALTS)

Late in the Pitts Head Tuff Formation–Lower Rhyolitic Tuff Formation interval, basaltic volcanism was extensive and is represented by the basalts, hyaloclastites and basaltic tuffs high in the sequence (e.g. Figure 45). Collectively, this basaltic volcanism is referred to here as the sub-Lower Rhyolitic Tuff Formation basalts. These are particularly prominent in the outcrop on the south-east limb of the Cwm Idwal syncline (Figure 51) into the Llanberis Pass, on the north side of the Snowdon massif, and in Cwm Tregalan on the south side (Plate 21).

In Llanberis Pass a basalt sheet, about 18 m thick, outcrops approximately 20 m below the base of the Lower Rhyolitic Tuff Formation. The massive central part of the flow is columnar-jointed, and in places its top is blocky. The basalt comprises few foliated, albitised feldspar laths and few microphenocrysts, up to 1.8 mm, in a groundmass of chlorite,

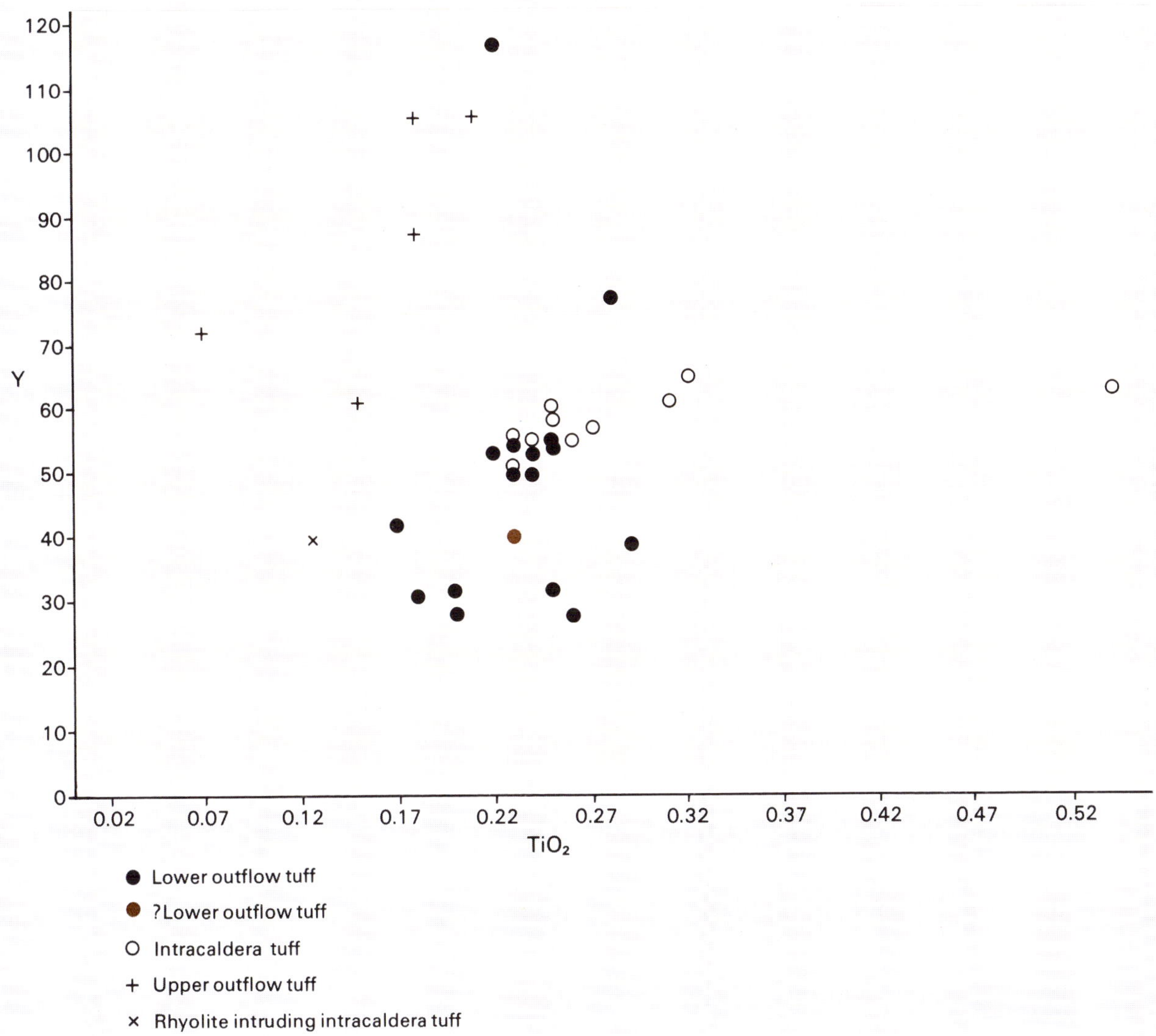

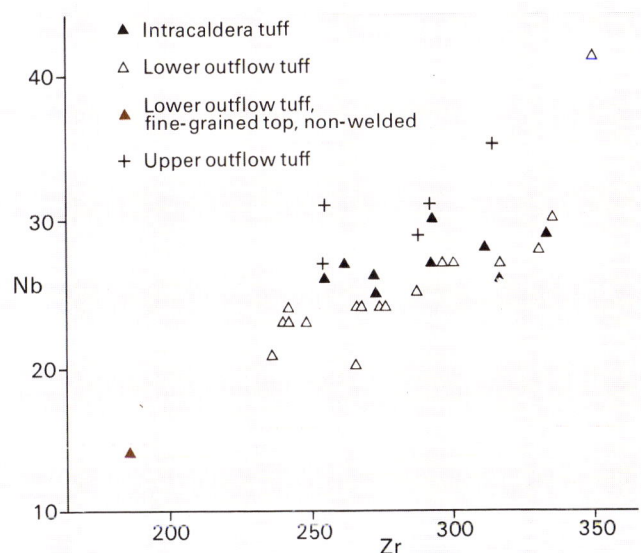

Figure 42 Pitts Head Tuff Formation, Y vs. TiO$_2$ diagram showing the geochemical similarity of the intracaldera and lower outflow tuff and their discriminance from the upper outflow tuff.

Figure 43 Pitts Head Tuff Formation, Nb/Zr diagram.

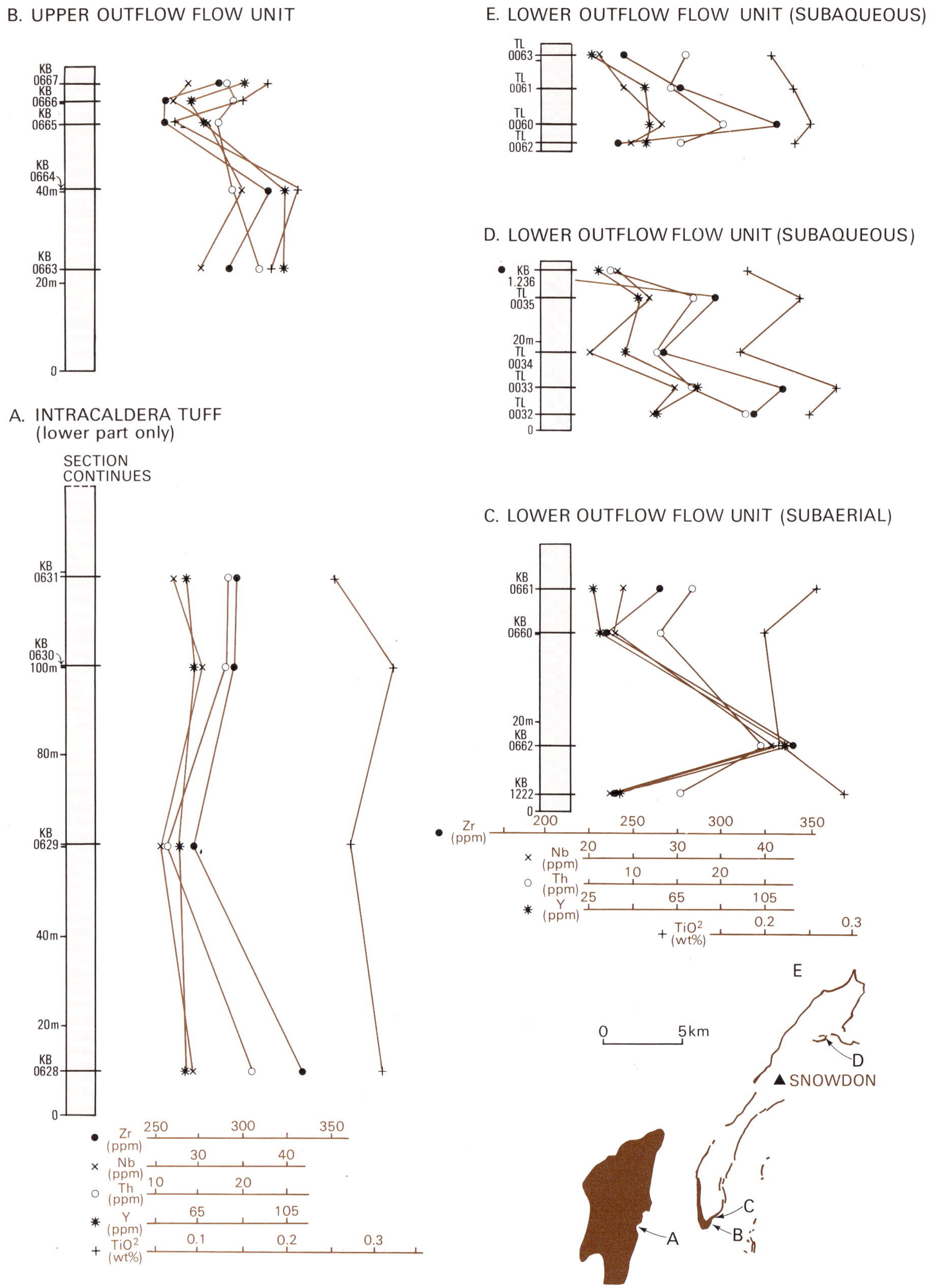

Figure 44 Pitts Head Tuff Formation, immobile trace element variations in serial sections of the intracaldera tuff (A), upper outflow tuff (B) and lower outflow tuff (C), subaerial, and (D) and (E), subaqueous. Location of Sections A–E indicated on outcrop map.

Plate 20   Trough cross-bedding and parallel lamination in sandstones cropping out above the lower outflow tuff of the Pitts Head Tuff Formation in its most distal expression, Llyn Cwm-y-ffynnon [SH 646 564].

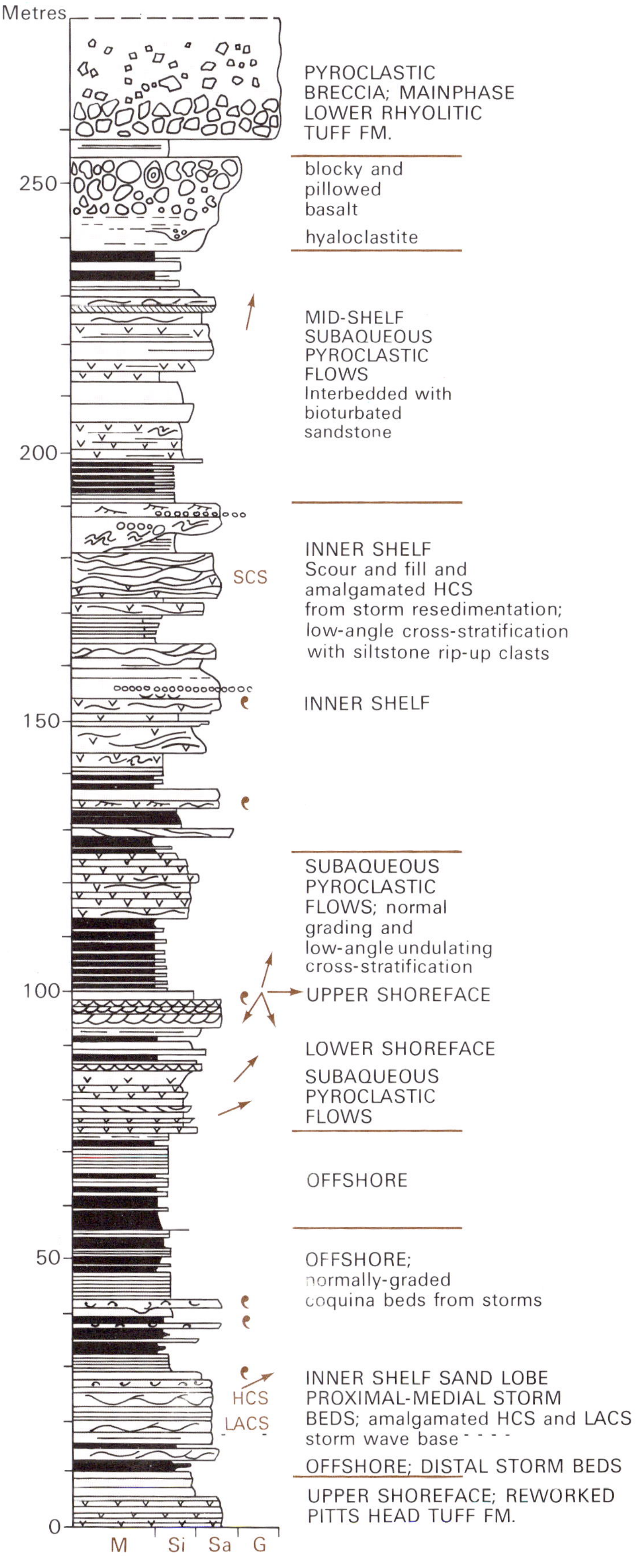

Figure 45   Graphic log and environmental interpretation of sequence above Pitts Head Tuff Formation and below Lower Rhyolitic Tuff Formation, Cwm Bochlwyd (SH 6438 5806 to 6425 5828) (after Orton, 1988a). Key as for Figures 22, 33 and 34.

PYROCLASTIC BRECCIA; MAINPHASE LOWER RHYOLITIC TUFF FM.

blocky and pillowed basalt

hyaloclastite

MID-SHELF SUBAQUEOUS PYROCLASTIC FLOWS Interbedded with bioturbated sandstone

INNER SHELF Scour and fill and amalgamated HCS from storm resedimentation; low-angle cross-stratification with siltstone rip-up clasts

SCS

INNER SHELF

SUBAQUEOUS PYROCLASTIC FLOWS; normal grading and low-angle undulating cross-stratification

UPPER SHOREFACE

LOWER SHOREFACE

SUBAQUEOUS PYROCLASTIC FLOWS

OFFSHORE

OFFSHORE; normally-graded coquina beds from storms

INNER SHELF SAND LOBE PROXIMAL-MEDIAL STORM BEDS; amalgamated HCS and LACS storm wave base - - - -

HCS
LACS

OFFSHORE; DISTAL STORM BEDS

UPPER SHOREFACE; REWORKED PITTS HEAD TUFF FM.

Metres

250

200

150

100

50

0

M   Si   Sa   G

Plate 21   The base of the Snowdon Volcanic Group exposed in the western wall of Cwm Tregalan. PT—Pitts Head Tuff Formation; B—basalt; WLRT—basal welded unit of the intracaldera facies of the Lower Rhyolitic Tuff Formation. [SH 606 535].

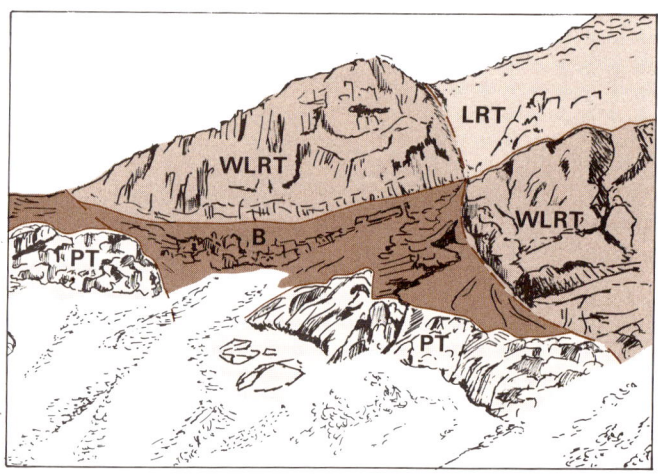

carbonate, disseminated ilmenite and fine acicular actinolite crystals. The few vesicles are infilled with carbonate and chlorite.

To the north, extensive outcrops of pillowed basalt, basaltic breccias and basic tuffs underlie the Lower Rhyolitic Tuff Formation on the south-east limb of the Cwm Idwal syncline (Locality 14, Figure 6; Figure 45). At two localities, basaltic breccias intrude the base of the Lower Rhyolitic Tuff Formation. These are interpreted as vents which initially were the source of the sub-Lower Rhyolitic Tuff Formation basalts and were later reactivated in Bedded Pyroclastic Formation times.

South-east and east of Snowdon, two small accumulations of basaltic breccia, hyaloclastite and associated basaltic sandstones are unconformably overlain by the Lower Rhyolitic Tuff Formation, indicating further basaltic vents, active prior to the main Lower Rhyolitic Tuff Formation eruptions.

South of Snowdon, in Cwm Tregalan (Plate 21), the interval between the lower Pitts Head outflow tuff and the Lower Rhyolitic Tuff Formation is occupied by up to 100 m of pillowed basalt and pillow breccia with lenses of basaltic sandstones. In the eastern part of the outcrop, the basalts overlie sandstones containing coarse beds consisting almost entirely of clasts of welded Pitts Head Tuff. However, south-eastwards they transgress on to much lower siltstone horizons, possibly several hundred metres below the Pitts Head Tuff. To the west and south-west, the pillowed basalts are replaced by massive and flow-banded basalt representing an intrusive and partially extrusive complex between the lower Pitts Head Tuff and the base of the intracaldera tuffs of the Lower Rhyolitic Tuff Formation (Plate 21). Because of their general concordance with these overlying acidic tuffs of the Lower Rhyolitic Tuff Formation, in contrast to their strongly discordant relationship with underlying sediments in the south-east, the Cwm Tregalan basalts have previously been included in the Lower Rhyolitic Tuff Formation (British Geological Survey, 1989). However, because of their distinctive geochemistry they are here included in the general category of 'sub-Lower Rhyolitic Tuff Formation' basalt. The nature of their basal contact is important in constraining the age of pre-Lower Rhyolitic Tuff Formation tectonism, related to the early development of the Snowdon Centre.

This extensive expression of basaltic volcanism prior to the eruption and emplacement of the ash-flow tuffs of the Lower Rhyolitic Tuff Formation is petrogenetically significant. The disposition of the basaltic centres is difficult to determine because of limited outcrop, but a crude north-east–south-west alignment of the basaltic accumulations can be distinguished (Figure 49), suggesting deep-seated fracture control.

GEOCHEMISTRY   Selected analyses of the sub-Lower Rhyolitic Tuff Formation basalts are presented in Appendix 1, Table 7. On a plot (Figure 46) of Zr/TiO_2 vs. Nb/Y the basalts form a coherent group within the field of andesite/basalt. Compared with all other compositions within the Snowdon Volcanic Group (Figure 37), they have markedly lower Nb/Y ratios, principally reflecting their very low relative contents of Nb. Chondrite-normalised REE patterns are generally flat (Figure 47), with a slight LREE enrichment. Eu anomalies are present in some samples though this, to

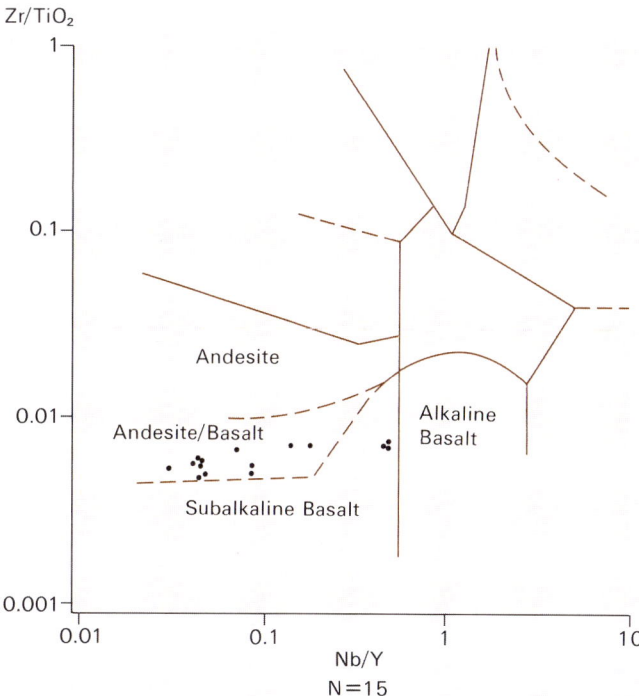

Figure 46   Basalts below the Lower Rhyolitic Tuff Formation ('sub-LRTF basalts'), Zr/TiO_2 vs. Nb/Y diagram.

some extent, may reflect alteration since both positive and negative anomalies occur. They are strongly depleted in Nb, Zr, TiO_2, P_2O_5, LREE and, to a lesser extent, Y and HREE, and hence represent a separate magmatic composition from later basalts. The only exception to these generalities is provided by the basalt below the base of the Lower Rhyolitic Tuff Formation in Llanberis Pass. In all respects, it is comparable to other basalts within the Snowdon Volcanic Group. Consequently, it is not considered to be related to the sub-Lower Rhyolitic Tuff Formation basalt magmatic group and may in fact not be extrusive as was previously thought (Howells et al., 1985a), but an intrusion related to later basaltic magmatism.

In terms of an N-MORB normalised multi-element plot (Figure 48), the sub-Lower Rhyolitic Tuff Formation basalts are particularly characterised by a negative Nb anomaly, lesser negative Yb and P anomalies, and a comparatively flat HFS element profile, slightly depleted relative to N-MORB. There is also a significant positive Ni and Cr anomaly. Enrichment in Rb to Sr relative to MORB is related to secondary alteration (no data for Th is available). In terms of immobile elements, the sub-LRTF basalts are compositionally comparable to some of the dolerite and basaltic dolerites which intrude the sequence below the level of the Lower Rhyolitic Tuff Formation, with the exception of those dolerites to the south-east of Snowdon. They differ markedly from all dolerites which intrude the sequence above the base of the Lower Rhyolitic Tuff Formation. In general terms, the N-MORB normalised profiles are comparable with those of volcanic-arc basalts and specifically with island-arc tholeiites (cf. South Sandwich Islands; Pearce, 1982b).

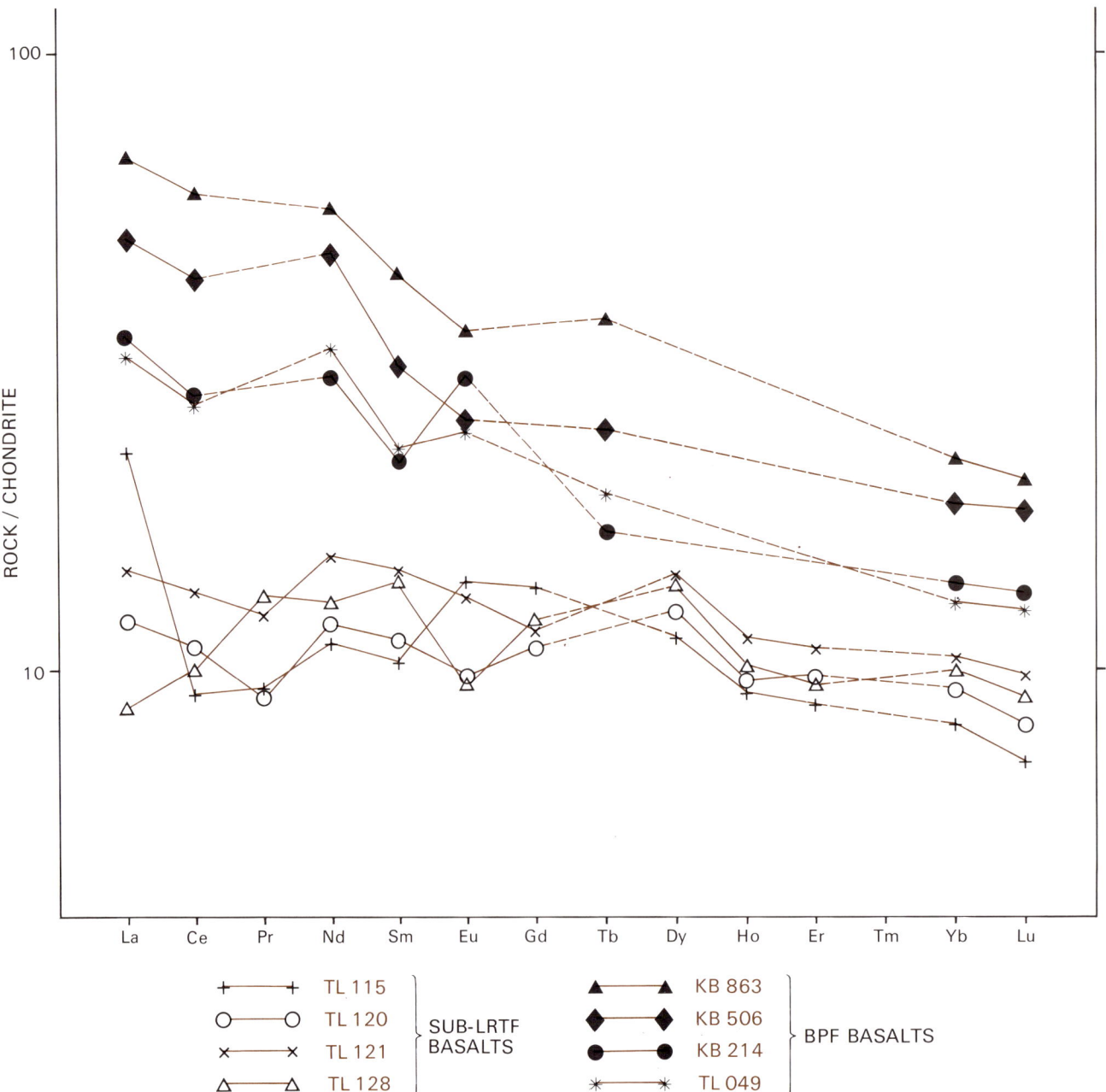

Figure 47   Sub-LRTF basalts and basalts of the Bedded Pyroclastic Formation (BPF), Chondrite-normalised REE patterns (normalising factors after Thompson.et al., 1984).

SNOWDON CENTRE

The deposits produced during the eruptive activity at the Snowdon Centre dominate the geology of the Snowdon massif itself and its periphery. A precursor minor acidic phase (Yr Arddu Tuffs) on the south-east side of the Snowdon massif was followed by a major acidic caldera-forming phase (Lower Rhyolitic Tuff Formation), a subsequent phase of basaltic volcanism (Bedded Pyroclastic Formation) and a late, possibly caldera-forming, acidic phase (Upper Rhyolitic Tuff Formation). The development of the centre was complex and

vent locations, largely controlled by the pre-existing tectonic framework, migrated with time.

Tectonic framework

Four north-east–south-west to north-north-east–south-south-west-trending fracture zones repeatedly influenced the development of the Snowdon Centre. One, in the north-west, extended from Moel Hebog to Cwm Idwal, another, in the central area, from Moel Ddu through Beddgelert and Cwm Llan, and two, in the south-east, through Nantmor and

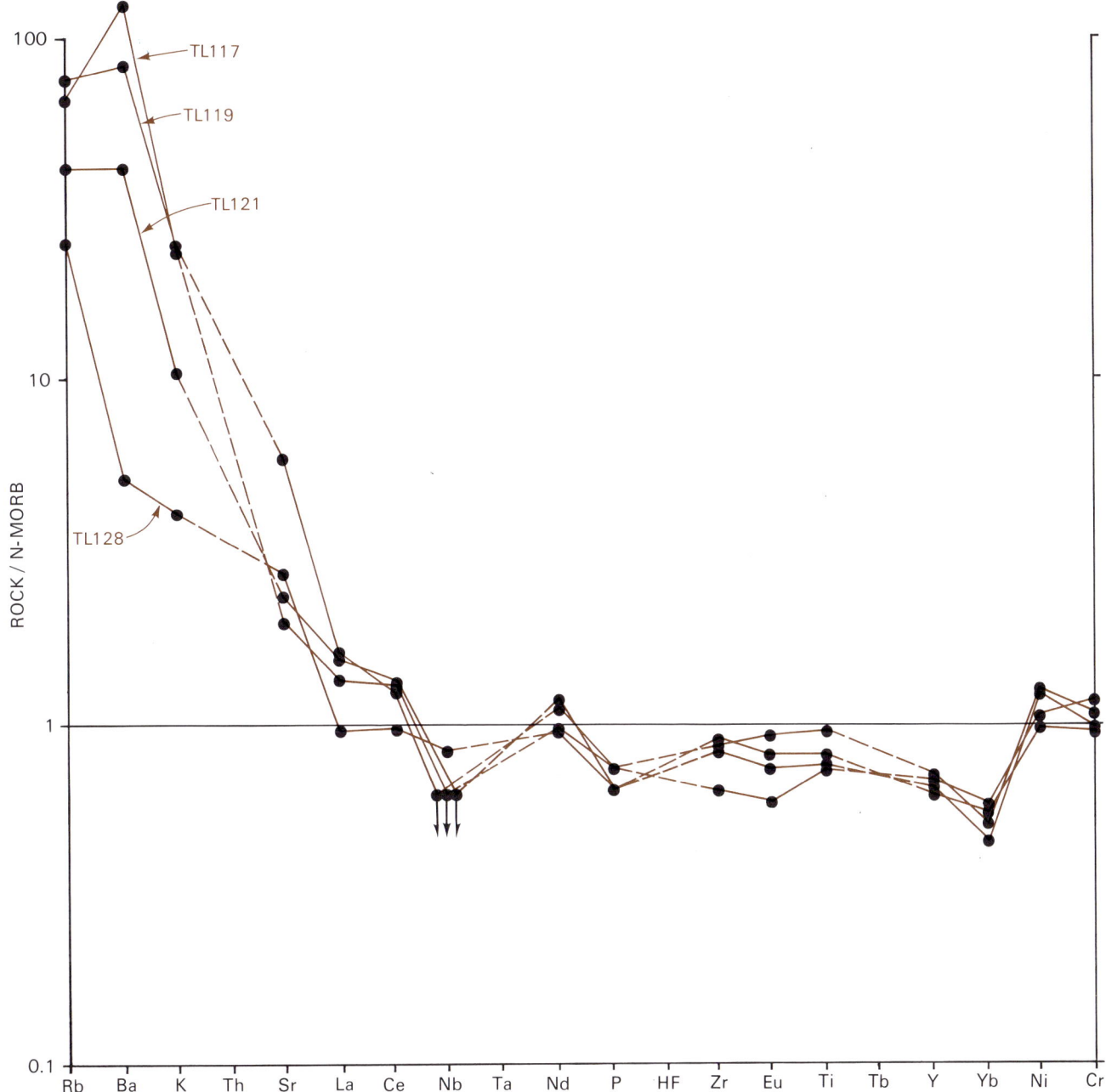

Figure 48   Sub-LRTF basalts, N-MORB normalised multi-element diagram (normalising factors after Pearce, 1982a). Arrows below data for Nb indicate values below the analytical detection limit as plotted.

Yr Arddu respectively (Figure 49). Each of these zones reflects the upward propagation of pre-existing basement faults into the cover sequence. To what extent the original basement fault motions were dominantly normal or strike-slip is largely a matter for conjecture. They controlled magma movement at depth, the eventual sites of eruption and, together with subsidiary east–west to east-north-east–west-south-west-trending faults, the form of the Lower Rhyolitic Tuff Formation caldera.

The earliest expression of tectonism related to the development of the Snowdon Centre is seen in an area about the Beddgelert Fault Zone where there is a marked angular discordance between the Lower Rhyolitic Tuff Formation and its substrate. The area is defined by a periclinal anticline (Figure 50) in the substrate, and beyond it the contact is generally conformable, though in places disconformable. The contrast between the anticline in the substrate and the broadly synclinal structure, with a subparallel axis, in the overlying tuffs indicates that the former developed prior to the initiation of volcanism. Both structures were eventually tightened by early Devonian folding.

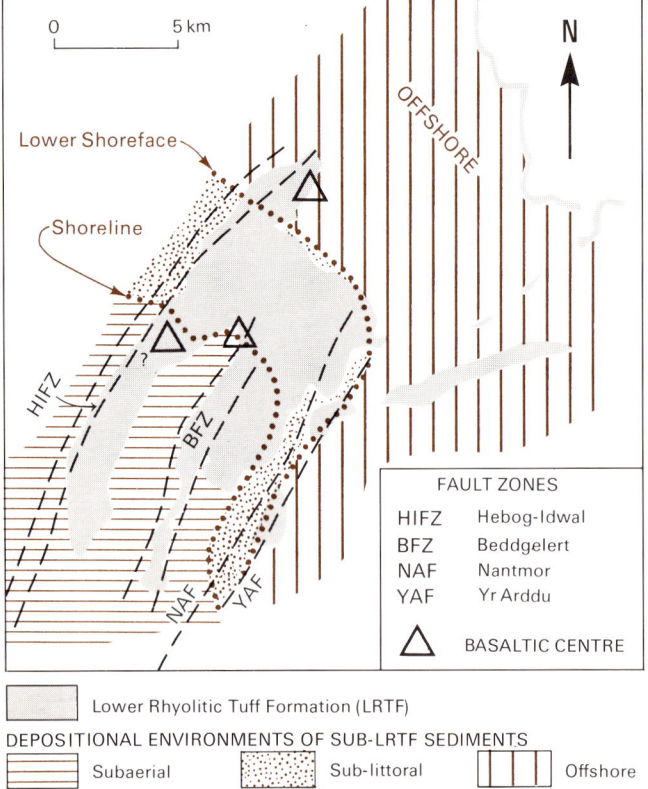

Figure 49 Sketch map showing the disposition of fault zones and the shoreface prior to the onset of the eruptive cycle at the Snowdon Centre and the positions of sub-LRTF basalt eruptive centres.

This pre-Lower Rhyolitic Tuff Formation pericline, here referred to as the Beddgelert pericline, is unique in the outcrop of the Caradoc volcanic sequence and, in the regional extensional setting, is difficult to explain by conventional folding. The structure developed in unlithified sediments; locally, where the basal tuffs lie with marked discordance on the sediments, the tuffs form lobe-like, downward protrusions into the substrate (Beavon, 1963, fig. 3b). Also, in the area of marked discordance, there is little evidence of reworked sediments beneath the tuffs. All the sediments exposed within the pericline are of Caradoc age and most are probably no older than the Soudleyan Stage. There was insufficient time following sedimentation to allow more than limited dewatering prior to folding and the subsequent emplacement of the Lower Rhyolitic Tuff Formation during the Longvillian Stage. The most plausible explanation of the structure is that it represents near surface deformation of the sedimentary cover caused by the upward movement of magma, in the vicinity of the Beddgelert Fracture Zone.

Similar structures have been distinguished from seismic profiles about salt diapirs in the Gulf of Mexico (Bally, 1983) and have been generated by experimentally produced diapirs (Jackson and Talbot, 1989). With such a mechanism the structure would develop incrementally and its surface expression would be limited by outward lateral movement of slumped sediments off the crest, faulting and local erosion. The difficulties of establishing a consistent stratigraphy in the

sequence involved in the Beddgelert pericline, and of distinguishing clearly reworked sediments, would support such an explanation.

The lack of a stratigraphical marker in the folded sequence also inhibits recognition of precise time constraints in its development. The only firm evidence is provided in Cwm Tregalan, on the north-west limb of the anticline, where the basalt at the base of the Lower Rhyolitic Tuff Formation is markedly discordant on the substrate and hence postdates the structure. Also the restriction of the Pitts Head Tuff to the west of the structure may indicate that the latter had an even earlier topographic expression.

However, the earliest acidic eruptions at the Snowdon Centre were not in the immediate vicinity of the Beddgelert pericline but occurred along the parallel-trending Yr Arddu Fracture (Figure 49).

Early activity: Yr Arddu Tuffs

The Yr Arddu Tuffs (Howells et al., 1987) crop out in a synclinal outlier (Figures 50 and 51) on the east side of the Nantmor Fracture, on the south-east side of the Snowdon Centre. They comprise a sequence of welded ash-flow tuffs more than 180 m thick which accumulated close to their eruptive centre, now marked by later rhyolite intrusive and extrusive domes aligned along a presumed north-north-east–south-south-west-trending fissure (Howells et al., 1987). The tuffs are locally discordant on shallow-marine sandstones and siltstones. They are lithologically distinctive in their block and pumice content, and locally grade into block and ash-flow tuff and pyroclastic breccias. The pumice fragments are typically elongate, up to 35 cm long, with splayed and frayed terminations. The blocks, up to 1.5 m, are dominantly of acid tuff and rhyolite, of subangular to subrounded equant shape, and fewer, smaller clasts of siltstone, sandstone, basalt and dolerite. Siliceous nodules up to 40 cm in diameter are common, in places concentrated in zones near the base or top of flows. Although individual flow units are up to 40 m thick, they can, despite excellent exposure, only be traced for limited distances. Locally, thinner, reworked, cross-laminated and flaggy-bedded tuffs occur. Disarticulated brachiopods and echinoderm debris in cross-laminated tuffs near the base of the sequence indicate shallow-marine reworking of the top of the lowest ash-flow.

The heterogeneity of the Yr Arddu tuffs, with patchy concentration of blocks, suggests that they accumulated close to their eruptive centre. With increasing transport from the vent, the block component would be concentrated at the bases of individual flows and the latter would be more clearly defined than the impersistent units occurring in the outlier. It suggests that the eruptions were characterised by low, or suppressed (boiling-over) eruption columns.

The restricted development of reworked tuffs and the absence of deep erosion and intercalated epiclastic sediments within the sequence suggest that the eruptions were fairly continuous, without prolonged intervening breaks. The close geochemical similarity of the tuffs to the associated rhyolite intrusions suggests that the latter represent magma emplaced non-explosively into the vent area during the waning phases of the eruption. The rhyolites occur just north of the centre of the Yr Arddu outlier, close to the axis of the syncline.

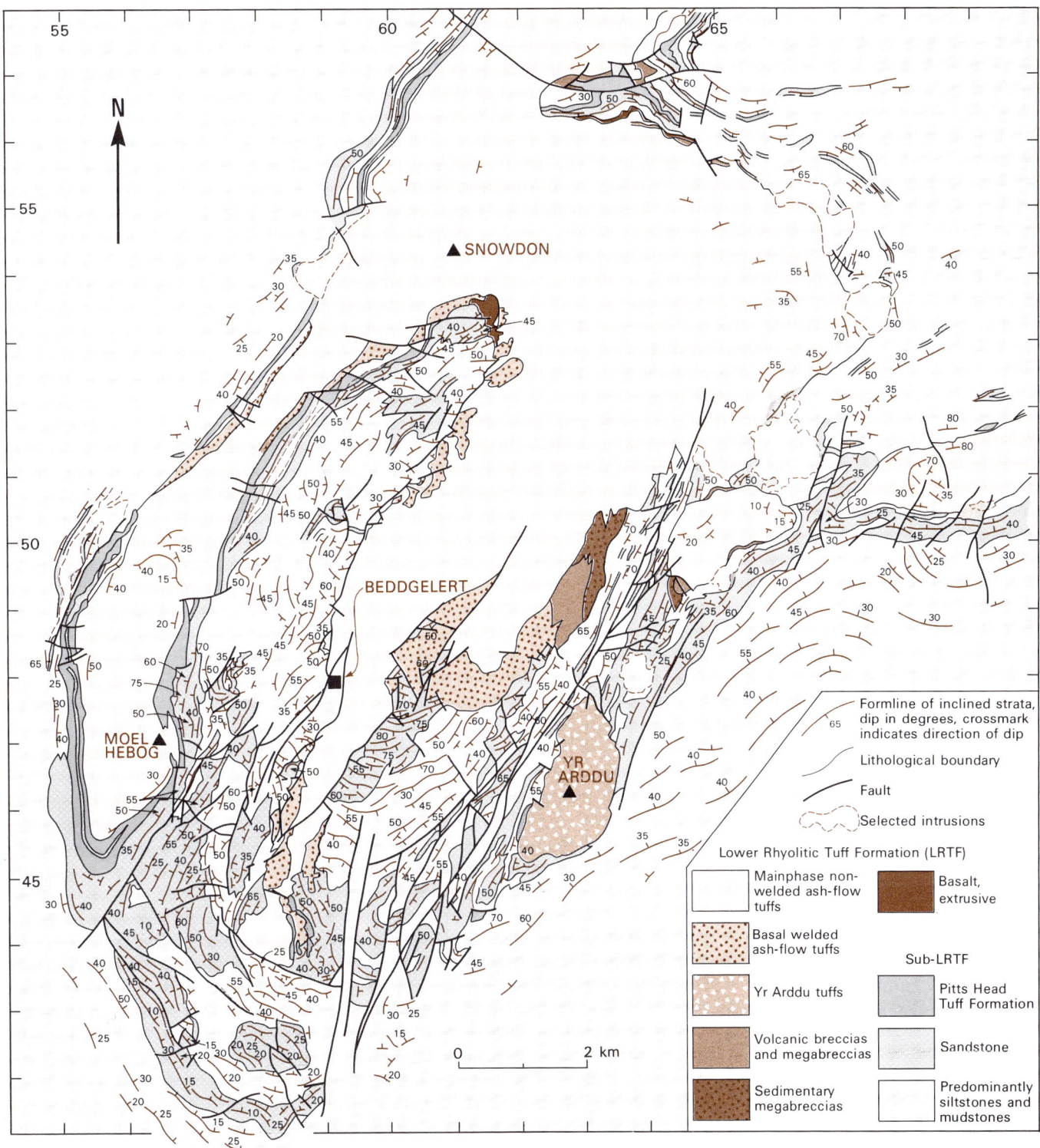

Figure 50 Sketch map, Snowdon Centre, showing the structure of the substrate to the outcrop of the Lower Rhyolitic Tuff Formation. The formlines of subjacent sediments define a periclinal anticline centred north-east of Beddgelert.

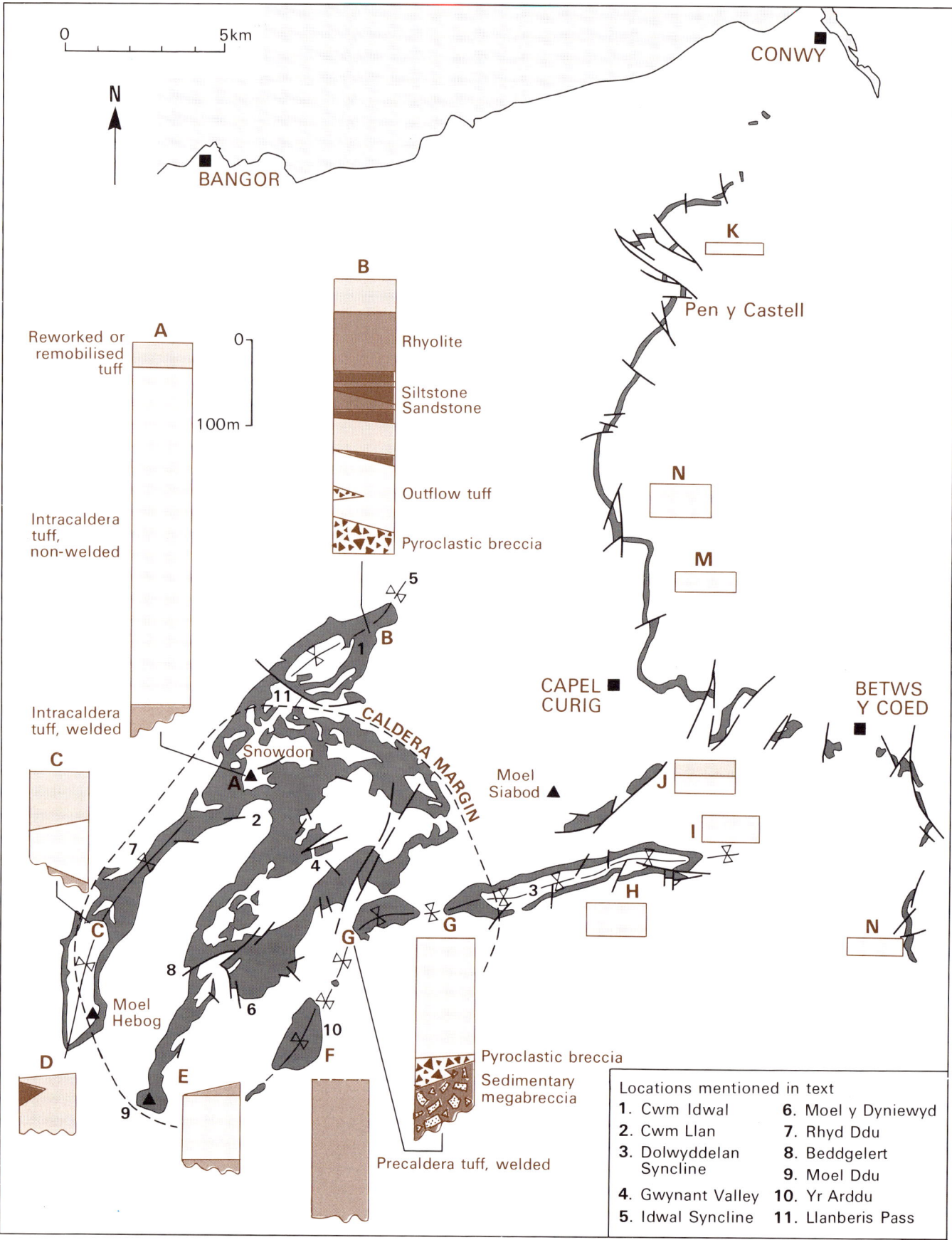

Figure 51   Outcrop and measured sections of the Lower Rhyolitic Tuff Formation (after Howells et al., 1985b). The distal outflow tuff sections, H–N, are of the lowest member (No. 1 tuff) of the Lower Crafnant Volcanic Formation.

Although this structure has been accentuated by later tectonism, it is interpreted as being, in part, primary, with the local discordance at the base of the tuffs being volcano-tectonic and the tuffs ponded in a downsag about the vent area. The character of the underlying sedimentary rocks and the marine reworking of the lower part of the tuff sequence indicate that the centre developed in a shallow-marine environment (Figure 49).

Close to Moel y Dyniewyd, north-west of Yr Arddu, a nonwelded acidic ash-flow tuff, about 40 m thick, overlying bedded accretionary lapilli tuffs (about 30 m thick), is locally exposed below the main Lower Rhyolitic Tuff Formation sequence. These tuffs are geochemically closely comparable with the Yr Arddu tuffs and are interpreted as outflow and air-fall tuffs derived from the Yr Arddu centre. To the north-east of Yr Arddu, a partly extrusive rhyolite dome (Figure 68), also geochemically comparable with and predating the main Lower Rhyolitic Tuff Formation volcanism, represents another early acidic centre along the Yr Arddu fracture.

PETROGRAPHY   The Yr Arddu tuffs comprise albite-oligoclase feldspar crystals and very rare quartz crystals, together with lithic clasts, in a matrix of devitrified recrystallised shards and vitric dust. Although all these components are invariably present, their proportion in serially collected samples is highly variable.

Feldspar crystals are either euhedral, slightly rounded or fragmented. Locally they form up to 70 per cent of the rock but elsewhere they are sparsely distributed. Clasts of acidic tuff are closely comparable with the host tuff and common rhyolite clasts are characterised by spherulitic recrystallisation of silica crowded with iron oxide grains and sericite flakes. Pumice clasts are distinctive both as collapsed elongate fragments and as chloritised irregular vesicular fragments.

The tuff matrix is characteristically heterogeneous as a result of variable states of recrystallisation, although only rarely is the original fabric obliterated. Shards are typically well preserved with eutaxitic to parataxitic textures, although occasionally nonwelded fabrics also occur. The devitrification of the original glass mainly resulted in a quartzose aggregate but the devitrified dust of the matrix also includes some sericite and chlorite. In places spherulites of radially arranged fibrous quartz, intergrown with sericite and chlorite, are well developed (Plate 17F).

The geochemistry of the Yr Arddu Tuffs is described below, together with the remainder of the Lower Rhyolitic Tuff Formation.

Main caldera activity: the Lower Rhyolitic Tuff Formation

The bulk of the Lower Rhyolitic Tuff Formation, together with its northern correlative, the lowest ash-flow tuff member of the Lower Crafnant Volcanic Formation (Figure 51), represent a major period of acidic ash-flow tuff eruption and caldera collapse in central Snowdonia (Howells et al., 1986). The formation, up to 600 m thick, comprises (Figure 51) sedimentary and pyroclastic breccias, acidic ash-flow tuffs, reworked and remobilised tuffs, rhyolites, rhyolite breccias and intercalated sediments. It rests mainly (Figure 49) on marine siliciclastic sedimentary rocks and locally on basaltic lavas, the latter showing evidence, in pillows and associated pillow breccias and hyaloclastites, of their subaqueous emplacement. The contact between the Lower Rhyolitic Tuff Formation and its substrate varies from conformable to disconformable and unconformable (Figure 50).

The thickness of the primary ash-flow tuff component of the Lower Rhyolitic Tuff Formation varies considerably across its outcrop (Figure 51) and the variations are broadly coincident with internal facies changes. The thickest sequence occurs in the central area, about the Snowdon massif, and from here it thins markedly in all directions. An estimated volume of 30 km^3 (minimum) occurred within the 200 m isopach (Figure 52) in the central area and, outside it, a minimum of 20 km^3.

From the evidence of thickness variations of the primary ash-flow tuffs, the character of their basal contacts, their internal facies and the distribution of associated rhyolites (Figure 52), it is considered that the central area represents a caldera, about 12 km in diameter.

A palaeogeographic reconstruction (Figure 49) indicates that, during the interval following the eruption of the Yr Arddu Tuffs and the commencement of the major caldera-forming eruptions of the Snowdon centre, a shoreline, with a low coastal plain in the south-west and deepening to the north and east, lay in the area where the caldera subsequently developed. The early caldera-forming eruptions probably commenced subaerially in the south but later migrated northwards beyond the shoreline.

EARLY FISSURE-CONTROLLED ERUPTIONS

To the north-west of Yr Arddu welded, acidic, ash-flow tuffs occur at the base of the Lower Rhyolitic Tuff Formation sequence over much of the south-west sector of the caldera (Figures 51 and 52) (Howells et al., 1986). They are distributed about the area of marked basal discordance, across the previously formed Beddgelert pericline, and in close proximity to the Beddgelert Fault Zone. Their distribution suggests that the vent occurred along the latter.

The current level of erosion through the tuff sequence above the folded substrate has exposed tuffs within the narrow downfaulted north-east–south-west graben of the Beddgelert Fault Zone (Figure 49) on the south-west side of the caldera. The relationships suggest that fault activity along the zone occurred pre-, syn- and post-eruption and further suggest that the zone was the site of an active fissure vent.

To the east of the Beddgelert Fault Zone, thick wedges of sedimentary and volcanic megabreccias accumulated at the base of the tuff sequence (Beavon, 1963; Howells et al., 1986). The sedimentary megabreccias contain rafts, up to 50 m long, of locally derived sediment in a mudstone to sandstone matrix. They appear to interdigitate with and be partly overlain by the volcanic megabreccias containing large rafts and blocks of acid tuff and sedimentary rocks in a tuffaceous matrix. The breccias probably developed by collapse associated with movement of the Beddgelert Fault Zone during enhanced uplift and instability immediately prior to the onset of volcanism (Beavon, 1963; Howells et al., 1986), and then during the construction of a flared vent along the fault zone, with the eruption of the basal welded acidic ash-flow tuff. In these ways they differ from the breccias directly associated with caldera formation as described by Lipman (1976) in the San Juan Mountains.

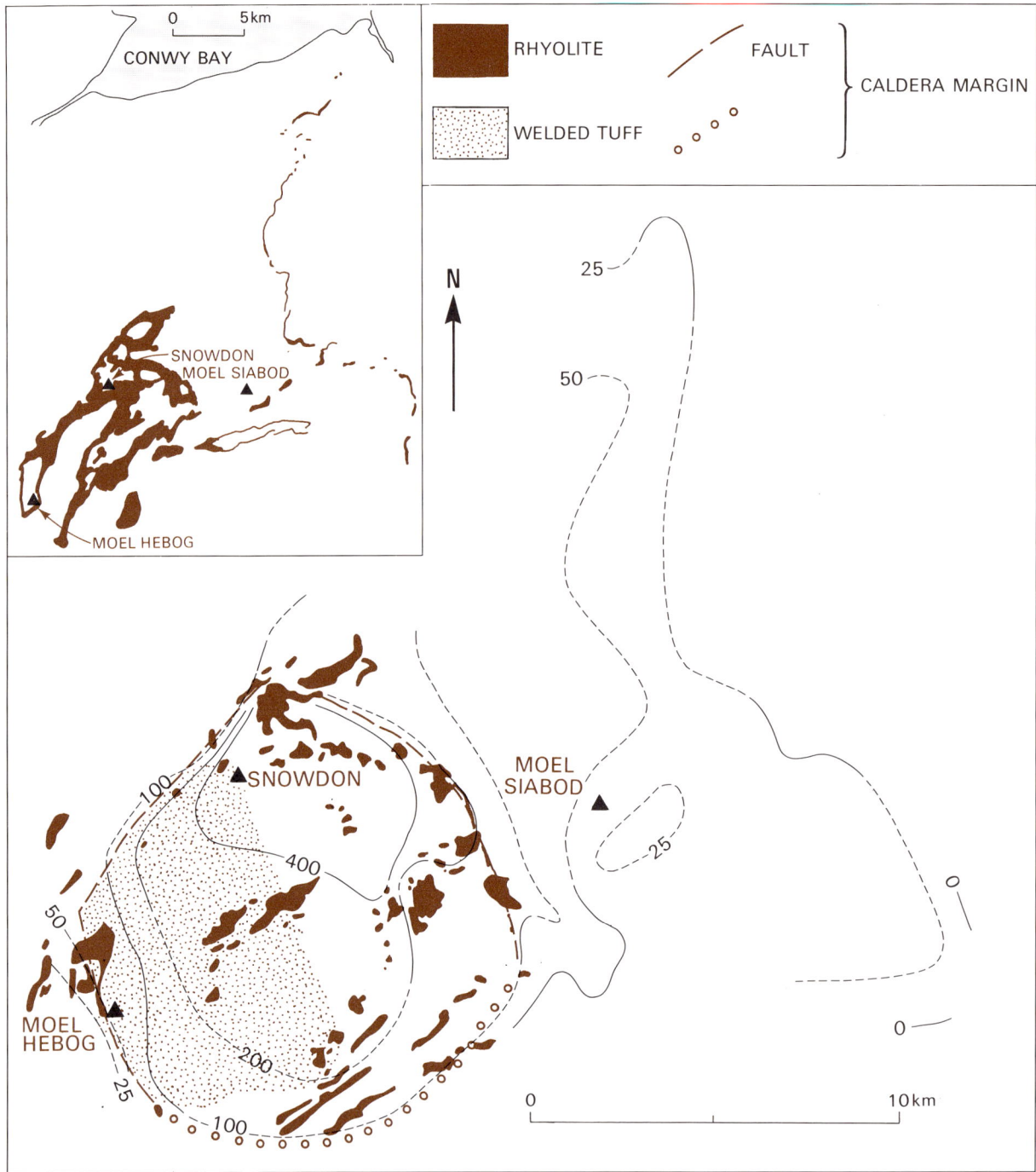

Figure 52   Lower Rhyolitic Tuff Formation, isopachs (metres) of the primary ash-flow tuffs and the distribution of welded tuff and associated rhyolites, palinspastically restored (assuming concentric folding). The current outcrop pattern of the LRTF is inset (after Howells et al., 1985b).

Plate 22   View north-westwards from near Yr Arddu. Moel y Dyniewyd [SH 612 478], in middle distance, formed of the basal welded tuff of the intracaldera facies of the Lower Rhyolitic Tuff Formation. The near-horizontal tuffs overlie steeply dipping sediments on the south-east limb of the Beddgelert pericline.

The basal welded tuff of the south-west sector is well exposed about Moel y Dyniewyd (Plate 22), Moel Ddu and between Cwm Tregalan and Moel Hebog (Figures 51 and 52). It is typically white-weathered and intensely jointed with, in places, well-developed columnar joint sets. At Moel y Dyniewyd its basal zone, up to about 8 m thick, is crowded with siliceous nodules. The overlying tuff, up to about 150 m thick, is fine-grained and silicified with a distinctive even, planar welding foliation. In its outcrop between Beddgelert and Moel Ddu, in the southern end of the Beddgelert Fault Zone, the welded tuff overlies and encloses large slide blocks and rafts of the Pitts Head Tuff (Howells et al., 1986; Reedman et al., 1987a). It is itself unconformably overlain and overstepped to the south by nonwelded acidic tuffs.

To the north-west, in the syncline between the west side of Cwm Llan and Moel Hebog, the basal welded tuff (Plate 23A) forms a prominent feature, up to 30 m thick, on the south-east limb and thins to less than 10 m on the north-west limb. Locally, the welded tuff is underlain by a thin sequence (<60 cm) of bedded nonwelded tuffs. The main body of the tuff is silicified, well foliated and locally columnar-jointed,

with siliceous nodules mainly concentrated at its base. In places the top is not clearly defined because the silicification extends up and grades into the overlying nonwelded, ash-flow tuffs.

On Moel Hebog the welded tuff is generally <20 m thick. It wedges out to the south and is overlain by nonwelded, lithic-rich tuff, locally with a 1 m thick, clast-supported breccia at its base. Welding to the upper contact suggests that the upper part of the flow unit was eroded prior to deposition of the overlying breccia.

The basal tuff, in its distribution, lithological characters, geochemistry, and locally its clearly defined upper contact with the overlying tuffs, indicates that it represents a distinct primary, welded, ash-flow tuff unit rather than the basal welded zone of a much thicker ash-flow deposit. Its vent may have lain in the Beddgelert Fault Zone, north-west of Moel y Dyniewyd. For much of its outcrop in the south-west sector the internal features of the ash-flow tuff, its locally well-developed basal nodular zone and its even foliation, suggest that the tuff was emplaced in a subaerial environment. In the vicinity of Cwm Tregalan, its relationship to pillowed basalts

A

B
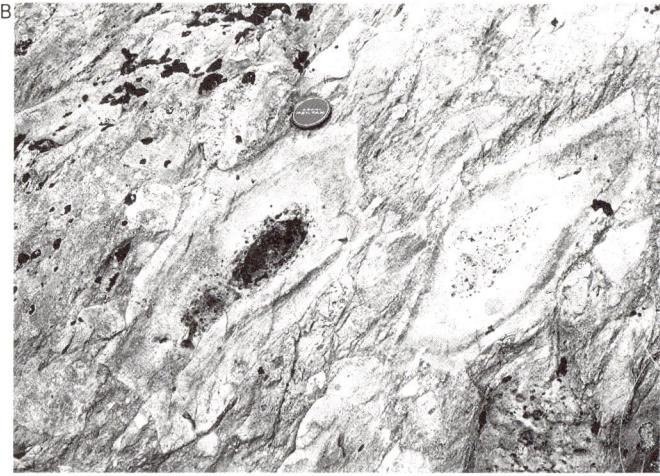

Plate 23

A   Variable silicification in the basal welded tuff of the intracaldera facies of the Lower Rhyolitic Tuff Formation cropping out in Cwm Clogwyn, west of Snowdon. In the lower part (A) irregular, silica-rich nodules are deformed in the cleavage. In the central, more evenly silicified zone (B), both cleavage and the welding foliation are displayed, and in the upper zone (C) the welding foliation dominates. [SH 594 542].

B   Clasts of pillowed basalt in pyroclastic breccia near the base of the intracaldera facies of the Lower Rhyolitic Tuff Formation, Cwm Clogwyn. [SH 555 473].

in the substrate indicates that it extended marginally into a marine environment. However, the exact configuration of the shoreline is difficult to establish because of the restriction of outcrop.

The basal welded tuff unit cannot be traced in continuity in the Lower Rhyolitic Tuff Formation outcrop south of Cwm Tregalan, on the east side of Cwm Llan. Here the basal tuffs are impersistently welded but extensive 'pods' of silicified tuff, up to 500 m across, with an even-banded foliation occur which, at a few localities, can be shown to be a relict welding fabric. These large silicified tuff pods are geochemically similar to the basal welded unit (Figure 56) but the relationship remains unclear. Their occurrence is restricted to the vicinity of the Beddgelert Fault Zone, close to the postulated fissure vent, which suggests that they formed in response to localised hydrothermal activity.

PETROGRAPHY   The basal welded tuffs on the south-west side of the caldera comprise a very fine-grained aggregate of quartz, sericite and chlorite with a foliation defined by segregation of coarser quartz recrystallisation (Plate 17C). In most of the tuffs the only indications of the original texture are isolated sericitised and chloritised feldspar phenocrysts. In rare instances, however, there is clear evidence (Plate 17D) that the foliation reflects a welded shard fabric that is over-

printed (Plate 17E) by areas, or 'clouds' of recrystallisation. The contacts between the two fractions are gradational.

MAIN PHASE CALDERA-FORMING ERUPTIONS

The main phase of the Lower Rhyolitic Tuff Formation eruptions emplaced up to 500 m of almost entirely massive, nonwelded acidic ash-flow tuff within the caldera about the Snowdon massif (Figure 51). The north-western and northern margin of the caldera is clearly defined by the pronounced thickness changes which occur about the 200 m isopach of the tuffs (Figure 52) in the north sector of the caldera, which probably lies close to the eruptive centre. The latter was probably sited on the Beddgelert Fault Zone, to the north-east of the earlier fissure vents, and reflects a north-eastern shift in the locus of activity soon after, and probably in response to the eruption of the early welded ash-flow tuff.

On the south and south-east sides of the caldera the thickness variations are more gradual and here the form of the caldera is more clearly distinguished by the distribution of the associated intrusive and extrusive rhyolites (Figure 52). A single outflow tuff, up to 55 m but generally <35 m thick, can be traced some 25 km to the north-east and east of the caldera.

The basal zone of the thickest accumulation of intracaldera tuffs (Plate 3), close to the north edge of the caldera,

Plate 24   Textures in ash-flow tuffs of the Pitts Head Tuff and Lower Rhyolitic Tuff formations

A   Brecciated welded ash-flow tuff of the lower flow unit of the Pitts Head Tuff Formation. Lenticular ribs of silicification accentuate the welding foliation [SH 568 467].

B   Siliceous nodules overprinting the welding fabric in the lower outflow tuff of the Pitts Head Tuff Formation, Moel Hebog. The central nodule encloses dark-coloured fiamme. [SH 568 467].

C   Clasts of welded tuff of the Pitts Head Tuff Formation in the basal part of the intracaldera facies of the Lower Rhyolitic Tuff Formation, Moel Hebog [SH 567 461].

D   Siliceous nodules concentrated along anastomosing joints in welded tuff of the intracaldera facies of the Lower Rhyolitic Tuff Formation, Moel Ddu [SH 586 456].

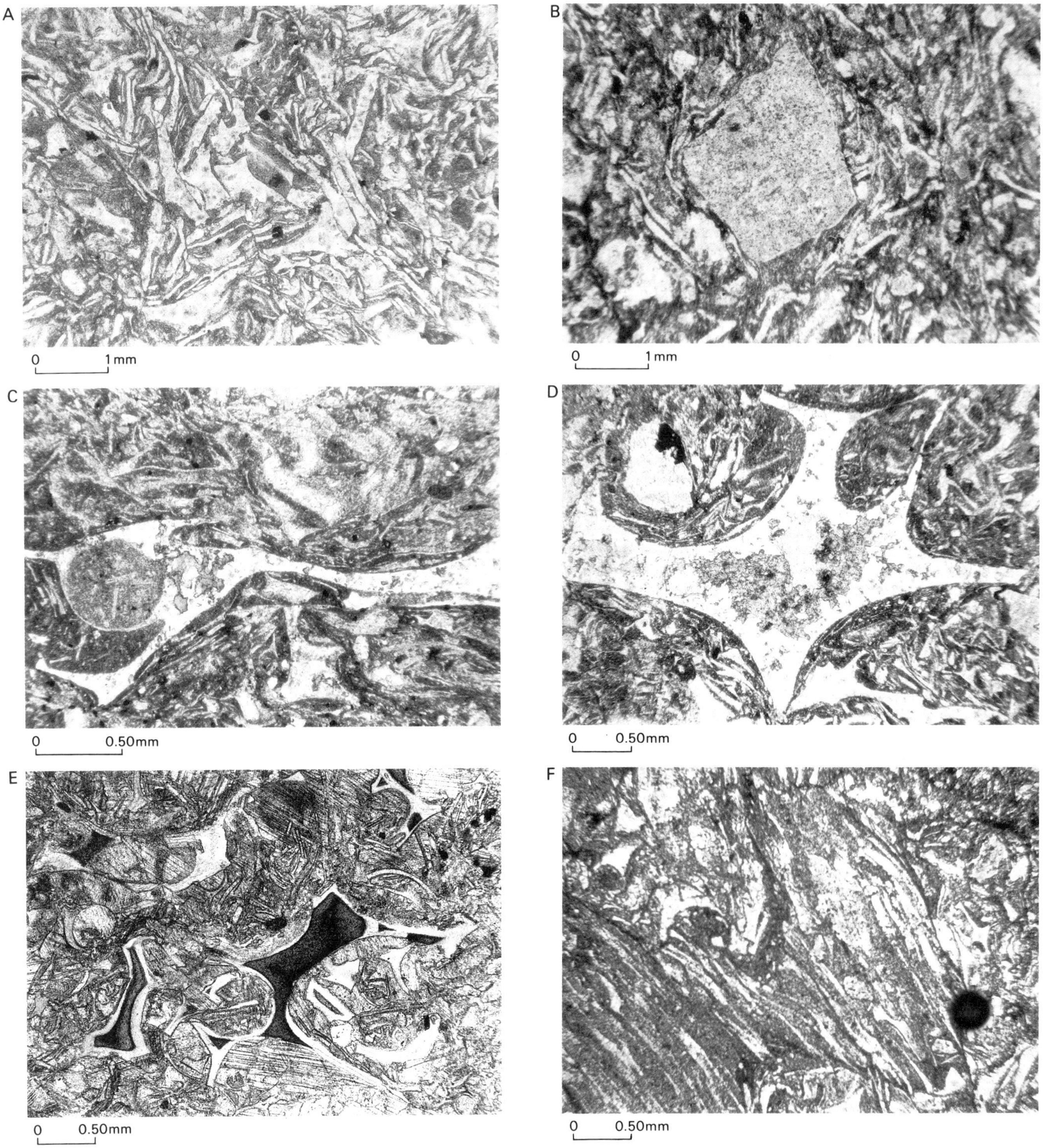

Plate 25 Photomicrographs, nonwelded ash-flow tuffs of the Lower Rhyolitic Tuff Formation

A   MH 883. Devitrified shards, a few with cuspate shapes and numerous rods and spikes, recrystallised to a fine quartzose mosaic, in a matrix of sericite and chlorite shreds. Lower Rhyolitic Tuff Formation, ppl. [SH 6340 5583].
B   MH 1215. Sericitised albite feldspar phenocryst with some moulding of adjacent devitrified shards. Lower Rhyolitic Tuff Formation, ppl. [SH 6250 4935].
C   MH 1203. Bimodal size distribution of shards, including an elongate cuspate fragment, with complete bubble form, in a matrix of finer rods and spikes. Lower Rhyolitic Tuff Formation, ppl. [SH 6279 4993].
D   MH 1244. Bimodal size distribution of shards, including a blocky cuspate fragment in a matrix of rods and spikes. Lower Rhyolitic Tuff Formation, ppl. [SH 6191 4939].
E   E 38967. Uncompacted and undeformed vitroclastic fabric, replaced by carbonate and chlorite, within a carbonate nodule. Lower Crafnant Volcanic Formation, No. 1 tuff (= outflow tuff of Lower Rhyolitic Tuff Formation), ppl. [SH 7270 5843].
F   E 38997. Tubular pumice clast in a fine shardic matrix. Outflow tuff, Lower Rhyolitic Tuff Formation, ppl. [SH 7330 6207].

comprises a massive, coarse, lithic breccia, up to 40 m thick (Figure 51; Plate 23B). It is clast supported at its base but grades up through a matrix-supported zone into the main body of the ash-flow tuff. The blocks, up to 0.7 m long, are mainly of basalt, acid tuff and rhyolite, with rare sandstone and siltstone. The matrix comprises shardic crystal tuff with little epiclastic debris. The breccias are interpreted as co-ignimbritic lag breccias (cf. Druitt and Sparks, 1982) emplaced close to the vent.

The overlying tuffs in the main accumulation are lithologically remarkably uniform through about 500 m thickness. They are massive, nonwelded, generally well cleaved and local indications of bedding foliation are weak and impersistent. Towards the south-west caldera margin, near Moel Ddu, two welded ash-flow units up to 25 m thick have been distinguished near the base of the formation but restriction of outcrop prohibits their lateral relationships being interpreted. Their geochemistry is similar to the basal welded tuffs. They probably represent local deposits from a vent in the southern part of the Beddgelert Fault Zone.

Within the thickest nonwelded tuff accumulation, in the north, narrow zones (< 2 m) of welding occur in the proximity of quartz-filled fractures, suggesting that circulation of hydrothermal fluids during cooling of the intracaldera sequence promoted welding in the adjacent wall rock. A similar mechanism, probably on a much larger scale, is tentatively suggested for the formation of very large 'pods' of silicified welded tuff near the early fissure vent on the Beddgelert Fault Zone.

The lithological uniformity of the intracaldera tuffs is not reflected in the geochemical analyses of serially selected samples (Figure 57) which suggest that their accumulation involved at least two eruptive pulses.

PETROGRAPHY   Microscopic examination of the intra-caldera tuffs corroborates the lithological uniformity determined mesoscopically. Typically they comprise a variable admixture of shards and feldspar crystals in a matrix of sericite and chlorite (Plate 25A and B). Locally, the shards are tightly packed but elsewhere the matrix component is greater, and here the tuffs are commonly well cleaved and the shards are tectonically distorted. The shards are generally < 0.2 mm, occurring mainly as fragmented rods and spikes, although multicuspate shards are common and complete bubble forms also occur (Plate 25C). Locally, a distinctly bimodal shard population can be distinguished (Plate 25C and D) with larger, thicker, cuspate and tabular fragments scattered in the densely packed finer shards of the matrix. The shards are devitrified and recrystallised, consisting of a fine quartz mosaic. They contrast sharply with the matrix of sericite and chlorite which is presumed to represent devitrified (argillised) and recrystallised vitric dust.

Crystals are of albite oligoclase feldspar and less abundant quartz, although locally, both are of rare occurrence. Typically, the crystals are of indented subhedral and hollow form, reflecting rapid, quenched crystallisation. Above the basal zone, clasts are almost entirely of small tubular pumice (Plate 25F) up to 4 mm, with extremely ragged terminations.

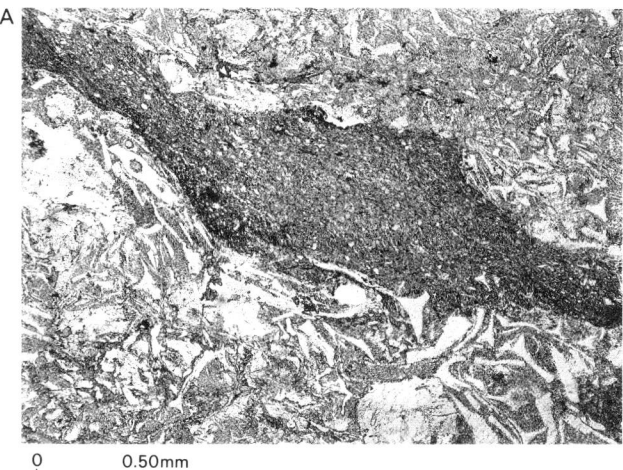

0   0.50mm

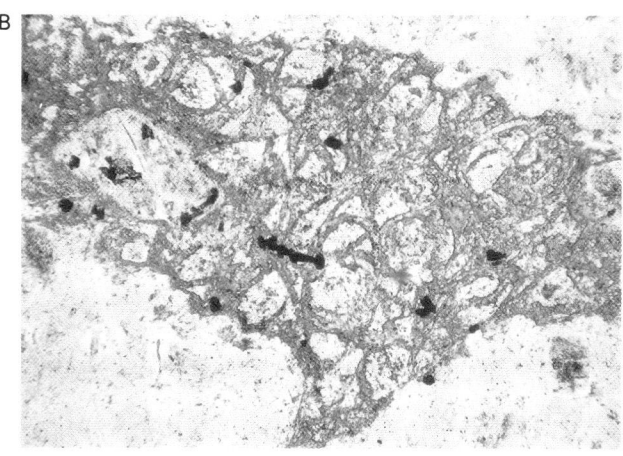

0   0.50mm

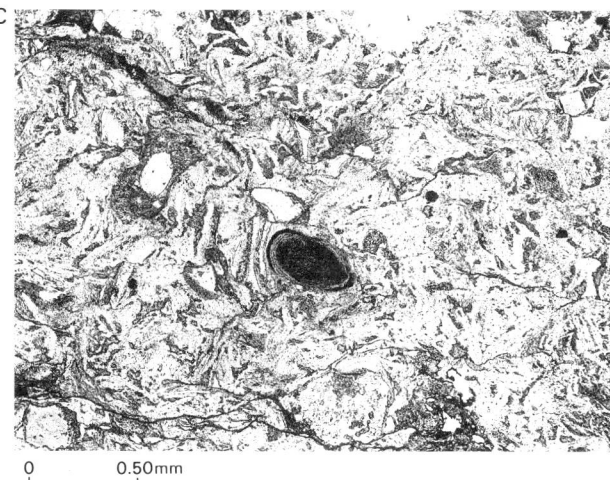

0   0.50mm

Plate 26   Photomicrographs, lithic clasts within the Lower Rhyolitic Tuff Formation

A   E 3992a. Siltstone clast in a vitroclastic matrix near base of ash-flow tuff. The irregular clast shape and included shards at its margins indicate an unlithified state on incorporation. Outflow tuff, Lower Rhyolitic Tuff Formation, ppl. [SH 7580 5775].
B   MH 1197. Clast of perlitic, fractured, devitrified glass in welded, crystal lithic tuff. Yr Arddu Tuffs, ppl. [SH 6213 4539].
C   E 39362. Ash-flow tuff. Vitroclastic fabric with an isolated oolith incorporated during flow. Outflow tuff, Lower Rhyolitic Tuff Formation, ppl. [SH 7188 5248].

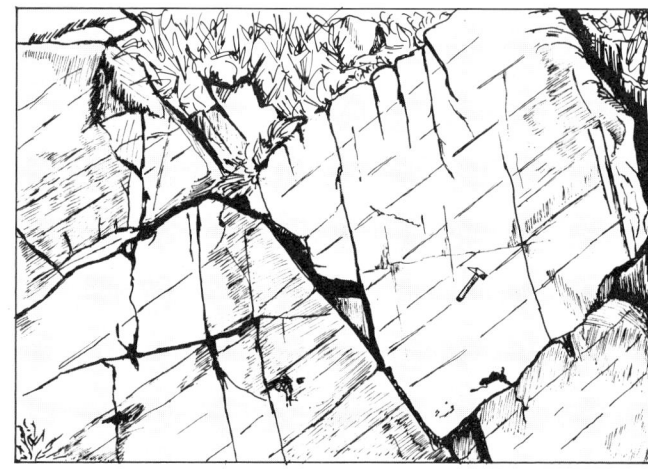

Plate 27   Weakly developed lamination in nonwelded outflow tuff of the Lower Rhyolitic Tuff Formation caldera (= No. 1 tuff of the Lower Crafnant Volcanic Formation), north-east of Capel Curig [SH 7270 5843].

PRIMARY OUTFLOW TUFFS

To the north and east of the caldera margin, the primary tuffs of the Lower Rhyolitic Tuff Formation thin abruptly and markedly. Close to the margin, at Cwm Idwal (Figure 51), the tuffs are about 110 m thick. Here they contain lenses of pyroclastic breccia, up to 20 m thick, which pass laterally into trails of isolated blocks and clasts and which are identical to the intracaldera lag breccias (see above). The breccias are generally, but not invariably found at the base and in the lowest 30 m of the tuff. They pass into and are overlain by poorly defined, massive beds, up to 1.5 m thick, of primary ash-flow tuffs with no intercalated sediments. These represent repeated pulses of ash-flows from a single eruptive phase (Howells et al., 1986).

Further to the east and north-east, the Lower Rhyolitic Tuff Formation is represented by a single outflow tuff which lies conformably in the siltstone sequence. The tuff, up to 55 m thick, can be traced some 25 km through north-eastern Snowdonia (Howells et al., 1973; Howells et al., 1985a). It comprises a crystal- and clast-rich base, a uniform central zone and a fine-grained top. Clasts in the basal zone include irregular blocks of siltstone, which were incorporated while unlithified (Plate 26A), and well-preserved brachiopods, trilobites and rare ooliths (Plate 26C), which indicate that the flow was emplaced in a marine environment. The fine-grained top, 1–6 m thick, comprises devitrified and recrystallised vitric dust which settled from the water column following emplacement of the body of the flow. It probably represents fine material elutriated mainly from the flow head.

Siliceous nodules are locally common in the outflow tuff and carbonate nodules are sparsely distributed. Welding is extremely localised, and has been recognised only at the west end of the Dolwyddelan syncline and near Pen y Castell (Figure 51) (Howells et al, 1981a). In both instances, it is considered that welding developed as a result of local insulation of the ash-flow, after emplacement.

Weakly developed bedding is a characteristic feature of the outflow tuff. In north-eastern Snowdonia, this develops from a crude internal bedding foliation (Plate 27) within the massive tuff to more well-defined beds, possibly reflecting intraflow units, away from source. The foliation probably reflects the development of laminar shear horizons within the flow. Additionally, with increasing transport, flow transformation (Fisher, 1983, 1984) due to the ingestion of water at the flow head and waning energy of the flow may be an important factor.

At the western end of the Dolwyddelan syncline (Figure 51) the main outflow tuff directly overlies a well-bedded, lithic and crystal-rich ash-flow tuff, up to 40 m thick, which locally includes shell debris near its base. To the east and north-east the lower tuff wedges out into shallow-marine siltstones and mudstones, and the ill-sorted beds incorporate increasing amounts of epiclastic debris which includes well-rounded rhyolitic pebbles (Howells et al., 1973). The relationships suggest that both tuffs transgressed a shoreline in this vicinity and that the lower tuff represents the distal accumulation of the early Beddgelert Fault Zone fissure vent eruption, and its reworking.

PETROGRAPHY

The main body of the Lower Rhyolitic Tuff Formation outflow tuff is petrographically similar to the massive facies of the intracaldera tuffs, in shard types, albite-oligoclase, feldspar crystal content and variable proportions of matrix. Shards in the basal zone are replaced by sericite and in these instances the cleavage is well developed. Above the basal zone there is a slight upwards decrease in abundance and size of the crystal fraction (Howells et al., 1973). Shard shapes within carbonate nodules (Plate 25C) are completely undistorted, suggesting that the nodules developed soon after emplacement and protected the overprinted fabric from subsequent modification.

GEOCHEMISTRY OF THE PRIMARY ASH-FLOW TUFFS

Representative analyses of the Lower Rhyolitic Tuff Formation are presented in Appendix 1, Table 7. On a plot (Figure 53) of Zr/TiO_2 vs. Nb/Y, most analyses of the Lower Rhyolitic Tuff Formation fall within the rhyolite field with a subsidiary cluster within the field of rhyodacite/dacite. A small cluster of points lies in the rhyolite field but close to the comendite/pantellerite field boundary. The few scattered points in the subalkaline basalt, basalt/andesite and andesite fields are samples of the intracaldera tuffs obtained adjacent to mineral veins, indicating that in such environments the so-called immobile trace elements are themselves susceptible to mobility. Analyses of the distal outflow facies, to the north-east of the main Lower Rhyolitic Tuff Formation outcrop, plot within the field defined by the main body of the Lower Rhyolitic Tuff Formation, substantiating their correlation with the proximal outflow and intracaldera tuffs.

The intracaldera basal welded tuffs, main phase tuffs and reworked tuffs, and the outflow tuffs, are all very similar with

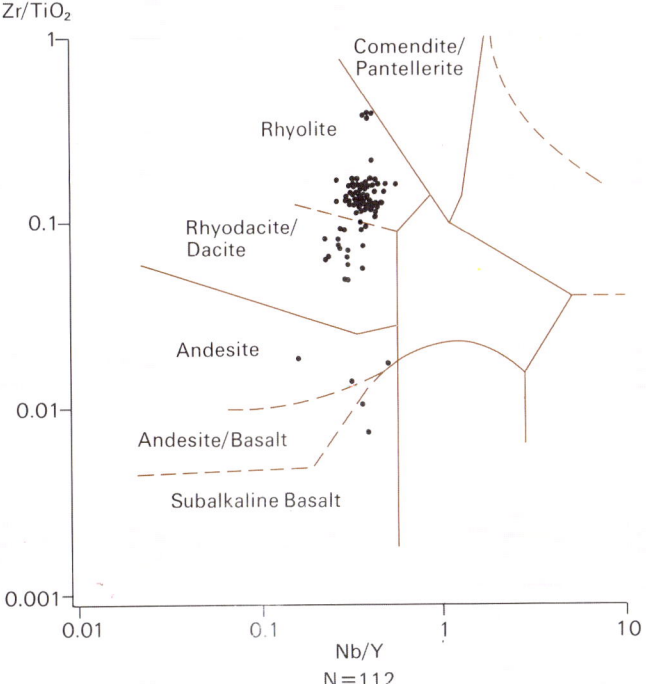

Figure 53 Lower Rhyolitic Tuff Formation, Zr/TiO_2 vs. Nb/Y diagram.

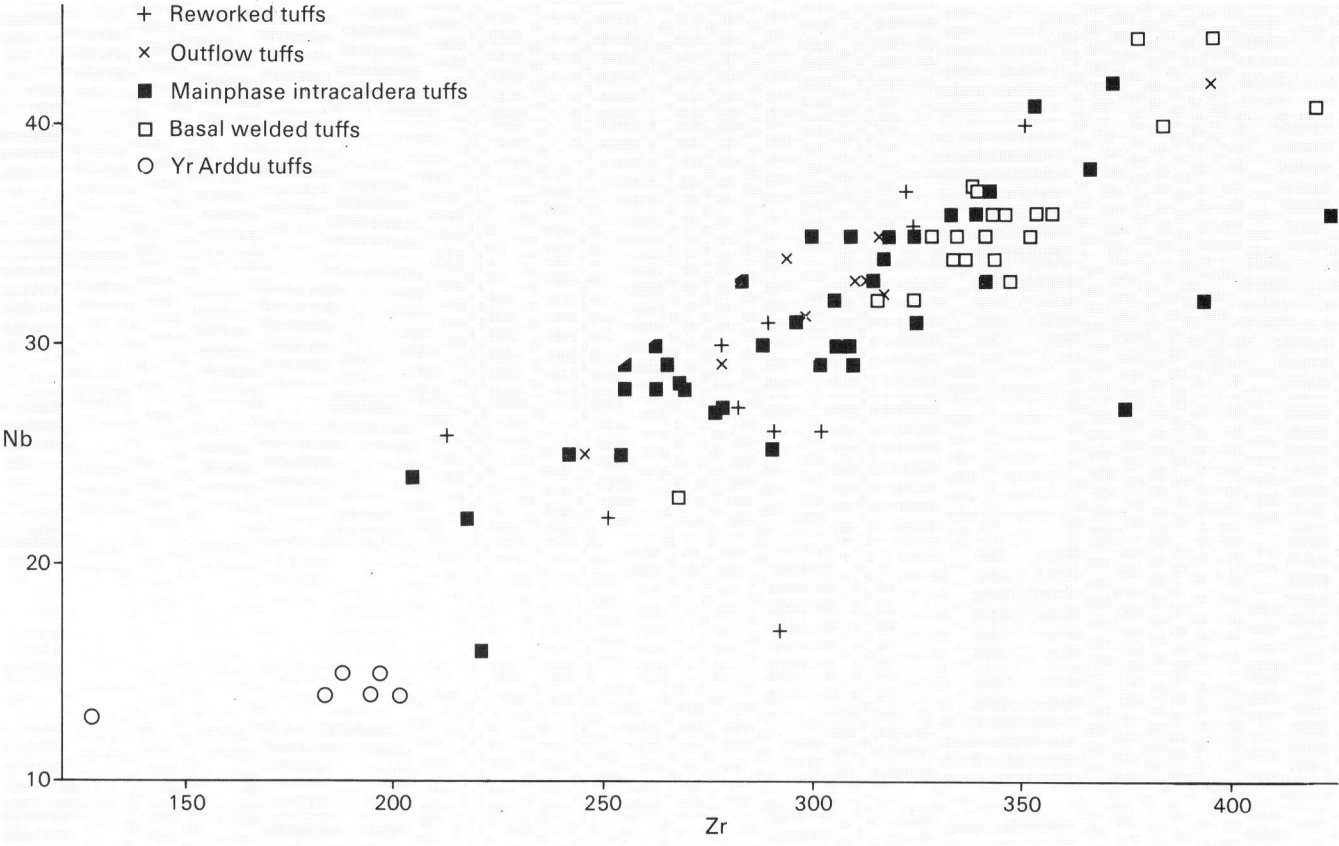

Figure 54   Nb vs. Zr diagram showing the various facies of the Lower Rhyolitic Tuff Formation.

respect to their immobile trace element contents and relative ratios. They can all be discriminated, however, from the Yr Arddu Tuffs in terms of their significantly higher concentrations (Appendix 1, Table 7) of Nb, Zr, Y, La and Ce, and to a lesser extent Ga, and particularly their lower concentrations of Th. Furthermore, although Nb/Zr ratios of the Yr Arddu Tuffs (Figure 54, Appendix 1, Table 7) and several other immobile element ratios are similar to the remainder of the Lower Rhyolitic Tuff Formation, they differ significantly in their lower Nb/Th (Figure 55), Nb/TiO$_2$ and Y/TiO$_2$ ratios, and higher La/Zr and Th/Zr ratios (Appendix 1, Table 7). Isolated exposures of nonwelded tuff and accretionary lapilli-bearing tuffs, underlying the basal welded facies of the Lower Rhyolitic Tuff Formation to the north of Yr Arddu, are very similar in terms both of immobile element concentrations and respective ratios to the Yr Arddu Tuffs (Figure 55, Appendix 1, Table 7), strongly indicating their correlation.

The basal welded facies, while similar in most respects to the overlying main phase intracaldera and outflow tuffs of the Lower Rhyolitic Tuff Formation, clearly differs on a plot (Figure 56) of TiO$_2$ vs. Zr, reflecting its enrichment of Zr relative to TiO$_2$. In this respect, the isolated pods of welded tuff are indistinguishable from the remainder of the welded facies. V/Zr ratios are also significantly lower in the basal welded facies.

In serial sections (Figure 57) through the intracaldera and the proximal outflow facies to the north of the caldera, the vertical profiles of the concentrations of Zr, Nb, Th and TiO$_2$ are remarkably consistent, reflecting the constancy of their relative abundances, although TiO$_2$ and Th are slightly more variable. Within the caldera, Y is yet more variable, although to the north of the caldera little relative variation is seen.

Variations within each of the serial sections suggest that more than one pulse of volcanism is represented in the accumulation of the primary tuffs. Near the centre of the caldera, at Cwm Tregalan (Figure 57), a distinct change is seen in the concentrations of Zr, Nb, Th and TiO$_2$ approximately at the top of the basal welded facies. The main body of the overlying nonwelded tuff is thereafter relatively homogeneous. A similar anomaly, approximately 100 m above the base of the Lower Rhyolitic Tuff Formation, is seen in the profile at Pen y Pass, close to the northern margin of the caldera. This suggests, therefore, that the lower part of the section in the northern part of the caldera may correlate with the basal welded facies further south. However, no welding is seen at Pen y Pass and TiO$_2$/Zr ratios are broadly similar throughout the section. The remainder of the exposure shows more variability than the Cwm Tregalan section, perhaps indicating yet further pulses of volcanism, though more detailed sampling would be needed for this to be substantiated.

A section through the outflow facies to the north of the caldera, in Cwm Idwal, again suggests the possibility of a geochemical break in the section approximately 60 m above its base and close to the top of the main body of the pyroclastic breccias. The onset of deposition of the remobilised

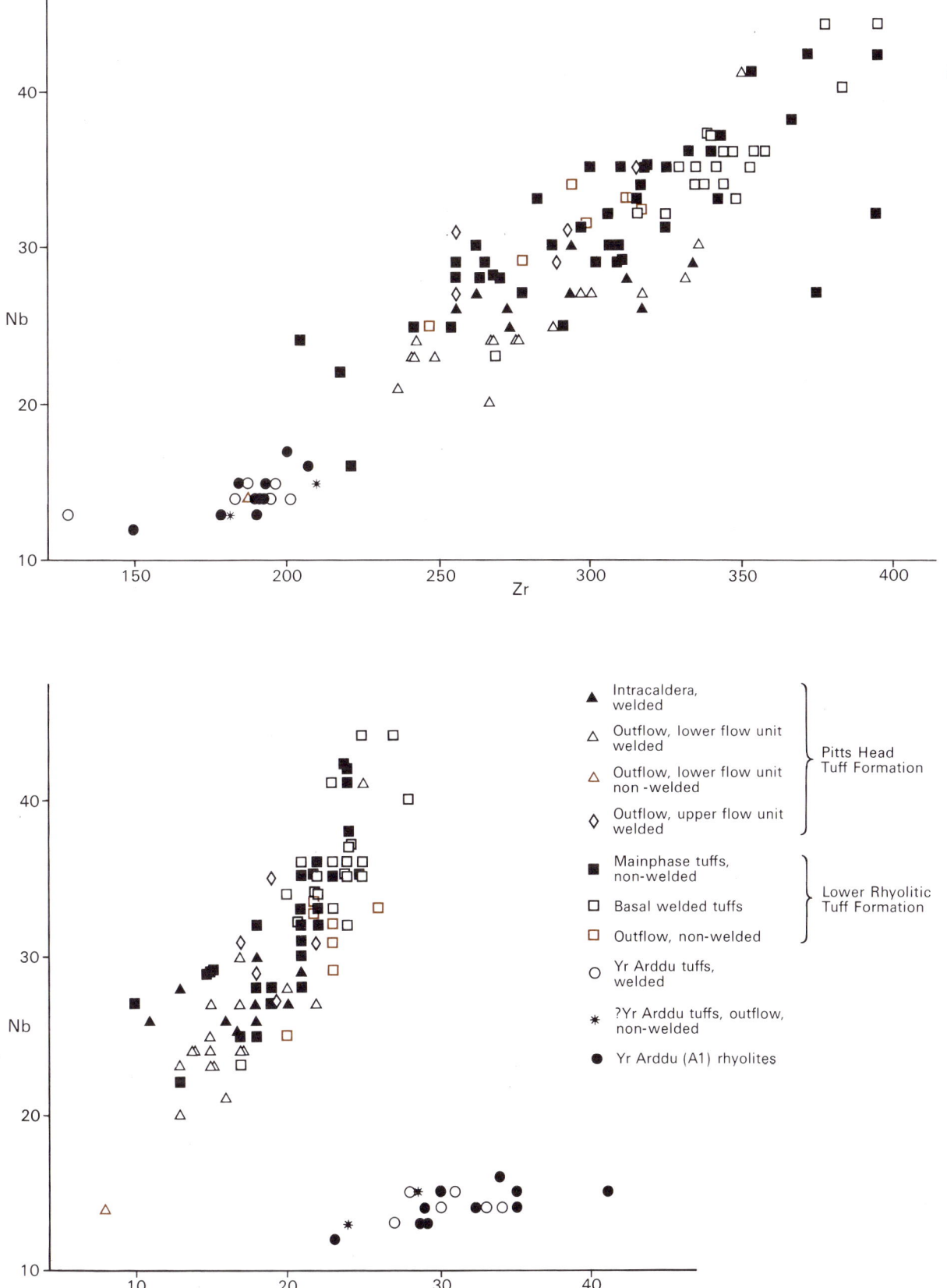

Figure 55   Lower Rhyolitic Tuff Formation and Pitts Head Tuff Formation, (A) Nb vs. Zr diagram (B) Nb vs. Th diagram, indicating similarities of the Yr Arddu tuffs, their possible outflow equivalents and the A1 rhyolites and their dissimilarity from other facies of the Lower Rhyolitic Tuff and Pitts Head Tuff formations.

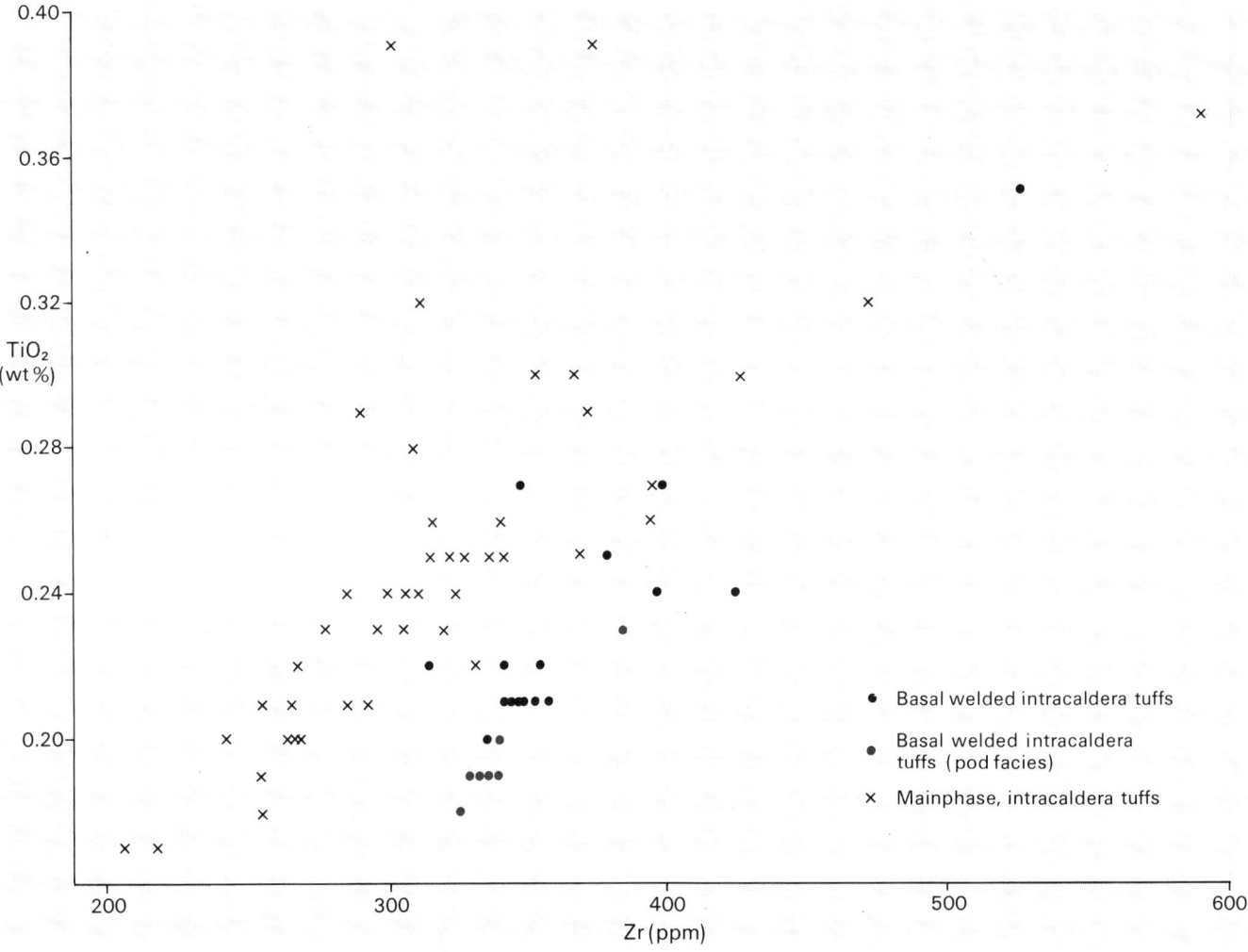

Figure 56   Lower Rhyolitic Tuff Formation, TiO$_2$ vs. Zr diagram, showing the discriminance between the basal welded intracaldera tuffs and the main-phase intracaldera tuffs.

tuffs and volcaniclastic sandstones above the primary tuffs is reflected by the geochemistry. Towards the top of the section, reworked tuffs and tuffites, which overlie a thick, locally extrusive rhyolite, are greatly enriched in Zr (1780 ppm maximum), Nb, Th and Y relative to the remainder of the section. They closely mirror the composition of the underlying rhyolite which is geochemically distinct from magma erupted during the main caldera-forming phase (see below). Elsewhere, analyses of the reworked facies (Figure 54, Appendix 1, Table 7) are broadly similar to those of the primary Lower Rhyolitic Tuff Formation.

In terms of discriminance between the Lower Rhyolitic Tuff Formation and the Pitts Head Tuff Formation, few clear distinctions can be drawn since the main-phase intracaldera and outflow tuffs of the Lower Rhyolitic Tuff Formation are very similar to the intracaldera and lower outflow tuff of the Pitts Head Tuff Formation (Appendix 1, Tables 6 and 7). The Pitts Head tuffs are, however, clearly resolved from the Yr Arddu Tuffs with respect to their higher Nb/Th and Zr/Th ratios, and their higher concentrations of Nb and Zr (Figure

55). Also, they have lower ratios with respect to TiO$_2$ than the basal welded facies of the Lower Rhyolitic Tuff Formation.

REWORKED TUFFS

Reworked tuffs overlying the primary ash-flow tuff sequence occur both within and outside the Lower Rhyolitic Tuff Formation caldera. The bed forms of these tuffs and their contained faunas identify the environment of reworking as marine. The processes involved can broadly be subdivided into either local wave-reworking or remobilisation of previously emplaced pyroclastic debris into sediment-gravity flows as a result of instability. Most sequences display a complex interdigitation of the full spectrum covered by these two types of reworking.

Intracaldera

The asymmetrical character of the caldera, defined by the thickness variation of the intracaldera ash-flow tuff sequence, is inversely reflected in the thickness variations of the intracaldera reworked tuffs. In the north, in the area of the

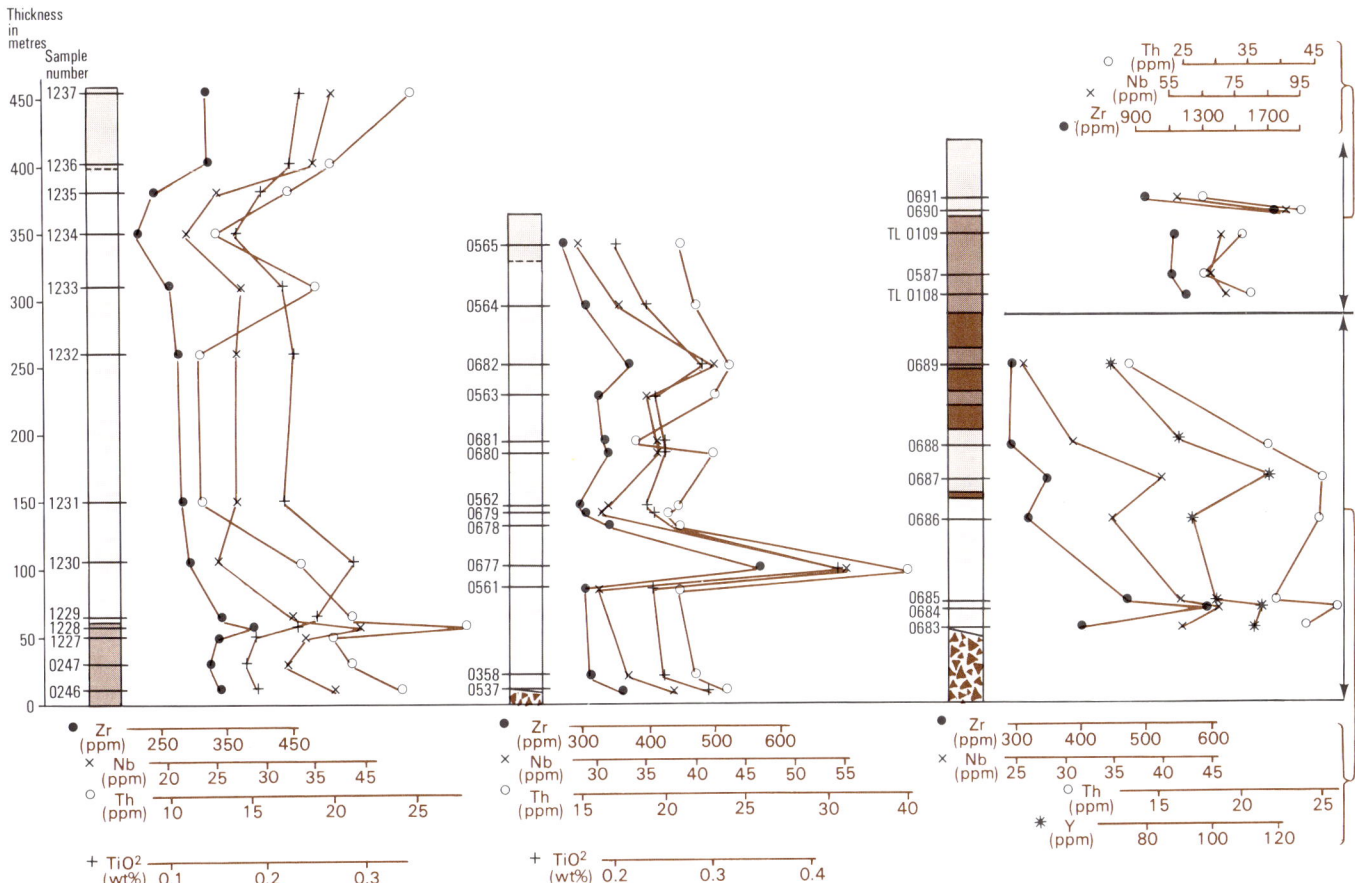

INTRACALDERA TUFFS
(Centre of caldera)

CWM TREGALAN
[SH 6155 5243 to 6141 5416]

INTRACALDERA TUFFS
(North margin of caldera)

PEN Y PASS
[SH 6474 5589 to 6451 5539]

OUTFLOW TUFF
(To north of caldera)

CWM IDWAL
[SH 6459 5900 to 6410 5878]

Figure 57   Lower Rhyolitic Tuff Formation, immobile trace element variations in serial sections of intracaldera tuffs (Cwm Tregalan and Pen y Pass) and outflow tuffs (Cwm Idwal). Key to facies as for Figure 51. All sample numbers prefixed by KB, except where marked TL.

thickest accumulation of primary tuffs, only the topmost 10–20 m of the sequence comprises shallow-marine, re-worked tuffs, whereas at the south-west margin, reworked tuffs extend down locally to the top of the basal welded tuff (Figure 58).

The palaeogeography in the vicinity of the caldera, and hence the processes of reworking, were effected by caldera resurgence which accompanied the intrusion of rhyolites following cessation of the main caldera-forming eruptions (Campbell et al., 1987). The extent and magnitude of resurgence related to this phase is difficult to distinguish because of the restriction of outcrop. However, it is apparent that wherever high-level intrusive/extrusive rhyolite domes were emplaced, they profoundly affected local sedimentation and cumulatively this effect was considerable.

Reworked tuffs at two intracaldera sections, at Moel Hebog, on the south-west side, and at Llyn Gwynant near the centre, have been examined in detail (Fritz et al., 1990). In the vicinity of Moel Hebog, about 90 m of the reworked facies

is well exposed. Large blocks and rafts of Pitts Head tuff occur near its base (Plate 19). These were derived by slumping off the caldera rim (Howells et al., 1986; Reedman et al., 1987a). Above, the sequence (Figure 58) comprises volcanogenic silt-stones, sandstones and conglomerates which in their weather-ing character reflect the primary ash-flow tuffs of the Lower Rhyolitic Tuff Formation as being the dominant source. Low in the sequence, thin (10–50 cm) massive beds of fine ash-grade material probably represent small flows of remobilised pyroclastic debris and suspension deposits. Above, the bed-forms are variably massive, structureless, plane-bedded, small-scale trough cross-stratified, herringbone cross-bedded, low-angle hummocky cross-stratified and straight-crested ripple cross-laminated. They are considered to reflect a com-bination of deposition from primary and secondary pyro-clastic flows and upper flow regime reworking, or tempestite storm currents in shallow water, possibly on the foreshore of 'beaches' along the rim of the caldera. Instability of the en-vironment is reflected in the many contorted beds and soft sediment deformational structures and by the mass flow deposits. The common development of wave-ripples and hummocky cross-stratification restrict maximum water depth to within storm wave-base and indicate that subsidence and sedimentation were largely in balance throughout the deposition of the 90 m-thick sequence.

A similar intracaldera section (38 m) at Llyn Gwynant has been interpreted (Fritz et al., 1990) to reflect sedimentation in shallow water above fair weather wave-base, possibly on a topographic high above a resurgent rhyolite dome. In both sections the shallow-marine influence is corroborated by the contained brachiopod-dominated faunas of the *Dinorthis* and *Nicolella* associations (cf. Pickerill and Brenchley, 1979).

Outside the caldera

Reworked tuffs are a distinctive feature of the Lower Rhyolitic Tuff Formation sequence outside, but within a limited distance of the caldera. Their outcrop is restricted by the subsequent folding and deep erosion, but their main distribution is clearly confined to a zone close to the north and east margins of the caldera. Further in these directions, the Lower Rhyolitic Tuff Formation is represented solely by the single outflow tuff emplaced in water depths below wave-base.

Two sections, at Cwm Idwal and Moel Siabod, have been examined in detail (Figure 58C and D) (Fritz et al., 1990) and they reflect markedly contrasting sedimentary environments. At Cwm Idwal, the primary tuffs are overlain by 2 m of laminated black mudstone, interpreted as background, non-volcanic sedimentation, below storm wave-base. The overlying laminated, tuffaceous siltstones and acid tuffites (60 m thick) with black mudstone intercalations reflect incursions of fine-grained volcanic ash. Beds with hummocky and swaley cross-stratification indicate that the environment eventually shallowed to above storm wave-base (possibly about 100 m). Particularly distinctive in this unit are large (< 1 m) carbonate concretions.

The upper 75 m of the Cwm Idwal section is dominated by graded volcaniclastic sandstone beds, 10–50 cm thick, with interbeds of tuffaceous siltstone interpreted as representing turbidity current deposition on the proximal to mid portions of a submarine fan complex (Orton, 1987; Fritz et al., 1990). The reworking of the tops of the graded sandstone beds into hummocks and swales resulted from the episodic influence of storm waves, and these beds resemble the tempestite sequence described by Aigner (1985). The dominantly volcaniclastic components of the fan sequence and its position above primary outflow tuffs suggest that it was fed from the caldera margin. The black mudstone bed between the outflow tuff and reworked tuffs suggests a significant time lapse between primary tuff emplacement and reworking from the caldera margin. The latter was possibly initiated by resurgence in the northern sector of the caldera. Towards the top of the sequence, the sandstones and siltstones are succeeded by a rhyolite lava flow, up to 75 m thick, at the edge of a thicker dome exposed to the south. The rhyolite, which is finely flow-banded, flow-brecciated and columnar-jointed, is in turn overlain by 35 m of laminated and cross-laminated, wave-rippled, rhyolitic, tuffaceous sandstones and tuffites. Geochemistry has confirmed that the rhyolite clasts in the latter are compositionally comparable (high Zr) to the underlying rhyolite.

At Moel Siabod (Figure 58D), the 20 m-thick section comprises coarse-grained volcaniclastic sandstones, with angular grains, interbedded with silicified tuffaceous mudstones (Plate 28). The latter are interpreted as containing a high proportion of suspension deposited ash which lithified quickly and consequently resisted resuspension in energy conditions that scoured and redeposited the sand. The sandstone contains plane beds, hummocky low-angle cross-stratification and small-scale wave-ripple cross-lamination. The section represents high-energy deposition in shallow water above normal wave-base (< 9–18 m), possibly a beach accumulation. The very thin sequence indicates deposition in a relatively stable area.

The evidence provided by these sections indicates the influence of the development of a major explosive acidic volcanic centre on the sedimentation patterns in a marine

Plate 28   Reworked facies of the Lower Rhyolitic Tuff Formation, Moel Siabod [SH 7172 5370]

A and B   Interbedded, laminated, volcaniclastic sandstones and fine-grained, variably silicified, tuffaceous siltstones. The sandstones erode into the siltstones and are characterised by plane-lamination, hummocky cross-stratification and small-scale wave-ripple cross-lamination.

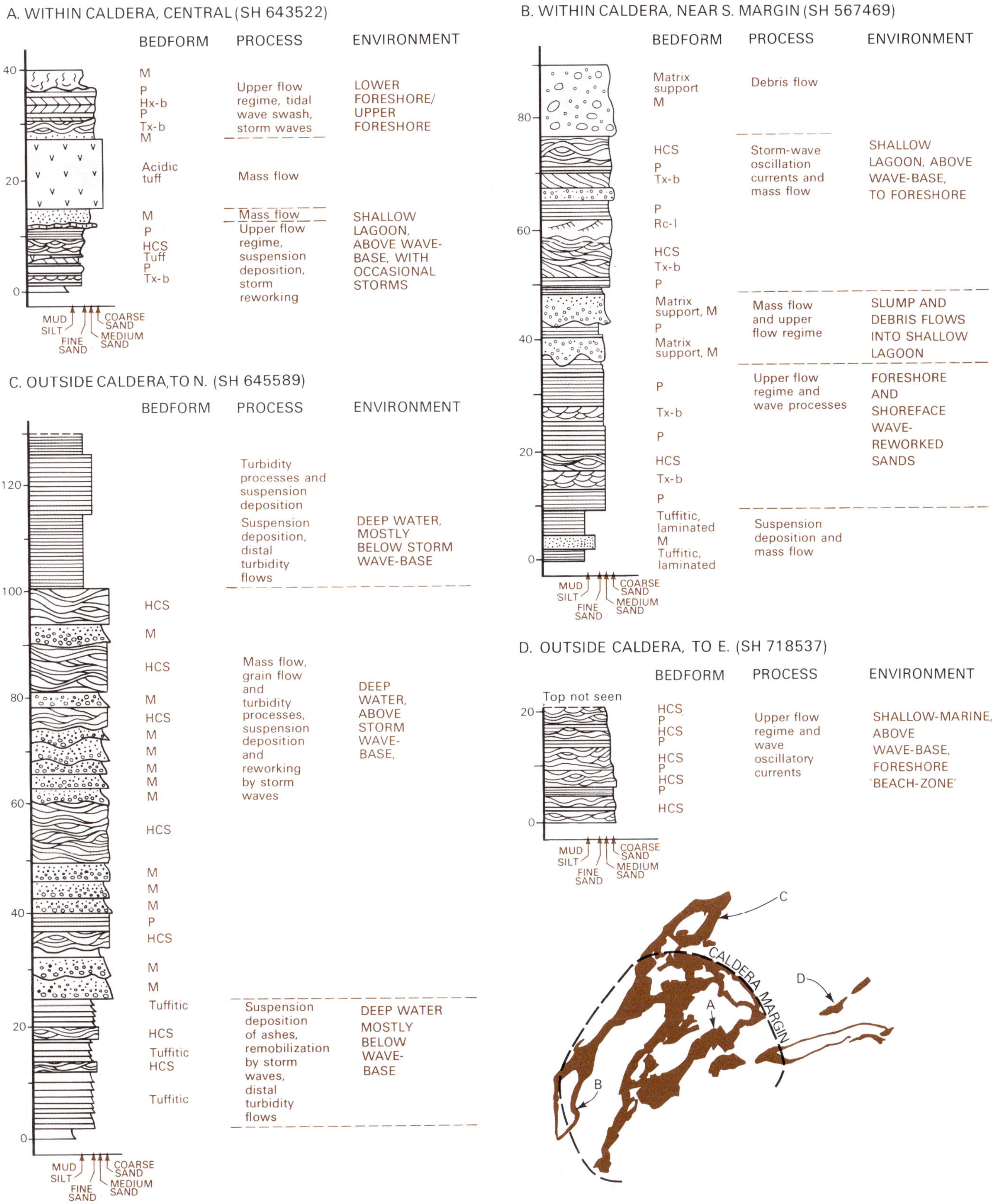
A. WITHIN CALDERA, CENTRAL (SH 643522)
BEDFORM
PROCESS
ENVIRONMENT
40
20
0
M
P
Hx-b
P
Tx-b
M
Upper flow regime, tidal wave swash, storm waves
LOWER FORESHORE/ UPPER FORESHORE
Acidic tuff
Mass flow
M
P
HCS
Tuff
P
Tx-b
Mass flow
Upper flow regime, suspension deposition, storm reworking
SHALLOW LAGOON, ABOVE WAVE-BASE, WITH OCCASIONAL STORMS
MUD
SILT
FINE SAND
COARSE SAND
MEDIUM SAND
B. WITHIN CALDERA, NEAR S. MARGIN (SH 567469)
BEDFORM
PROCESS
ENVIRONMENT
80
60
40
20
0
Matrix support
M
Debris flow
HCS
P
Tx-b
Storm-wave oscillation currents and mass flow
SHALLOW LAGOON, ABOVE WAVE-BASE, TO FORESHORE
P
Rc-l
HCS
Tx-b
P
Matrix support, M
P
Matrix support, M
Mass flow and upper flow regime
SLUMP AND DEBRIS FLOWS INTO SHALLOW LAGOON
P
Tx-b
P
HCS
Tx-b
P
Upper flow regime and wave processes
FORESHORE AND SHOREFACE WAVE-REWORKED SANDS
Tuffitic, laminated
M
Tuffitic, laminated
Suspension deposition and mass flow
MUD
SILT
FINE SAND
COARSE SAND
MEDIUM SAND
C. OUTSIDE CALDERA, TO N. (SH 645589)
BEDFORM
PROCESS
ENVIRONMENT
120
100
80
60
40
20
0
Turbidity processes and suspension deposition
Suspension deposition, distal turbidity flows
DEEP WATER, MOSTLY BELOW STORM WAVE-BASE
HCS
M
HCS
M
HCS
M
M
M
M
M
HCS
M
M
M
P
HCS
M
M
Mass flow, grain flow and turbidity processes, suspension deposition and reworking by storm waves
DEEP WATER, ABOVE STORM WAVE-BASE,
Tuffitic
HCS
Tuffitic
HCS
Tuffitic
Suspension deposition of ashes, remobilization by storm waves, distal turbidity flows
DEEP WATER MOSTLY BELOW WAVE-BASE
MUD
SILT
FINE SAND
COARSE SAND
MEDIUM SAND
D. OUTSIDE CALDERA, TO E. (SH 718537)
BEDFORM
PROCESS
ENVIRONMENT
Top not seen
20
0
HCS
P
HCS
P
HCS
P
HCS
P
HCS
Upper flow regime and wave oscillatory currents
SHALLOW-MARINE, ABOVE WAVE-BASE, FORESHORE 'BEACH-ZONE'
MUD
SILT
FINE SAND
COARSE SAND
MEDIUM SAND
C
CALDERA MARGIN
D
A
B

Figure 58 Graphic logs and environmental interpretations of sections of the reworked facies of the Lower Rhyolitic Tuff Formation (after Fritz et al., in press).

Ornament is as for Figures 22, 33 and 34 and in addition; M/G: Massive and graded; P: Planar laminated; Tx-b: Trough cross-bedded; Rc-l: Ripple cross-lamination. Inset map shows outcrop of the Lower Rhyolitic Tuff Formation and position of sections.

Plate 29   Columnar-jointed intrusive rhyolite exposed on Crib Goch [SH 623 552], at the northern margin of the Lower Rhyolitic Tuff Formation caldera. Snowdon summit, Yr Wyddfa is on the skyline.

Plate 30   Flow-banding in intrusive rhyolite near the northern margin of the Lower Rhyolitic Tuff Formation caldera [SH 618 570].

environment. The effects are clearly two-fold, the rapid incursion of volcanic debris into the environment and the contemporaneous volcanotectonic modification of the sedimentary setting. However, the amount of reworked pyroclastic debris associated with the Lower Rhyolitic Tuff Formation caldera is small in amount and restricted in distribution when compared to the volume of primary pyroclastics preserved. It is this simple observation that is most persuasive in concluding that the caldera was contained, to a large extent, within the marine environment and that its subaerial expression was extremely limited. The erosion of a subaerial, resurgent caldera edifice of the scale of the Lower Rhyolitic Tuff Formation caldera would have contaminated Caradoc and probably subsequent sedimentation over the whole of north-west Wales and this is manifestly not the case.

RHYOLITE EMPLACEMENT RELATED TO THE LOWER RHYOLITIC TUFF FORMATION CALDERA

Intrusive and extrusive rhyolites are particularly abundant within and about the Lower Rhyolitic Tuff Formation caldera (Figure 52) and they also intrude the underlying sedimentary sequence. The rhyolites are massive, locally feldsparphyric, columnar-jointed (Plate 29), flow-banded (Plates 30, 31C) and autobrecciated (Plate 31D) (Figures 59 and 60). They occur as dykes, sills, small stocks and large domes. The latter are elliptical to subcircular in plan, with diameters from tens of metres to over 1 km.

The relatively deep-seated dykes and sills are generally massive, with sharp contacts and narrow thermal alteration zones, which suggest that the adjacent sediments were consolidated at the time of intrusion. At the south-west edge of the caldera, some of the sills in the Lower Rhyolitic Tuff Formation substrate are composite or multiple intrusions with basalt or dolerite margins and rhyolite cores. However, shallower intrusions, within or close to the base of the Snowdon Volcanic Group, form dome-like bodies with brecciated margins, locally intruded by flames of sediment (Plate 31B), indicating intrusion into unconsolidated, water-saturated sediment (Kokelaar et al., 1984; Campbell et al., 1987). The near-surface intrusions also caused updoming and slumping of unlithified sea-floor sediments and some broke surface, creating exogenous domes or flows with carapaces of autobreccia (Figure 60). Subsequent movement caused mass collapse at the dome or flow margin, with the development of a peripheral apron of poorly sorted and bedded rhyolite breccias, commonly interdigitated with extruded rhyolite and other sediments (cf. Kokelaar et al., 1985).

Topographic highs on the sea floor about shallow-level intrusions and extruded domes became areas of nondeposition of volcaniclastic sediments. In the shallow-marine environments, some high-level intrusions, as that at Cerrig Cochion (Figure 61), were exhumed and extruded domes were eroded. The eroded rhyolite debris was incorporated into parallel-laminated and trough cross-laminated rhyolitic sandstones across the bevelled tops of the rhyolites (Figure 61).

Three main phases of rhyolite extrusion and, by implication, intrusion have been determined (Campbell et al., 1987) during the evolution of the Lower Rhyolitic Tuff Formation caldera (Figure 68):

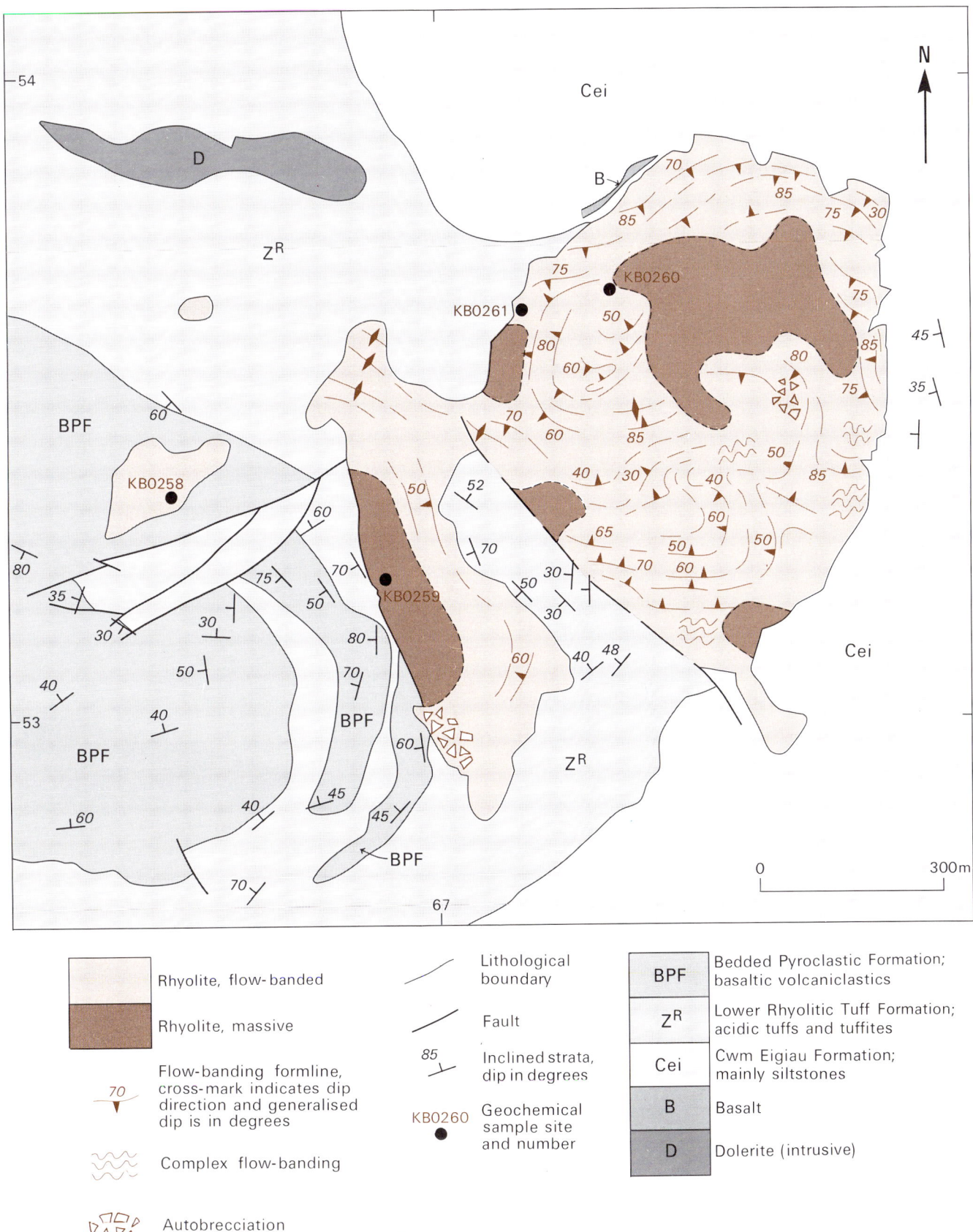

Figure 59   Map of partly exogenous rhyolite dome showing internal structure, facies variations and relationship to adjacent strata, Carnedd y Cribau.

A

B

C

D

Plate 31   Features of rhyolites

A   Rhyolite forming the conspicuous peak of Crib Goch [SH 624 552], intruding the Bedded Pyroclastic Formation at the northern margin of the Lower Rhyolitic Tuff Formation caldera. (BGS Photograph L2566).

B   Fluidised basaltic sandstones 'intruding' flow-banded rhyolite at Clogwyn y Person. (BGS Photograph L2599). [SH 6167 5555].

C   Contorted flow-banding, Llanberis Pass [SH 6119 5734]. (BGS Photograph L1942).

D   Autobrecciated rhyolite, Ysgolion Duon [SH 6661 6365]. (BGS Photograph L1873).

SHALLOW-LEVEL INTRUSION BREAKS
SURFACE CAUSING EITHER GRAVITATIONAL
COLLAPSE OF AUTOBRECCIATING CARAPACE
AND GENERATION OF MASS-FLOW OR LATERAL
BLAST ERUPTION

SHALLOW LEVEL INTRUSION IN WET
SEDIMENT CAUSES TOPOGRAPHIC
SWELL OF SEDIMENT COVER AND
FLUIDISATION OF ADJACENT SEDIMENTS

CONTINUED REWORKING AND
DILUTION OF OVERLYING SEDIMENTS BY
SEDIMENT GRAVITY FLOWS AND/OR
SHALLOW MARINE PROCESSES

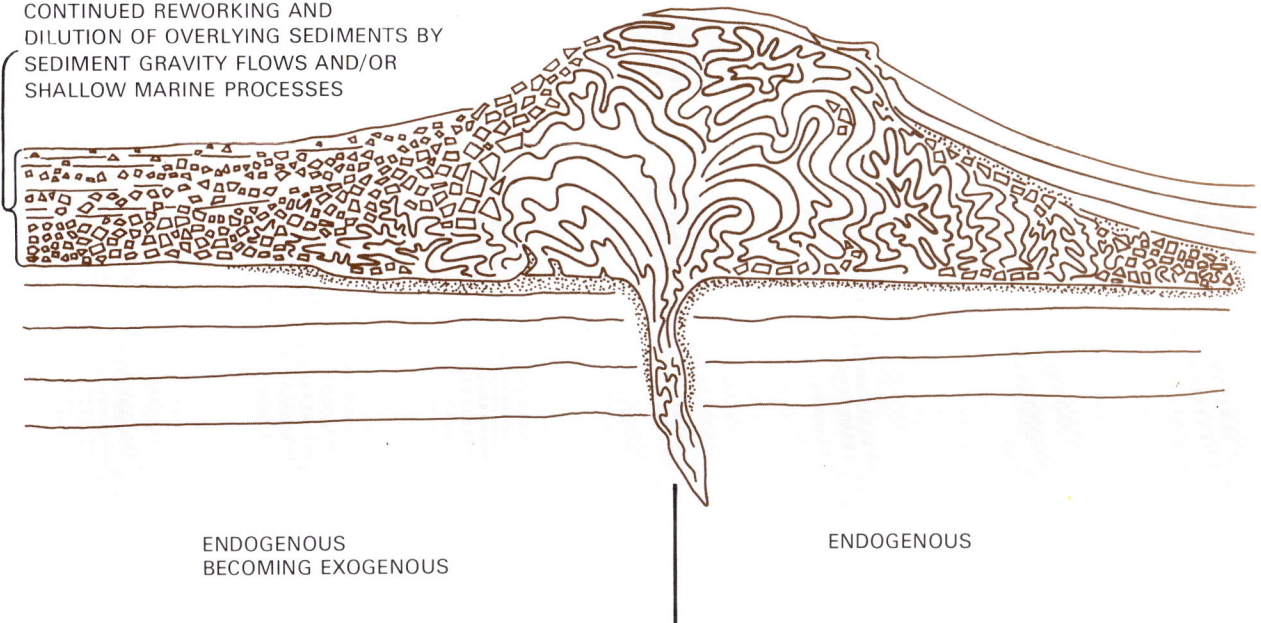

ENDOGENOUS
BECOMING EXOGENOUS

ENDOGENOUS

Figure 60   Composite schematic section through a shallow-level rhyolite dome.

1   Immediately preceding the main phase of caldera-forming eruptions of the Lower Rhyolitic Tuff Formation.
2   During resurgence and reworking of primary ash-flow tuffs of the Lower Rhyolitic Tuff Formation and emplacement of the overlying basaltic Bedded Pyroclastic Formation.
3   During the emplacement of the Upper Rhyolitic Tuff Formation.

This contradicts the postulated single phase in the waning stages of volcanism (Rast, 1969; Beavon, 1980; Leat et al., 1986) and accordingly, the geochemically based model of Leat et al. (1986) has been revised.

PETROGRAPHY   The rhyolites are sparsely microporphyritic with phenocrysts of albite, alkali feldspar and lesser quartz. The feldspar phenocrysts (<3.5 mm) are subhedral with rounded terminations and are slightly to completely sericitised. The isolated quartz phenocrysts (<1 mm) occur as intensely corroded relicts.

The groundmass is devitrified and variably recrystallised as a platy quartz-feldspar mosaic crowded with inclusions of sericite, chlorite and opaque dust, as spherulitic quartz-feldspar within a fine chlorite and sericite aggregate (Plate 32A and C), as a quartz-feldspar aggregate with a pervasive snowflake texture (Plate 32D) and as a perlitic texture with the arcuate fractures accentuated by chlorite segregation. Locally, flow-oriented feldspar microlites in a devitrified originally glassy mesostasis can be distinguished (Plate 32B).

Chlorite and iron-oxide pseudomorphs of ferromagnesian mineral are rare and, together with primary opaque oxides, form <2 per cent of the constituents. Apatite is a common accessory mineral.

A wide range of accessory minerals has been identified in the groundmass, by A T Kearsley (Department of Geology, Oxford Polytechnic), using back scattered electron imagery (Plate 33) and analysed using energy-dispersive techniques. These include aeschynite, allanite, lanthanite (-Ce?), larsenite-fayalite, monazite, parisite, synchisite, thorite, xenotime, zircon, sphene and gadolinite (or spencite). Of these, the apatite, zircon, sphene and some of the allanite are probably primary magmatic phases. All of the other phases occur as anhedral crystals and intergrowths, in veins and as secondary rims, and are clearly of secondary (?hydrothermal) origin, with the possible exceptions of monazite and xenotime.

GEOCHEMISTRY

Representative analyses of rhyolites spatially associated with the Lower Rhyolitic Tuff Formation caldera are presented in Appendix 1, Table 8. On a plot of Zr/TiO_2 vs. Nb/Y (Figure 62), two main groupings are observed. One of these straddles the boundary between the rhyolite and rhyodacite/dacite fields, while the other is concentrated on the boundary between the rhyolite and comendite/pantellerite fields. The representatives of the latter group have very high Zr contents (>700 ppm), Nb, Y, Hf, Ta and REE and have previously been interpreted as peralkaline rhyolites (Leat et al., 1986).

Figure 61   Map of exogenous rhyolite dome, Cerrig Cochion, showing distribution of rhyolitic sandstones associated with extrusion of the rhyolite.

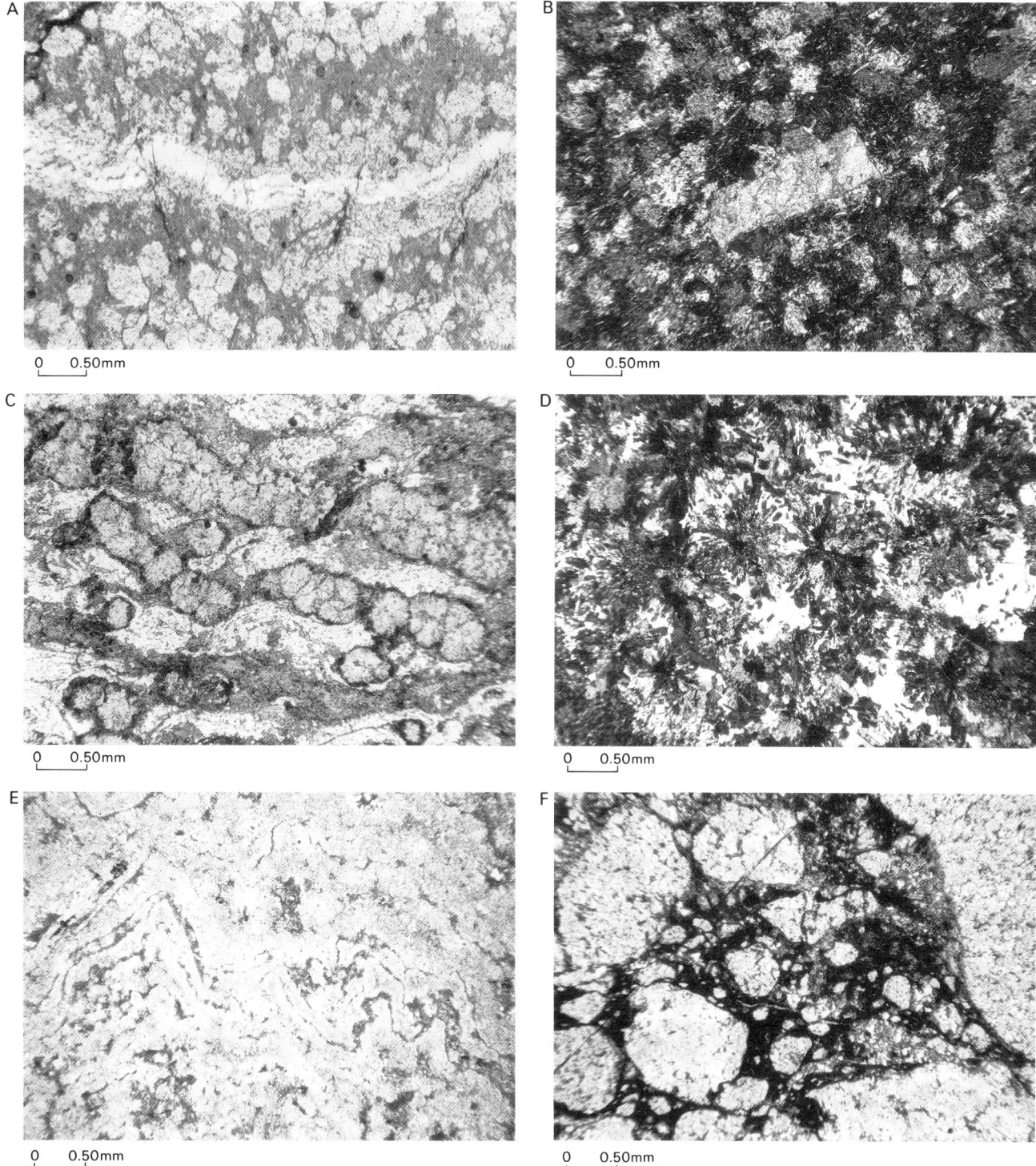

Plate 32 Photomicrographs, rhyolites

A   MH 1175. Small quartz-feldspar spherulites in a cleaved sericite matrix with quartz recrystallisation accentuating flow-banding, ppl. [SH 6030 5056].

B   KB 197. Feldspar microlites, crudely flow-orientated in a devitrified, glassy groundmass, recrystallised as a mosaic of interlocking quartz-feldspar patches. The microphenocryst is of albite feldspar. X polars. [SH 7550 7780].

C   E 44173. Spherulites of fibrous quartz-feldspar and recrystallised quartz accentuating flow-bands, ppl. [SH 6413 5812].

D   KB 250. Platy, spherulitic recrystallisation of glassy groundmass resulting in a 'snowflake' granophyric quartz and feldspar intergrowth. X polars. [SH 6390 5343].

E   KB 535. Contorted flow-lamination, ppl. [SH 5790 4410].

F   E 49021. Autoclastic brecciation. Clasts of devitrified rhyolite in a chloritic matrix with iron oxide grains, ppl. [SH 6383 5910].

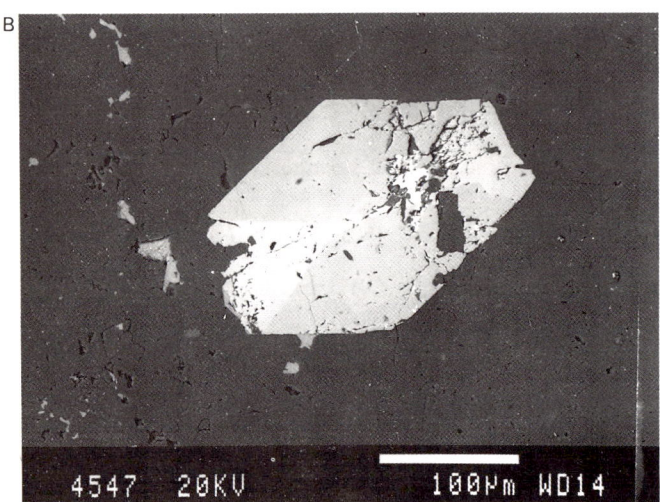

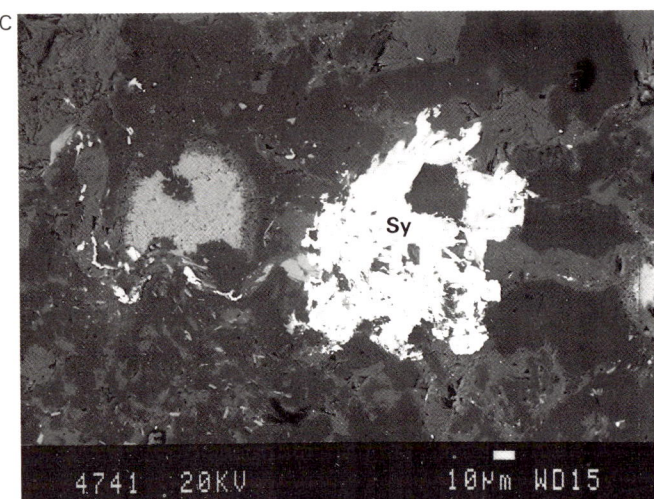

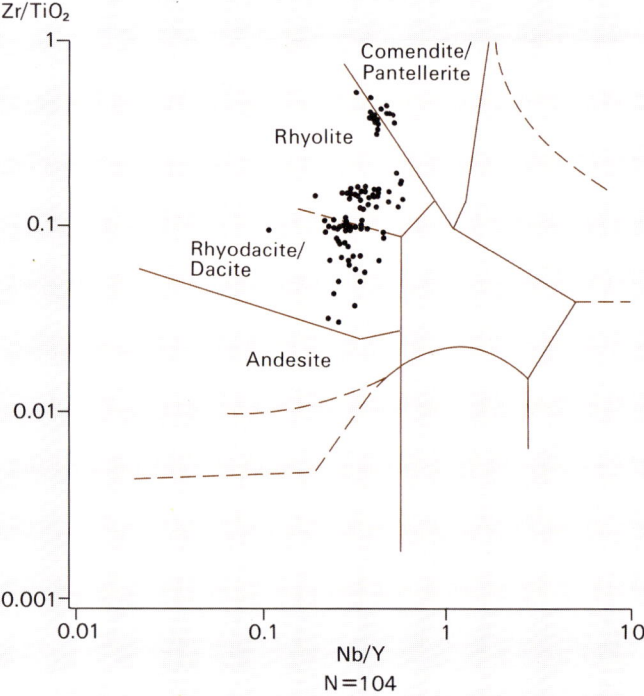

They fall predominantly within the comendite field on the Al$_2$O$_3$ vs. Fe$_2$O$_3$(T) diagram of Macdonald (1974). They also have lower Ta/Th ratios than pantellerites and, for example, are comparable to the Huckleberry Ridge and Lava Creek Tuffs of Yellowstone (western USA; see Hildreth, 1981), which are both comenditic. The other group of rhyolites have lower Zr contents (<550 ppm) and other trace element characteristics typical of subalkaline, high K-rhyolites (Leat et al., 1986). Similar subalkaline/peralkaline rhyolite bimodality has also been observed in the Caradoc on the Lleyn Peninsula (Leat and Thorpe, 1986) and in the Llandovery Skomer Volcanic Group in south-west Wales (Thorpe et al., 1989).

In terms of their immobile element chemistry, the rhyolites are further subdivisible. In particular, on a plot of Nb. vs. Zr, they can be subdivided into 5 groups (Campbell et al., 1987) (Figure 63, Appendix 1, Table 8): Groups A1 and A2 are characterised by Nb/Zr ratios of about 0.09 and Groups B1, B2 and B3 by ratios of about 0.05. This suggests that at least two fractionation trends were involved in their genesis. The population data and ranges of Nb and Zr (ppm) are shown in Table 1.

A plot of Nb vs. Th (Figure 64) also demonstrates these groups, although Groups A1 and B1 cannot be discriminated. However, on this plot group B2 can be further subdivided into two subgroups, B2i and B2ii. The different Th/Nb and Th/Ta ratios of the groups indicate that they cannot be derived from a common parent magma by direct fractional crystallisation. Groups A2 and B3 have similar Th/Ta ratios to many rhyolites erupted through continental crust (e.g. Hildreth, 1981; Macdonald et al., 1987). Groups A1 and B1 have very high Th/Ta ratios and those for the latter

Plate 33   Photomicrographs (back-scatter SEM; by Anton Kearsley, Oxford Polytechnic) of rhyolites from the Snowdon Centre

A   KB 454 (A1 rhyolite). Sphene (S) overgrown by allanite (All).
B   KB 454 (A1 rhyolite). Euhedral sphene.
C   KB 474 (A2 rhyolite). Synchysite (Sy) in radiating sheaves and along cracks.

Figure 62   Rhyolites of the Snowdon Centre, Zr/TiO$_2$ vs. Nb/Y diagram.

Table 1 Sample populations and Nb and Zr ranges of the rhyolite groups

Group	Nb range (ppm)	Zr range (ppm)	Number of intrusions sampled	Number of samples
Al	12 – 16	14 – 207	6	10
A2	30 – 51	265 – 530	6	10
B1	11 – 16	220 – 292	11	14
B2	16 – 25	334 – 510	15	28
B2a (composite sills)	15 – 25	510 – 515	15	31
B3	38 – 82	680 – 1384	26 (including 4 microgranite sills)	36
Total			84	129

are among the highest known for silicic magmas, being comparable with silicic lavas of Santorini (Mann, 1983; Huijsmans, 1985).

Chondrite-normalised patterns (Figure 65) for the five groups are broadly similar, implying genetic links. Significant Eu anomalies are characteristic of all of the groups, although those for the A1, A2 and B3 rhyolites are greater than those for the B1 and B2 rhyolites. The B1 rhyolites are the most REE-enriched, but their relative LREE/HREE enrichment is somewhat less than for the other groups.

Rhyolites from the multiple intrusions (sills) in the area south-east of Snowdon have geochemical similarities to the B2 rhyolites (Appendix 1, Table 9) and straddle the compositional interval between the B2 and B1 subgroups of the Nb vs. Zr plot (Figure 66). Because of their distinctive field relations, they are designated as a separate subgroup B2a.

The similarities between the various immobile HFS and LIL element concentrations and their relative ratios (e.g. Nb/Zr, Nb/Th) of the Yr Arddu Tuffs and the A1 rhyolites, and between those of the basal welded tuff of the Lower Rhyolitic Tuff Formation and the A2 rhyolites (Appendix 1, Tables 7 and 8) both imply a common parentage. The main phase of nonwelded intracaldera and outflow tuffs is compositionally

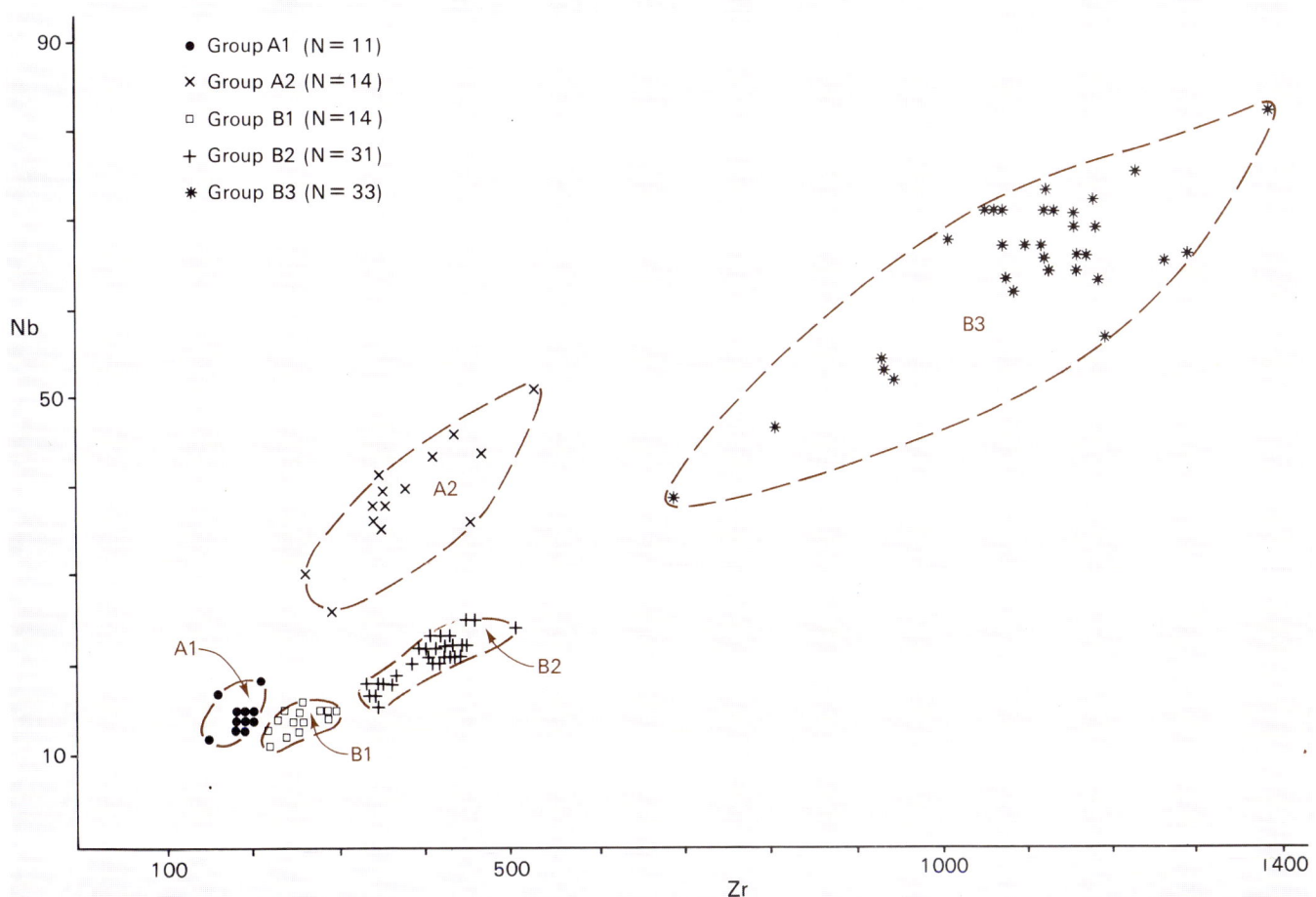

Figure 63 Rhyolites of the Snowdon Centre, Nb vs. Zr diagram (modified after Campbell et al., 1987).

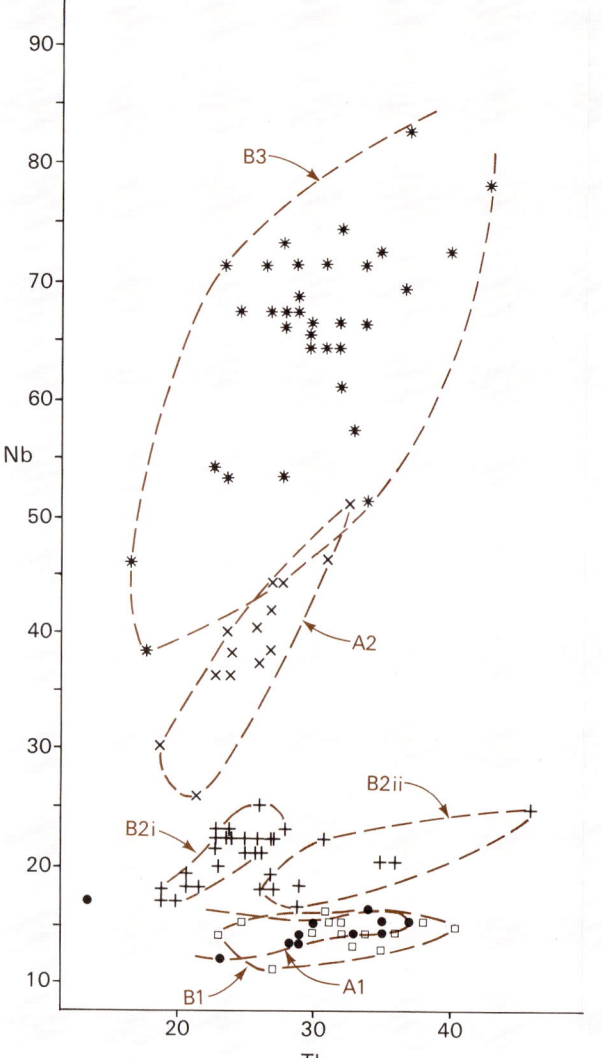

Figure 64 Rhyolites of the Snowdon Centre, Nb vs. Th diagram (modified after Campbell et al., 1987). Key as for Figure 63, except Group B2 subdivided (B2i, N = 22; B2ii, N = 9).

Despite their significant enrichment in Rb and Th, the B3 rhyolites have relatively flat patterns, although Hf, Zr, Sm, Y and Yb are somewhat depleted relative to the other elements. Hence, except for Rb and Th, the B3 rhyolite would reasonably approximate to ocean-ridge granite if a lower value of F (the proportion of residual liquid as a result of applying Rayleigh fractionation to N-MORB) was used to calculate the hypothetical ocean-ridge granite composition than was used by Pearce et al. (1984a; F = 0.25). The pattern for A2 rhyolite is similar to B3, albeit with less incompatible element enrichment, and relatively greater enrichment of Ce and Nb compared with the elements Hf to Yb, which approximate to ocean-ridge granite composition. The patterns of B2, B1 and A1 differ most markedly from B3 and A2 in their depletions of Ta and Nb, and additionally in B1 rhyolite of Hf and Zr, relative to Ce. Enrichment of Rb and Th relative to Nb and Ta is considered to reflect crustal involvement (cf. Pearce et al., 1984a; see also Thirlwall and Jones, 1983), a conclusion consistent with the isotopic evidence. The Ba negative anomaly, though smaller in the B2 rhyolite, is typical of all the rhyolites and is a feature of most within-plate granites, though not typically of those associated with attenuated continental lithosphere. In general, the A1, B1 and B2 rhyolites have patterns similar to within-plate granites associated with attenuated continental lithosphere such as the Mull and, to a lesser extent, Skaergaard granites (cf. Pearce et al., 1984a), although the Ba anomaly is more typical of other within-plate granites such as those which occur in the Oslo rift. The patterns also show some similarity to the volcanic-arc granite from Chile (cf. Pearce et al., 1984a), though the latter has a significant positive Sm anomaly. The A2 and B3 rhyolites have patterns either more comparable with volcanic-arc granites, for example from Newfoundland, though at significantly enriched concentrations of the incompatible elements, or are transitional in character to within-plate granites. As the relative enrichment of Rb and Th is least in B3 rhyolite, the relative crustal involvement is considered to be less than in the other rhyolites, whereas crustal involvement would be relatively greatest in the A1 rhyolites since it has the highest Th/Hf, Th/Zr and Th/Yb ratios. Isotopic data supports this generalisation.

STRATIGRAPHICAL AND SPATIAL RELATIONSHIPS OF THE RHYOLITE GROUPS

The five geochemically defined groups of rhyolites can be related (Campbell et al., 1987) (Figure 68) to:
a   The stratigraphic horizon at which they were extruded.
b   The stratigraphic range in which they were intruded.
c   The variation in form: domes reflect near surface intrusion or extrusion, whereas stocks, dykes and sills reflect deeper-seated intrusion.

From their extrusive ages, each of the groups represents a discrete phase of rhyolite emplacement during the evolution of the Snowdon Centre. The relationships of each group to caldera development are shown on the palinspastic diagram (Figure 68). The first phase (A1 rhyolites), immediately prior to the main caldera-forming Lower Rhyolitic Tuff Formation eruptions, was controlled by the north-east–south-west Yr Arddu fracture. The second phase (A2, B1 and B2, and some B3 rhyolites) postdated caldera subsidence and was associated with resurgence. Rhyolites were emplaced within the Bedd-

variable though, on grounds of the similarities of trace element ratios to those of the A2 rhyolites, derivation from the same magma is again implied. Similarly, the Upper Rhyolitic Tuff Formation ash-flow tuffs (Appendix 1, Table 10) are very similar to the B3 rhyolites.

The trace element discrimination diagrams for the tectonic interpretation of granites (Pearce et al., 1984a; O'Brien et al., 1985), generally suggest that the rhyolites are within-plate types, e.g. Nb vs. Y and Ta vs. Yb. Multi-element ocean-ridge-normalised granite diagrams (normalised to a hypothetical ocean-ridge granite) for each of the five rhyolite groups are shown in Figure 67. General enrichment of the incompatible elements K, Rb, Ba, Th, Ta, Nb and Ce is observed in each of the groups, but only B3 rhyolite is enriched in Hf, Zr, Sm, Y and Yb. Significant enrichment of Rb and Th relative to K and Ba is also a feature of all of the groups and their high Th/Ta ratios are demonstrable.

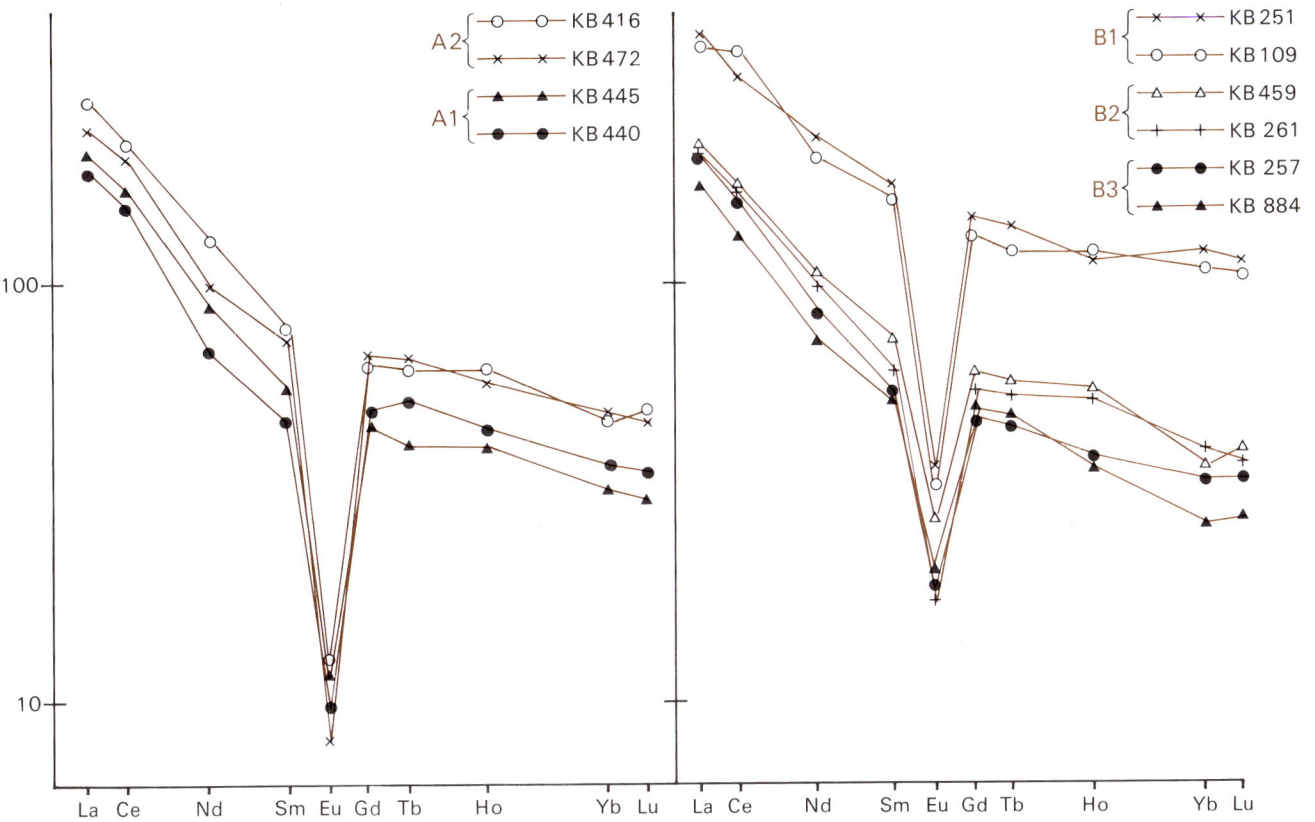

Figure 65 Rhyolites of the Snowdon Centre, chondrite-normalised patterns (after Campbell et al., 1987), (normalising factors after Thompson et al., 1984).

gelert Fault Zone and along associated north-east – south-west fractures. The concentration of intrusions along the fault zone suggests that at this time it had developed into an apical graben within the caldera. Intrusions also occurred in an arcuate array about the north and north-eastern margins of the caldera. In the final phase, during Upper Rhyolitic Tuff Formation times, most of the B3 rhyolites were emplaced mainly in an area within the northern margin of the caldera and as isolated intrusions elsewhere.

The geochemical variation of magma types related to the phases of caldera development may reflect a density-stratified magma chamber that was tapped at varying levels and which evolved with time. Alternatively, the variations could reflect tapping of different magma chambers, discrete magma pulses or periodic influxes of new magma into a single magma chamber. The emplacement of rhyolites during Bedded Pyroclastic Formation times, together with the presence of basaltic blebs in some of the B1 rhyolites, indicates the simultaneous availability of both basaltic and rhyolitic magma. The final phase of peralkaline (B3) rhyolites represents compositions enriched in HFS elements and REE. The similarity of their HFS ratios to those of the B1 and B2 rhyolites favours a genetic link, probably crystal fractionation rather than a fortuitous association with the earlier subalkaline rhyolites. The two arcs of B3 rhyolite domes may reflect nested caldera development, related to the eruption of the peralkaline Upper Rhyolitic Tuff Formation, which is a characteristic of peralkaline volcanic centres (Mahood, 1984).

Post-caldera basaltic volcanism: Bedded Pyroclastic Formation

Following resurgence and shallow-marine reworking of the main caldera phase at the Snowdon Centre there was an episode of basaltic volcanic activity. The deposits resulting from this activity are up to 450 m thick and are known as the Bedded Pyroclastic Formation. This is one of the principal elements of the geology of central Snowdonia, forming extensive outcrops about the Snowdon massif (see Williams, 1927; British Geological Survey, 1989) and in the cores of the Cwm Idwal, Moel Hebog and Dolwyddelan synclines (see Williams, 1930; Shackleton, 1959; and Williams and Bulman, 1931, respectively) (Figure 69). It was referred to as the 'Bedded Pyroclastic Series' by Williams (1927), but this was modified by Howells et al. (1983) to a formation to accord with modern stratigraphic nomenclature. However, the name is somewhat of a misnomer because genuinely pyroclastic elements form only a small part of the formation. Tuffaceous sediments of basaltic derivation dominate, with high-level intrusive and extrusive basalts, hyaloclastites and basic tuffites abundant in places. The sequence developed from a number of small eruptive centres, of which only a few are either preserved or recognised. Locally, thin beds of acid tuffs occur, which represent small expressions of explosive volcanism associated with synchronous effusion of rhyolite related to the Lower Rhyolitic Tuff Formation caldera. The base of the formation is variably conformable, disconform-

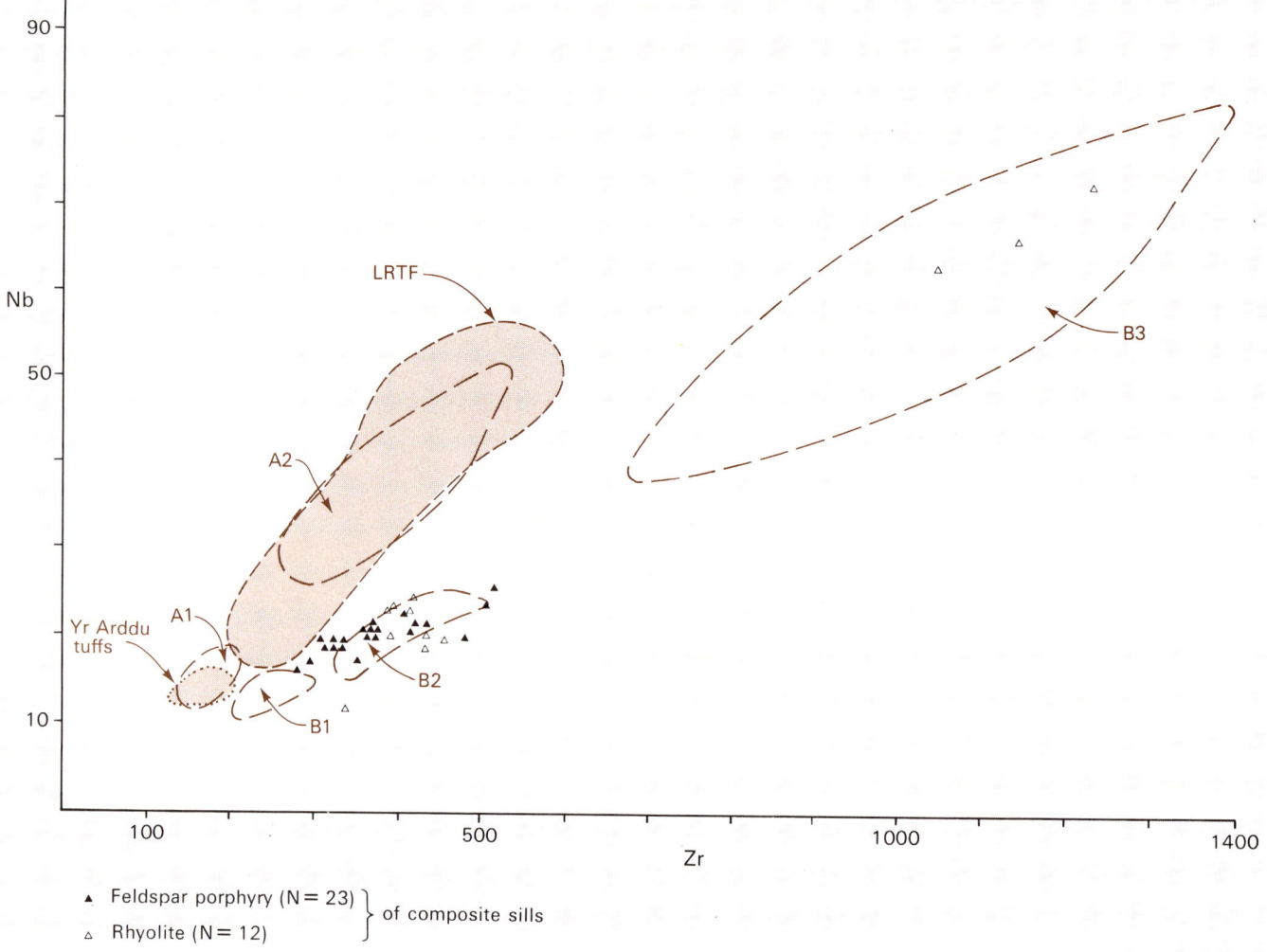

Figure 66   Rhyolites of the Snowdon Centre, Yr Arddu Tuffs and Lower Rhyolitic Tuff Formation,
Nb vs. Zr diagram.

able and unconformable with the underlying Lower Rhyolitic Tuff Formation, but only rarely is its upper contact preserved.

In spite of its excellent exposure, the sequences have not been examined in detail since the early work listed above, an omission probably caused by the generally high elevation and inaccessibility of the outcrop. During this project, the entire outcrop has been remapped at 1:10 000 scale. In addition, the sequence between Cwm Glas and Snowdon Summit, on the north side of the Snowdon massif, has been studied in detail by Dr Peter Kokelaar.

Cwm Glas to Snowdon Summit

The Bedded Pyroclastic Formation is well exposed in Cwm Glas (Plate 34) and Glaslyn, on the north and south sides of Crib y ddysgl ridge respectively, on the north side of the Snowdon massif. The sequence has been described in detail (Kokelaar, in press) and the following account is drawn from this work.

The sequence lies within the Lower Rhyolitic Tuff Formation caldera and overlies the thickest accumulation of intracaldera tuffs close to its northern margin. Here, in contrast to the general tectonic subsidence elsewhere in the graben,

repeated uplift due to resurgence caused periods of subaerial volcanism and several unconformities. Because of the latter and the restriction of outcrop caused by recent erosion, the sequence is incomplete. However, nine stages of development have been recognised within the sequence, each of which reflects specific phases of volcanism and sedimentation in a tectonically active shallow-marine environment.

STAGE 1   SUBSIDENCE AND SUBMARINE ERUPTIONS   Evidence of this stage is preserved in strata, up to 95 m thick, in the north-east face of Snowdon (Plate 36). The strata comprise basaltic tuffs, breccias, hyaloclastite and pillow lavas with 5–6 m of thin-bedded turbidites and basalt cobble conglomerate near the base, overlying the Lower Rhyolitic Tuff Formation. An abundance of vesicular scoriae and cuspate shards indicates that magmatic explosivity was not suppressed, but there is no evidence to suggest that the accumulation emerged from a marine environment. The deposits reflect sloughing of material directly from the vicinity of a vent (or vents) during eruption, but evidence of the late evolution of the volcanoes is missing because the deposits are truncated by a later (Stage 5) unconformity.

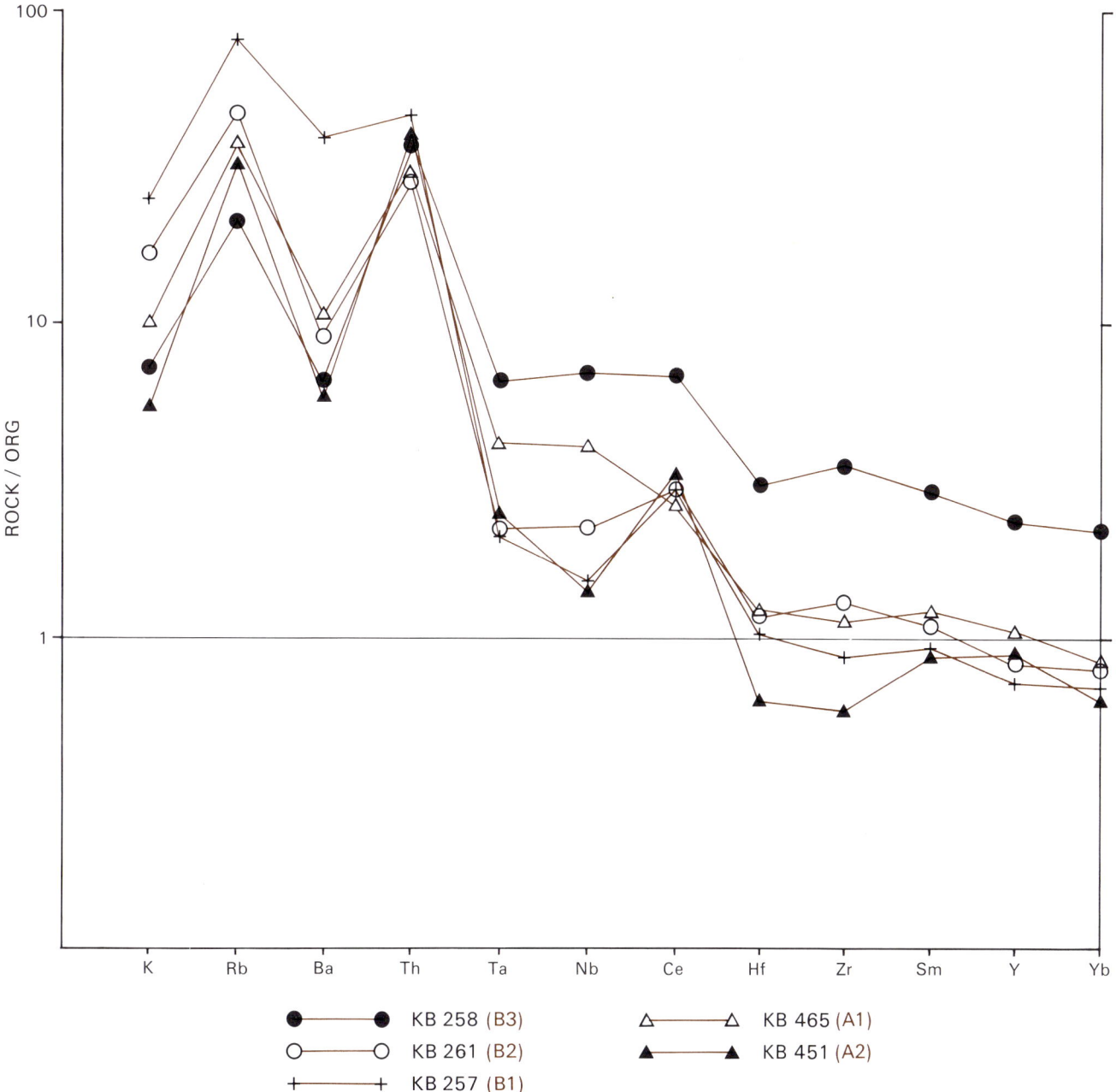

Figure 67 Rhyolites of the Snowdon Centre, multi-element ocean-ridge-normalised granite diagram (normalising factors after Pearce at al., 1984a).

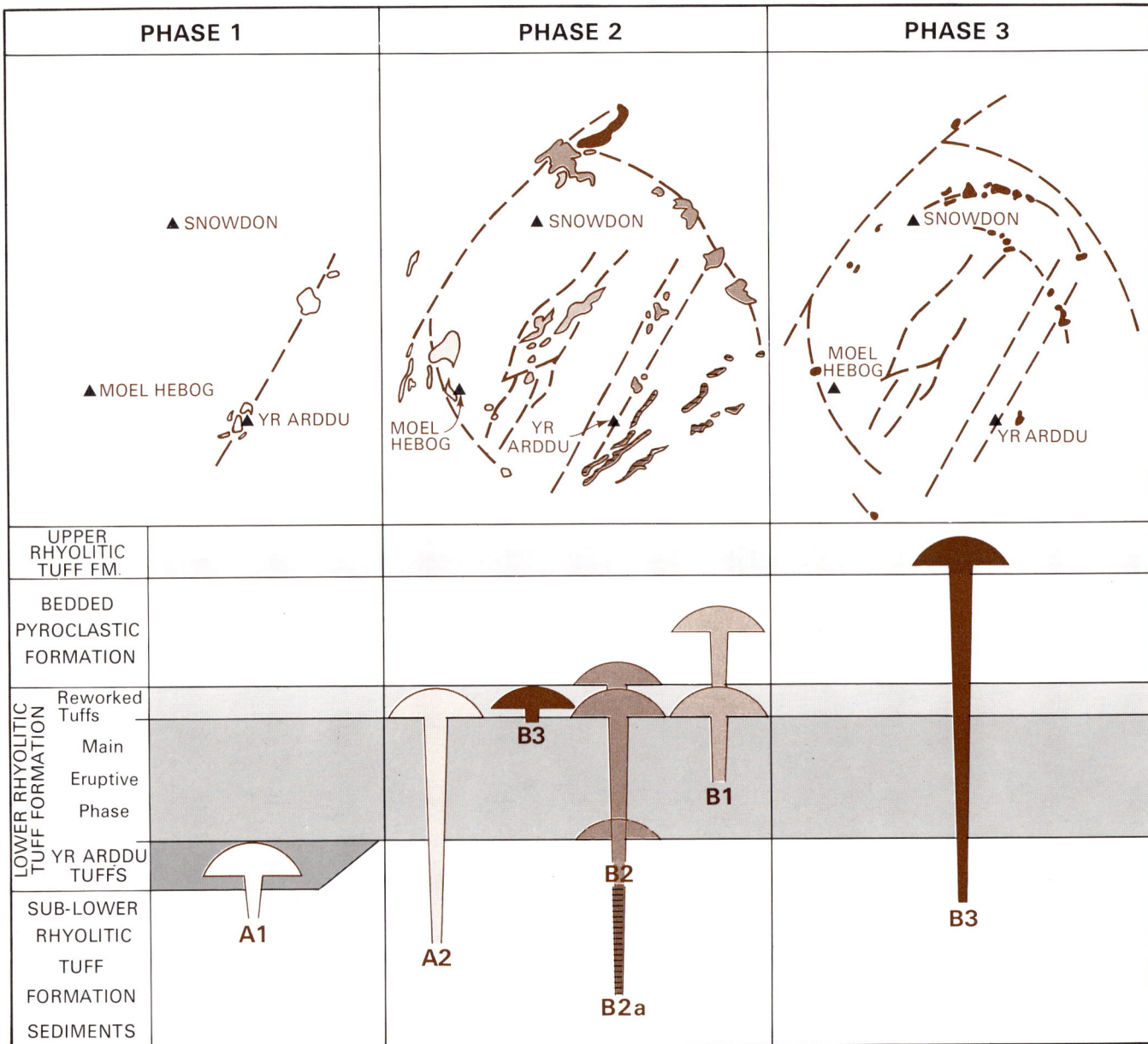

Figure 68 Rhyolites of the Snowdon Centre. Stratigraphical and spatial distribution of the groups (modified after Campbell et al., 1987).

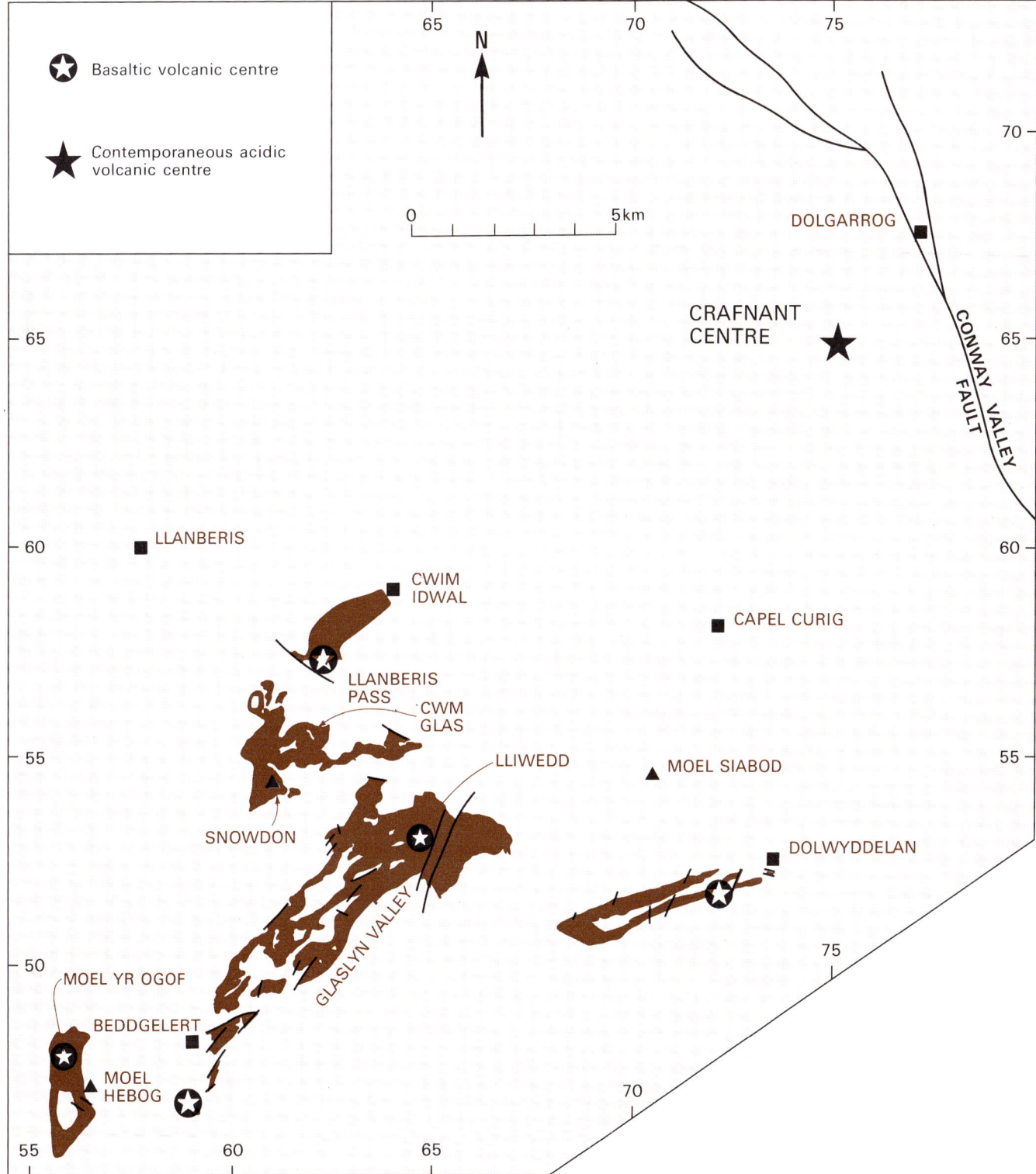

Figure 69 Outcrop of the Bedded Pyroclastic Formation showing main basaltic volcanic centres and contemporaneous acidic volcanic centre (Crafnant Centre).

Plate 34 Contact between the reworked facies of the Lower Rhyolitic Tuff Formation (below) and Stage 4 basaltic sandstones of the Bedded Pyroclastic Formation (above), exposed in the lip of Cwm Glas [SH 623 555].

STAGE 2 SUBAERIAL VENT CONSTRUCTION Two agglomerate-filled vents, cutting both the Lower Rhyolitic Tuff Formation and the Bedded Pyroclastic Formation Stage 1 deposits, have been distinguished on the north-east face of Snowdon (Plate 36). The Britannia vent (maximum diameter about 150 m) and the Glaslyn vent (minimum diameter 280 m) both contain a chaotic mixture of subangular to rounded lithic basaltic lapilli and blocks, up to 20 cm across. The largest clasts are non- to poorly vesiculated and the lapilli are moderately vesicular. The nature of the matrix is difficult to distinguish, but in both vents thin irregular basalt intrusions have been determined.

The height of the contemporary surface above the agglomerates is unknown as both structures are truncated by a later (Stage 5) unconformity. However, the character of the present infilling constituents suggests that the eruptions were dominantly Strombolian in character, although they reflect waning activity, possibly with late phreatic explosions. The lithological similarity, form and association of the two structures suggest that they developed penecontemporaneously.

STAGE 3 UPLIFT AND EROSION ASSOCIATED WITH RHYOLITE EXTRUSION This stage is distinguished by partial erosion of Stage 1 and 2 deposits to form a beach platform at or slightly below the level of the earlier sub-Bedded Pyroclastic Formation unconformity in the north and a cliff cut into Stage 1 deposits in the north-east face of Snowdon (Plate 36). Uplift of greater than 100 m is indicated.

In the north, in Cwm Glas, the stage is recognised in deposits which unconformably overlie the reworked top of the Lower Rhyolitic Tuff Formation (Plates 34, 35A). These comprise rhyolitic pebble and cobble conglomerates which are locally overlain by up to 6 m of flow-banded rhyolite lava. The relationships indicate that the lava was extruded on to a beach and locally eroded into small stacks while the syncline in the Lower Rhyolitic Tuff Formation substrate reflects contemporary emplacement of a large rhyolite intrusion. It is suggested that the rhyolites reflect resurgence which, together with the emplacement of similar rhyolites in the area, contributed greatly to the overall uplift distinguished during this stage of activity.

STAGE 4 STROMBOLIAN VOLCANISM AND SHALLOW–MARINE REWORKING OF ITS DEPOSITS The stage is represented by a complex sequence, up to 75 m thick, of turbidites, reworked turbidites, lithic vitric breccias and conglomerates (Plate 35) (Figure 70). These deposits overlie the Stage 3 rhyolite in Cwm Glas and abut the ancient cliff cut in Stage 1 deposits in

Plate 35 Sedimentary structures in the Bedded Pyroclastic Formation, Cwm Glas [SH 621 555]. Photographs by B P Kokelaar.

A   Basal bed of the BPF (Stage 4) resting with slight angular conformity on reworked acidic tuffs of the Lower Rhyolitic Tuff Formation.
B   Winnowed basaltic clasts in foreset lithic breccia of Stage 4.
C   Ripple lamination in turbidite apron deposits of Stage 4.
D   Dewatering ('dish' or 'tepee') structures in turbidite apron deposits of Stage 4.
E   Large-scale foresets (Stage 4) overlain by wave-worked deposits of Stage 5.

the north-east face of Snowdon. The sequence records littoral marine sedimentation during eruption and shoreline reworking of deposits from a nearby Strombolian volcano (Figure 71). Four cycles of breccia fan and turbidite apron progradation have been distinguished. Each cycle was followed by erosion and substantial vertical movements are indicated.

The deposits include beds dominated by basaltic pyroclasts which indicate minimal reworking during transport from a subaerial eruptive source to submarine depositional site. Interbeds with a preponderance of rounded clasts, diverse basalt types, shelly debris and rhyolite clasts suggest protracted nearshore storage and reworking of clastic material. It

is suggested that the fans and aprons grew during eruption episodes and that these were interspersed with periods when diminished sediment supply allowed organic burrowing of the turbidites and carbonate alteration. Wave-ripples and hummocky cross-stratification indicate deposition within wave-base. The geometry of the deposits and palaeocurrent data (Kokelaar, in preparation) indicate their derivation by coastline-parallel currents from a volcanic source to the south or south-east.

The lithologies and lithofacies of deposits of the four cycles are closely similar although successive deposits are thinner. Unconformities between the cycles reflect uplift and tem-

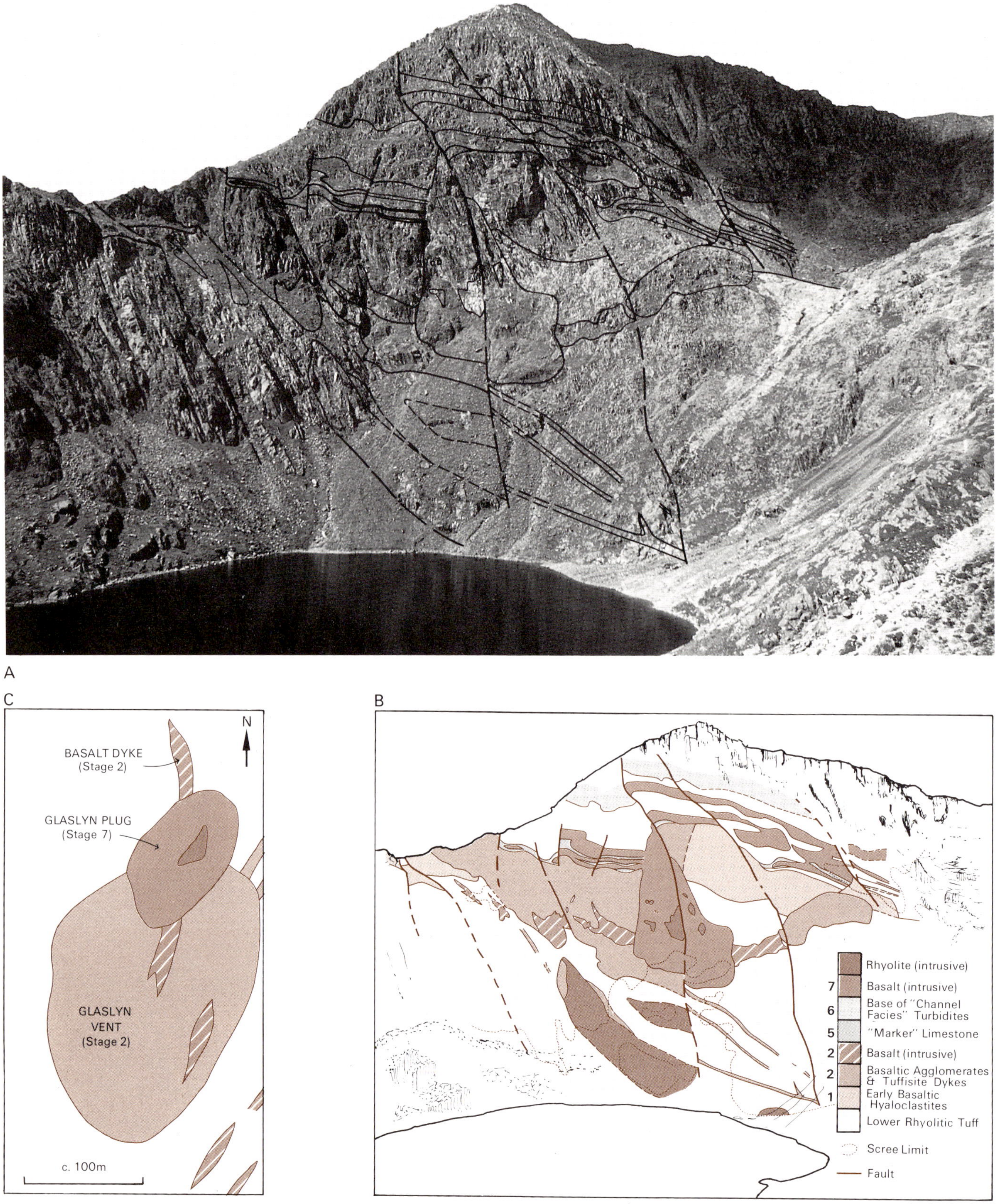

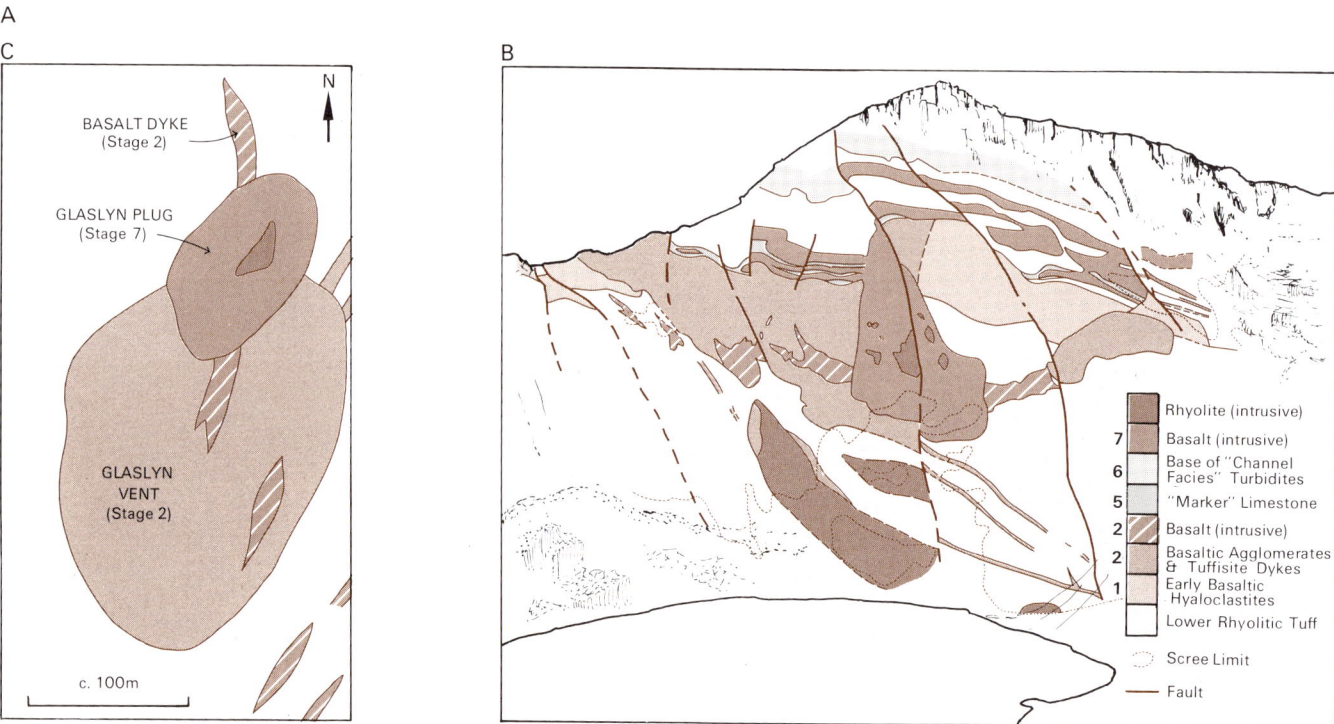

A

C   B

BASALT DYKE
(Stage 2)

GLASLYN PLUG
(Stage 7)

GLASLYN
VENT
(Stage 2)

N

c. 100m

	Rhyolite (intrusive)
7	Basalt (intrusive)
6	Base of "Channel Facies" Turbidites
5	"Marker" Limestone
2	Basalt (intrusive)
2	Basaltic Agglomerates & Tuffisite Dykes
1	Early Basaltic Hyaloclastites
	Lower Rhyolitic Tuff

⟋ Scree Limit

— Fault

Plate 36   Details of the Bedded Pyroclastic Formation cropping out on the north-east face of
Snowdon above Llyn Glaslyn.

(A) Photograph of north-east face with geological boundaries added. (B) Key to the geological units exposed
in A. Numbers in the legend refer to depositional stages of the Bedded Pyroclastic Formation. (C) Sketch map
of the Glaslyn Vent Complex. (All information supplied by B P Kokelaar).

WEST

EAST

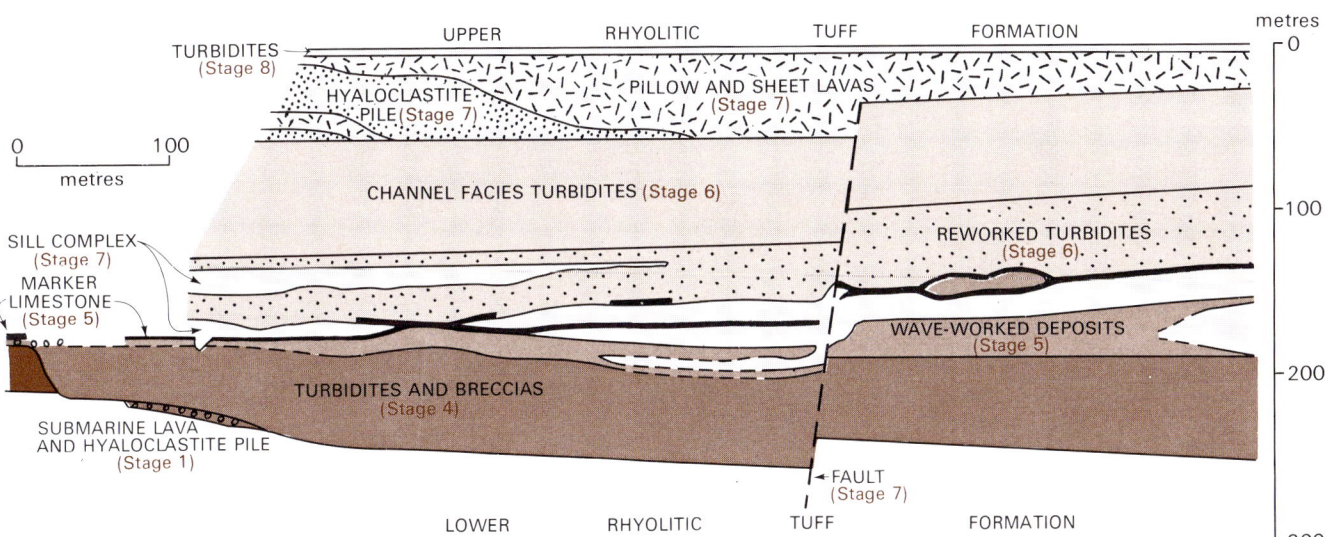

Figure 70 Bedded Pyroclastic Formation, generalised cross-section of south face of Crib y ddysgl [SH 611 551] (after Kokelaar, in press).

porary cessation of the volcanic activity, possibly with uplift, while progradation indicates its renewal. Overall, the stage reflects total subsidence up to about 90 m. The Glaslyn and Britannia vents (Stage 2) cannot represent the source of volcanoes because the deposits wedge out in their direction and abut the cliff cut in Stage 2 agglomerates.

STAGE 5 UPLIFT, SUBSIDENCE, TRANQUILLITY This stage is represented throughout the area by heterolithic sediments, up to 50 m thick, which overlie an irregular unconformity (on Stage 1, 2 and 4 deposits). The unconformity reflects development of a north-facing shoreline of platforms and small cliffs and is overlain by up to 3.5 m of cross-stratified gravel to boulder-grade conglomerates. Basalt clasts predominate, but clasts of rhyolite and lithified Stage 4 deposits also occur. The conglomerates are generally matrix-supported and suggest accumulation from density-modified grain flows rather than in-situ beach accumulation.

Succeeding strata are mainly turbiditic, coarse- to fine-grained sandstones, granule conglomerates and siltstones. Planar and trough cross-lamination, hummocky cross-stratification and current- and wave-rippled surfaces are common. Palaeocurrent data indicate along-shore east–west currents and influx of material from a southerly retreating coastline.

The uppermost 4 m comprises fine- to medium-grained sandstone, with brachiopod and crinoid debris, which has suffered intense carbonate alteration. The latter is interpreted to be the result of sea-floor weathering rather than the original lithological character. These strata, which include acidic tuffaceous beds, form a distinctive stratigraphic marker throughout the Snowdon outcrop and reflect a protracted period of relatively quiescent sedimentation.

STAGE 6 DEVELOPMENT OF TURBIDITE APRONS The turbidites, up to 140 m thick, can be subdivided into a lower,

thin- to thick-bedded, extensively reworked sequence and an upper, less reworked, coarser, more proximal and channelled sequence. The deposits, as those developed in Stage 5, are distinctly heterolithic and suggest shoreline erosion reworking, rather than direct volcanic input.

The substantial and variably trending thickness changes are more probably the result of contemporary faulting, differential subsidence and ponding in sea-floor topography than of proximal to distal thinning on submarine fans. A source to the south is, however, indicated and the contrasting sequences suggest activation of a new volcanic and sedimentary centre. Variable clast populations suggest that the turbidites reflect a multisourced apron complex.

STAGE 7 REACTIVATION OF GLASLYN VENT AND SILL AND LAVA EMPLACEMENT The stage is represented in the vicinity of Glaslyn vent (Plate 36) by up to 85 m of basaltic tuffs, hyaloclastites and lavas which prograde away from the centre into Cwm Glas. Clastic deposits form the bulk of the sequence and represent both explosive and non-explosive fragmentation. The accumulation is interpreted to reflect submarine fountaining of magma, with limited steam explosivity and contemporary sloughing of deposits away from the vent area.

To the south-east, the pile interdigitates with basalt lavas and is overlain by about 32 m of basalt lava with local lenses, up to 1.5 m thick, of heterolithic sandstone turbidites. The lavas comprise thin, variably vesicular sheets and pillowed units, columnar-jointed sheets up to 4 m thick, lava breccias and hyaloclastites. Plagioclase phenocrysts, up to 1 cm, are a distinctive component.

A sill complex, at or about the level of the Stage 5 carbonate horizon (Figure 70), comprises petrographic varieties of basalt similar to those occurring in the extrusive pile and it is suggested that they represent part of the same episode of magma movement. The sills show bulbous pillowed contacts

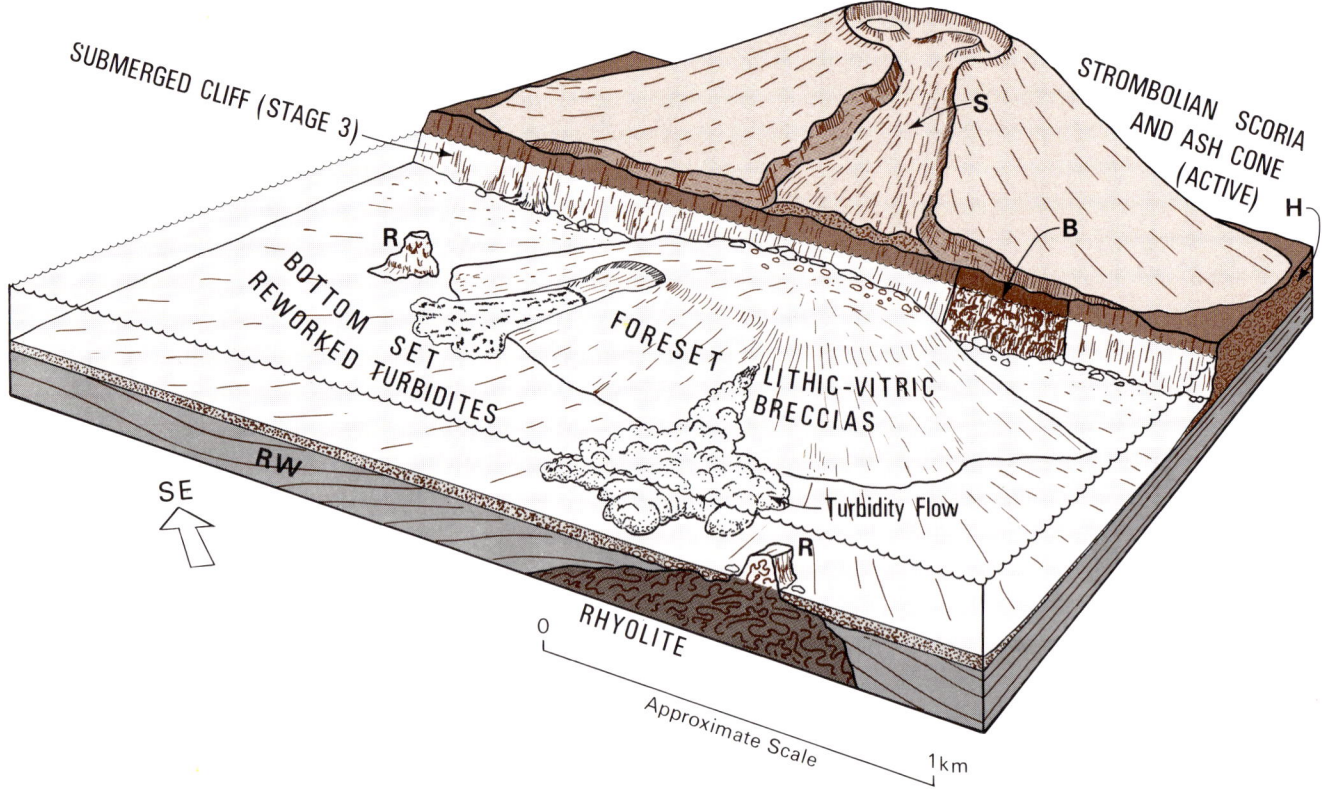

S Scoria and ash chute

H Hyaloclastite pile (Stage 1)

B Brittania vent (Stage 2),
       exposed in cliff

R Submerged and partly buried rhyolite stacks (Stage 3)

RW Reworked tuffs of the Lower
       Rhyolitic Tuff Formation

Figure 71 Bedded Pyroclastic Formation, model for evolution of Stage 4 deposits (after Kokelaar, in press).

and finger-like brecciated terminations, indicating emplacement in unlithified sediments. An irregular, complexly interconnected stack of up to four sheets can be traced into a plug-like multiple intrusion which cuts the Glaslyn vent agglomerate (Stage 2) and the adjacent Lower Rhyolitic Tuff Formation in the north-east face of Snowdon. The sites of the stack transgress the host deposits along faults which ponded erupted lavas, thus indicating that intrusion, effusion and faulting were contemporaneous.

STAGE 8 BURIAL OF EXTRUSIVE PILES   Throughout the area, the extrusive lavas and hyaloclastites (Stage 7) are discontinuously covered by up to 6 m of turbiditic and pebbly sandstones, which in turn are overlain by 1–3 m of silicic siltstones. The turbidites are intensely carbonated (as in Stage 5) and slow sedimentation rates are inferred.

In addition to scoria fragments, the beds contain an abundant derived shelly fauna, including brachiopods of a *Dinorthis* assemblage indicating derivation from a shallow sublittoral environment, in the range 0–10 m water depth (cf. Pickerill and Brenchley, 1979).

STAGE 9 UPLIFT AND RHYOLITE EMPLACEMENT   The stage is reflected in a slightly undulose and generally only slightly downcutting unconformity which developed across the deposits of Stages 7 and 8. However, local substantial uplift and faulting are evident where Stage 7 deposits are deeply eroded and missing across faults. Locally, rhyolitic pebbly sandstones resting on the unconformity indicate local emergence of rhyolite and littoral erosion. The unconformity marks the top of the Bedded Pyroclastic Formation.

North of Llanberis Pass

The Bedded Pyroclastic Formation to the north of Llanberis Pass, up to 220 m thick, is well exposed in the core of the Cwm Idwal syncline (Figure 69) and its base is exposed in the northern fold closure where 24 m of thin- and flaggy-bedded, light green, basic tuffites and tuffaceous sediments conformably overlie massive and thick, flaggy-bedded acid tuffites at the top of the Lower Rhyolitic Tuff Formation. The basic tuffites and tuffaceous sediments are coarse to fine grained, with lapilli and block-rich beds. Normal grading, cross-lamination and wave-ripple marked surfaces occur throughout. The

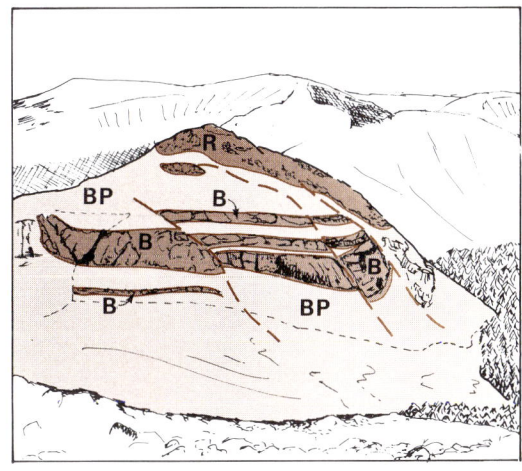

Plate 37   View, generally northwards, from Moel Hebog, showing broad features of geology on Moel yr Ogof [SH 556 478]: basaltic tuffs, hyaloclastites and volcaniclastic sediments (BP) and pillowed or massive basalts (B) of the Bedded Pyroclastic Formation, intruded by rhyolite (R).

lithologies comprise variable proportions of highly altered, vesicular and non-vesicular basaltic glass and basalt clasts, a few albitised and sericitised plagioclase crystals, epidote, chlorite, ilmenite, carbonate and aggregates of fine quartz and feldspar. In some beds, the clearly determinable volcanic components are sparse and they grade into tuffaceous sediments. The beds have yielded a shelly fauna (Dean, 1965) reinterpreted here as of possibly Upper Longvillian age.

These basal tuffites are conformably overlain by about 70 m of pillowed basalt. The pillows are well formed, up to 1.5 m, but generally 0.5 m in largest dimension. The basalt is overlain by basic tuffites and, to the south-west, interdigitates with two sequences of basic tuffites, indicating that the basalt in the north comprises several flows. To the south-west, the flows are locally massive, although pillows do occur in places. The cross-lamination, wave-ripples and shelly faunas in the bedded tuffites indicate a shallow-marine environment for their accumulation.

At the south-west closure of the syncline (Figure 69) a vent agglomerate crops out. It transects the upper rhyolite of the Lower Rhyolitic Tuff Formation and the lower, bedded, basic tuffites and contains abundant rounded, basalt clasts, up to 0.5 m in diameter, and a few angular, acidic fragments. Its matrix comprises ragged fragments of chloritised, vesicular, basaltic glass and rounded clasts of vesicular basalt in a fine aggregate of feldspar, chlorite, carbonate and ilmenite.

The agglomerate is unbedded. Detailed clast distribution has not been determined and the internal structure of the vent has not been defined. It is likely that the vent was the source of both the extrusive basalts within the outlier and the pyroclastic elements which were reworked into the tuffite and tuffaceous sedimentary sequence. Its present infill represents the last eruptive phase, the extrusive equivalents of which have been removed by erosion.

The basalts in this sequence are highly chloritised and carbonated, locally vesicular, containing scattered feldspar phenocrysts and, more rarely, augite. The feldspars, occurring both as microlites in the groundmass and as phenocrysts, are albitised and sericitised, and the phenocrysts include grains of clinozoisite and carbonate. Augite is wholly or partly replaced by actinolite and magnetite. Ilmenite occurs in small grains throughout the groundmass. Olivine has not been determined, and any original glassy mesostasis has been devitrified and replaced by a chloritic aggregate. Small rounded vesicles, less than 1.3 mm diameter, are infilled with chlorite and small prisms of epidote which commonly occur about the peripheries.

East of Lliwedd and the apical graben

To the east of Lliwedd (Figure 69), the Bedded Pyroclastic Formation conformably overlies the Lower Rhyolitic Tuff Formation. The contact is gradational, with the reworked acidic tuffs at the top of the Lower Rhyolitic Tuff Formation gradually being overwhelmed by the influx of marine, reworked basic tuffites, tuffaceous sandstones and siltstones at the base of the Bedded Pyroclastic Formation. The latter, up to 200 m thick, are mainly thin, flaggy-bedded, with common cross-lamination, ripple-marked bedding planes, and locally contain marine shelly faunas. A few thicker beds, about 1 m thick, of poorly sorted admixtures of pyroclastic and epiclastic debris, suggest emplacement by debris flows.

The source of these bedded tuffites and basaltic sediments has not been determined. However, they represent the marine reworking of a basaltic centre. Within the well-exposed sequence, only a few thin, altered basalt flows are preserved.

The bedded sequence is overlain by a thick accumulation of basalt, which is predominantly massive, but locally pillowed. Estimates of its thickness are difficult to make because of folding and the prevailing dip slope, but it must be in excess of 150 m. The basalts are highly altered, locally distinctively vesicular, and porphyritic, with albitised feldspar phenocrysts, up to 9 mm. In the thickest part of the sequence, individual flows are difficult to determine, hyaloclastites are extremely restricted, and there is no evidence of extensive reworking. To the east and south, thin flows can be distinguished, locally wedging out into reworked basaltic sediments. This suggests that the thickest sequence represents constant and fairly rapid effusion and accumulation close to the eruptive centre. Specific vents have not been recognised, but it is possible that the local intrusive rhyolite (Group B3) ascended a pre-existing basaltic vent.

To the south-west, in the apical graben of the Lower Rhyolitic Tuff Formation caldera, the accumulation of the Bedded Pyroclastic Formation was affected by penecontemporaneous faulting and the emplacement of rhyolite domes. On the south-east side of the graben, the contact between the Lower Rhyolitic Tuff Formation and Bedded Pyroclastic Formation is conformable and gradational from marine-reworked acidic tuffs and tuffites, with few basaltic sandstone intercalations, into dominantly basaltic sandstones. The latter, up to about 200 m thick, are well bedded and closely similar to the sequence in the equivalent position east of Lliwedd.

On the north-west side of the graben, the contact between the Lower Rhyolitic Tuff Formation and Bedded Pyroclastic Formation is generally faulted, and the latter comprises massive beds of fine- to coarse-grained, poorly sorted admixtures of pyroclastic and epiclastic debris, with blocks of basalt, acidic tuff and rhyolite locally common near the faulted margin. Bedding is generally ill defined and internal bed forms are rarely developed. The sequence is interpreted as representing accumulation from sediment gravity flows, generally below wave-base, within the tectonically unstable graben.

The contrasting relationships of both the contact between the Lower Rhyolitic Tuff Formation and Bedded Pyroclastic Formation, and the Bedded Pyroclastic Formation bed forms and lithologies on the north-west and south-east sides suggest that the graben was profoundly asymmetric and is better described as a half graben.

Both endogenous and exogenous rhyolite domes (Group B1) were emplaced in the apical graben during Bedded Pyroclastic Formation times. Aprons of rhyolite debris developed around the exogenous domes, and the debris was incorporated into the basaltic sediment in the vicinity. In places, rhyolite pebble and cobble conglomerates overlie the domes, indicating shallow-marine erosion prior to burial beneath the accumulating sediment gravity flows of the Bedded Pyroclastic Formation. Local enclaves of siltstone, bedded acid tuffs, tuffites and rare accretionary lapilli tuff are spatially associated with some of the rhyolites. These probably accumulated on topographic highs, relative to the sea floor, caused by the dome intrusion and as such lay above the influence of the basaltic sediment gravity flows which dominated sedimentation within the half graben.

Basaltic eruptive centres within the apical graben have not been distinguished, but two possibilities occur in the peripheral areas. One lies at its south-west end near Moel Hebog, where a small vent structure intrudes the Lower Rhyolitic Tuff Formation (Figure 69). It is infilled with blocky basalt and hyaloclastite and possibly represents a deep section of a vent. The other possible centre lies on the south-east flank of the graben, between the sections east of Lliwedd and the western closure of the Dolwyddelan syncline (Figure 69). It is marked by a rectilinear array of approximately east–west and subordinate north–south-orientated, fine-grained, basalt dykes within the Lower Rhyolitic Tuff Formation. Some dykes can be traced upwards into the base of the Bedded Pyroclastic Formation and down into a rhyolite (Group A1) and sediments subjacent to the Lower Rhyolitic Tuff Formation. The dykes are interpreted as possible high-level feeders related to Bedded Pyroclastic Formation volcanism. Such an interpretation is supported by the occurrence along one dyke of an elongate vent infilled with block-rich basalt, hyaloclastite and bedded basaltic tuffite with clasts typically less than 1 cm and bedding on a scale of 1–2 cm.

Dolwyddelan

In eastern Snowdonia, the Bedded Pyroclastic Formation forms a well-developed feature within the east-north-east-trending Dolwyddelan syncline, west of Dolwyddelan (Figure 69). This is the only outcrop which demonstrates the lateral and vertical relationships between the Bedded Pyroclastic Formation and the acidic sequences of the partly coeval Crafnant acidic eruptive centre. The formation varies markedly in thickness and facies both along the length of the syncline and across its axis. The thickest sequence, up to 180 m, on the western end of the north limb of the syncline, comprises basaltic tuffites and volcaniclastic sandstones and siltstones, with individual beds generally from 1 to 10 cm thick but with a few more massive beds up to 1 m thick. Grading and cross-lamination are common and grain size varies from silt grade to coarse lapilli up to 15 cm. Individual lapilli are subangular to subrounded and consist mainly of devitrified basaltic glass, some vesicular, and basalt; clasts of chloritised tuffaceous siltstones, acid tuff and flow-banded rhyolite are less common. The matrix comprises a fine-grained aggregate of carbonate, chlorite, iron-oxide and feldspar. Thin beds of fine-grained acidic tuff and tuffite occur locally within the sequence. Shelly fossils were obtained, particularly in the lower part of the sequence, and these include dinorthid and dalmanellid brachiopods and occasional trilobites, most notably *Estoniops alifrons*. The dominant depositional mechanism in the accumulation was sediment gravity flow, as both tuff-turbidites and debris flows.

On the south-south-east side of the syncline, the sequence is generally much thinner (50–80 m thick) with well-developed bedding on the scale of 1–5 cm. The sequence fines upwards, with silt-grade material predominating towards the top. In addition to the pronounced thinning across the axis of the syncline, which suggests penecontem-poraneous fault control of deposition, the facies also vary sub-parallel to the synclinal axis. Silt- and fine sand-grade material increase in abundance eastwards, becoming predominant at the fold closure, indicating a palaeoslope from west to east. These relationships suggest that most of the debris was derived from centres to the west, in the vicinity of the Snowdon massif. The only exception to this pattern is a local accumulation of basaltic agglomerates and poorly bedded, coarse tuffs high in the sequence, close to the eastern closure of the syncline. This accumulation is interpreted as representing a small, short-lived, eruptive centre close to the south-east margin of the Snowdon graben (Figure 69).

The upward transition from the Bedded Pyroclastic Formation to the Upper Rhyolitic Tuff Formation on the north-north-west limb is marked by a fine-grained acidic tuff which overlies mainly tuffaceous siltstones. The tuff is succeeded by approximately 6 m of black siltstones and mudstones, with dispersed acidic shards and feldspar crystals, and then by the main body of the Upper Rhyolitic Tuff Formation.

Moel Hebog

In south-western Snowdonia, the Bedded Pyroclastic Formation is preserved in the core of the Moel Hebog syncline (Figure 69, Plate 37) close to the south-west margin of the Lower Rhyolitic Tuff Formation caldera. The formation overlies both a flow-banded rhyolite and the reworked tuffs at the top of the Lower Rhyolitic Tuff Formation, which locally contain a thin basic tuffite. The rhyolite is possibly an extrusive expression of the large intrusive rhyolite (Group A2) which occurs to the north-east of Moel yr ogof.

The lower part of the Bedded Pyroclastic Formation is dominated by extrusive basalts, variably pillowed, up to 120 m thick, (Plate 38), with pillow breccias and hyaloclastites. Above, the relative proportion of reworked and remobilised basaltic tuffites and volcaniclastic sandstones increases and the frequency of primary basalt flows decreases. The basalts are variably massive, with columnar jointing, and they are pillowed, particularly towards the base. They pass laterally and vertically into pillow breccias, hyaloclastites and well-bedded, fine- to medium-grained basaltic tuffites. The pillow breccias comprise vesicular pillow fragments up to 50 cm, in a matrix of coarse clastic and hyaloclastitic basaltic fragments and feldspar crystals. They are succeeded by up to 110 m of bedded basaltic tuffites and hyaloclastites, within which two relatively massive basalt flows (45 m and 10 m thick) and other thinner flows occur. The basaltic tuffites are flaggy-bedded, generally 1–10 cm thick, with plane parallel-lamination and locally low-angle cross-lamination. More massive beds up to 40 cm thick occur in places. These lateral facies relationships suggest that a basaltic vent lay in the region of Moel yr ogof, and this is supported by the numerous thin (up to about 5 m), fine-grained, transgressive basalt sills and dykes which occur within the basalts immediately to the north of Moel yr ogof (Plate 37). These intrusions pass down into a basalt and dolerite dyke system, orientated approximately north-north-east–south-south-west, which penetrates down to the Pitts Head Tuff substrate, almost to the level of a massive dolerite sill to which it is probably related.

The spatial association of a large, flow-banded rhyolite (B3 Group) intrusion (Figure 68) with the presumed locus of the basaltic centre suggests that the rhyolite may have been

Plate 38   Pillowed basalt of the Bedded Pyroclastic Formation, Moel Hebog [SH 568 469].

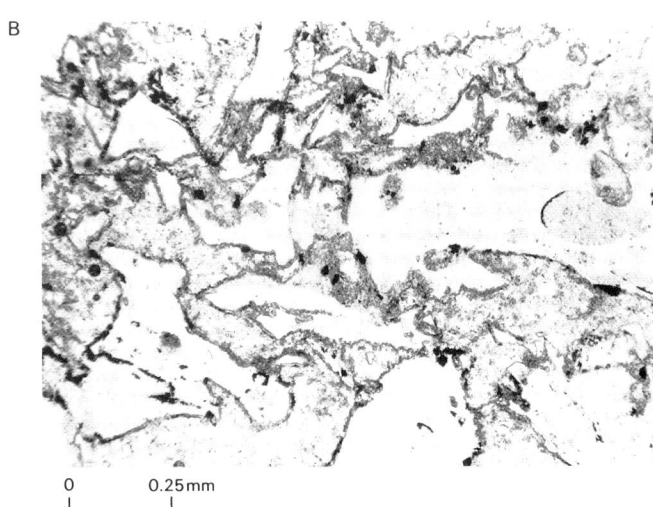

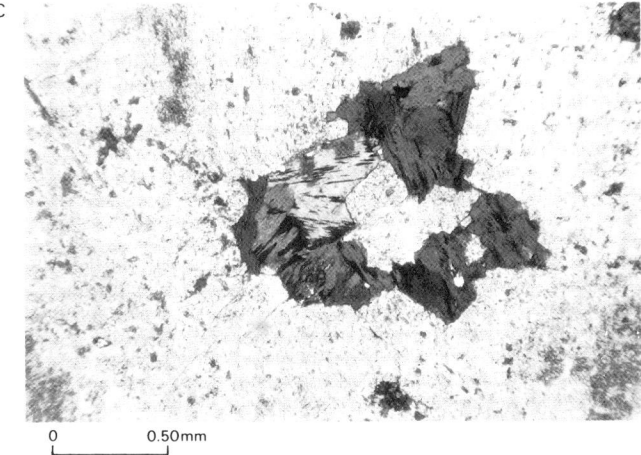

Plate 39

A   MH 906. Hyaloclastite. Chloritised fragments of basaltic glass and a prominent fragment of ferritised basalt. Bedded Pyroclastic Formation, ppl. [SH 6173 5560].
B   MH 1179. Hyalotuff. Chloritised fragments of vesicular basaltic glass in a dusky, recrystallised quartzofeldspar aggregate. Bedded Pyroclastic Formation, ppl. [SH 6131 5117].
C   KB 362. Tan y Grisiau Granite. Perthitic feldspar and quartz and chloritised biotite, ppl. [SH 6151 3413].

emplaced within the basaltic vent itself. A small outlier of bedded acidic tuffites succeeds the Bedded Pyroclastic Formation with apparent conformity. These tuffites are geochemically similar to the adjacent rhyolite intrusion on Moel yr ogof and a genetic relationship is presumed, thus implying only a short time between the cessation of Bedded Pyroclastic Formation volcanism and rhyolite emplacement.

GEOCHEMISTRY OF THE BEDDED PYROCLASTIC FORMATION Selected analyses of the Bedded Pyroclastic Formation are presented in Appendix 1, Table 10. On a plot of Zr/TiO_2 vs. Nb/Y (Figure 72) most basalt analyses ($N = 11$) cluster within the field of subalkaline basalt (Zr/TiO_2 ratios varying between 0.0062 and 0.0091). However, the volcaniclastic sediments ($N = 65$) are more variable (Zr/TiO_2 ratios varying between 0.0053 and 0.12) with most being basaltic but including compositions within the fields of andesite, dacite/rhyodacite and, in one case, even within the field of rhyolite. The more acidic compositions are of samples with a significant dilution of rhyolite and acidic tuff clasts, highlighting the contemporaneous local acidic volcanism and rhyolite dome emplacement during Bedded Pyroclastic Formation times. Despite the variable dilution, both volcaniclastics and

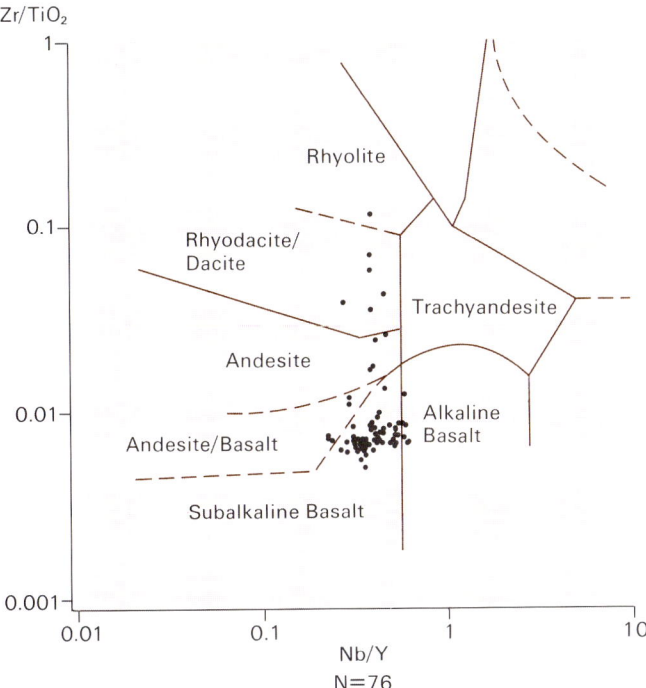

Figure 72 Bedded Pyroclastic Formation, Zr/TiO$_2$ vs. Nb/Y diagram.

basalts have similar Nb/Zr ratios (Figure 73) (basalts, 0.063–0.098; volcaniclastics, 0.0534–0.112). The basalts are relatively evolved in terms of the incompatible elements Zr (152–296 ppm), Nb (11–26 ppm) and Y (35–74 ppm), and are enriched in TiO$_2$ (1.98–4.79 wt%).

N-MORB normalised multi-element plots of the basalts (Figure 74) are characterised by enrichment in the LIL and HFS elements relative to N-MORB, with stronger enrichment of most of the LIL elements. However Sr and Ni show pronounced negative anomalies. Ce is consistently depleted, relative to La and Nb, with respect to N-MORB. The N-MORB normalised profiles are similar in terms of the HFS elements such as Hawaiian basalts (cf. Pearce, 1982b) and they have the general characteristics of E-type (enriched) MORB.

The basalts are compositionally similar to dolerites which intrude the Lower Rhyolitic Tuff Formation and Bedded Pyroclastic Formation, and to many of the dolerites which intrude the sequence below the Lower Rhyolitic Tuff Formation to the south-east of the caldera. They differ markedly from the sub-LRTF basalts (Figure 48) in their general LREE and HFS element enrichment and particularly with respect to N-MORB, in their negative Ni anomalies and in their enrichment rather than depletion of Nb, relative to Ce.

Compared with basalts and hyaloclastites of the Tal y Fan Volcanic Formation (Appendix 1, Table 12), the Bedded Pyroclastic Formation basalts are slightly enriched in HFS elements but are otherwise broadly similar. The hyaloclastites of the Dolgarrog Volcanic Formation (Appendix 1, Table 12) are very similar but are slightly enriched in Nb.

Peralkaline reactivation of the Snowdon Centre: Upper Rhyolitic Tuff Formation

In central Snowdonia the deposits of the basaltic activity are overlain by the Upper Rhyolitic Tuff Formation, a sequence of acidic ash-flow tuffs, bedded tuffs and tuffites and few tuffaceous siltstones. It also includes some basaltic beds. The sequence reflects the final activity of the Snowdon Centre and its outcrop is restricted to small outliers, the largest of which lie on the north side of the Snowdon massif and in the Dolwyddelan syncline. The last phase of the activity was of rhyolite intrusion (Group B3). Both intrusive and extrusive phases are of peralkaline composition.

North side of Snowdon

On the north side of Snowdon the formation, up to c.100 m thick, predominantly comprises a massive, acidic ash-flow tuff up to 35 m thick, bedded acidic tuffs and rhyolite and is well exposed on Clogwyn y Person (Plate 40). Here, it overlies a slightly undulose unconformity (the final phase, Stage 9, of the Bedded Pyroclastic Formation) and locally basaltic and rhyolitic pebbly sandstone occur at its base. Rounded rhyolite pebbles on the unconformity indicate local emergence and littoral erosion of rhyolite. In places the main ash-flow tuff oversteps these basal sediments and lies directly on basalt and hyaloclastite (Stage 7 of the Bedded Pyroclastic Formation). The ash-flow tuff is blue-grey, with bleached, weathered surfaces on which impersistent bands of small lithic clasts and carbonate nodules can be determined locally. The tuff comprises a microcrystalline aggregate of quartz, feldspar, sericite and chlorite, with dispersed shards, a few fragmental quartz and albite crystals and lithic clasts. It is poorly sorted, with markedly variable proportions of constituents. Shards are generally fine-grained, fragmental rods and spikes, although a few multicuspate forms are preserved. Lithic clasts include acid tuff, perlitic fractured, devitrified acidic glass, after rhyolite or welded tuff, ragged tubular pumice and, most distinctively, a concentration of chloritised basaltic fragments. The last include both rounded basalt and ragged ves-

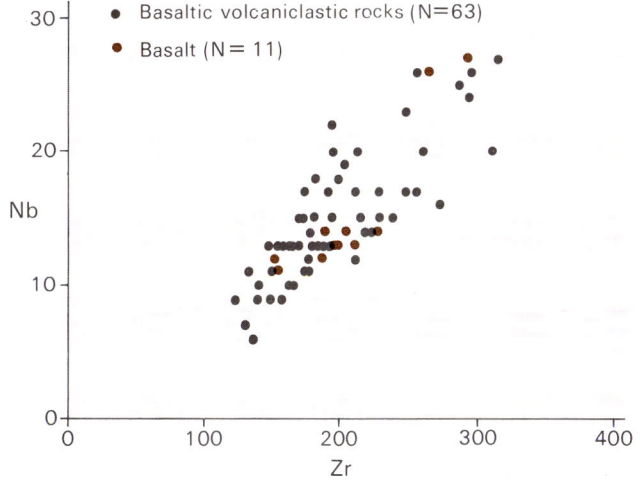

Figure 73 Bedded Pyroclastic Formation, Nb vs. Zr diagram.

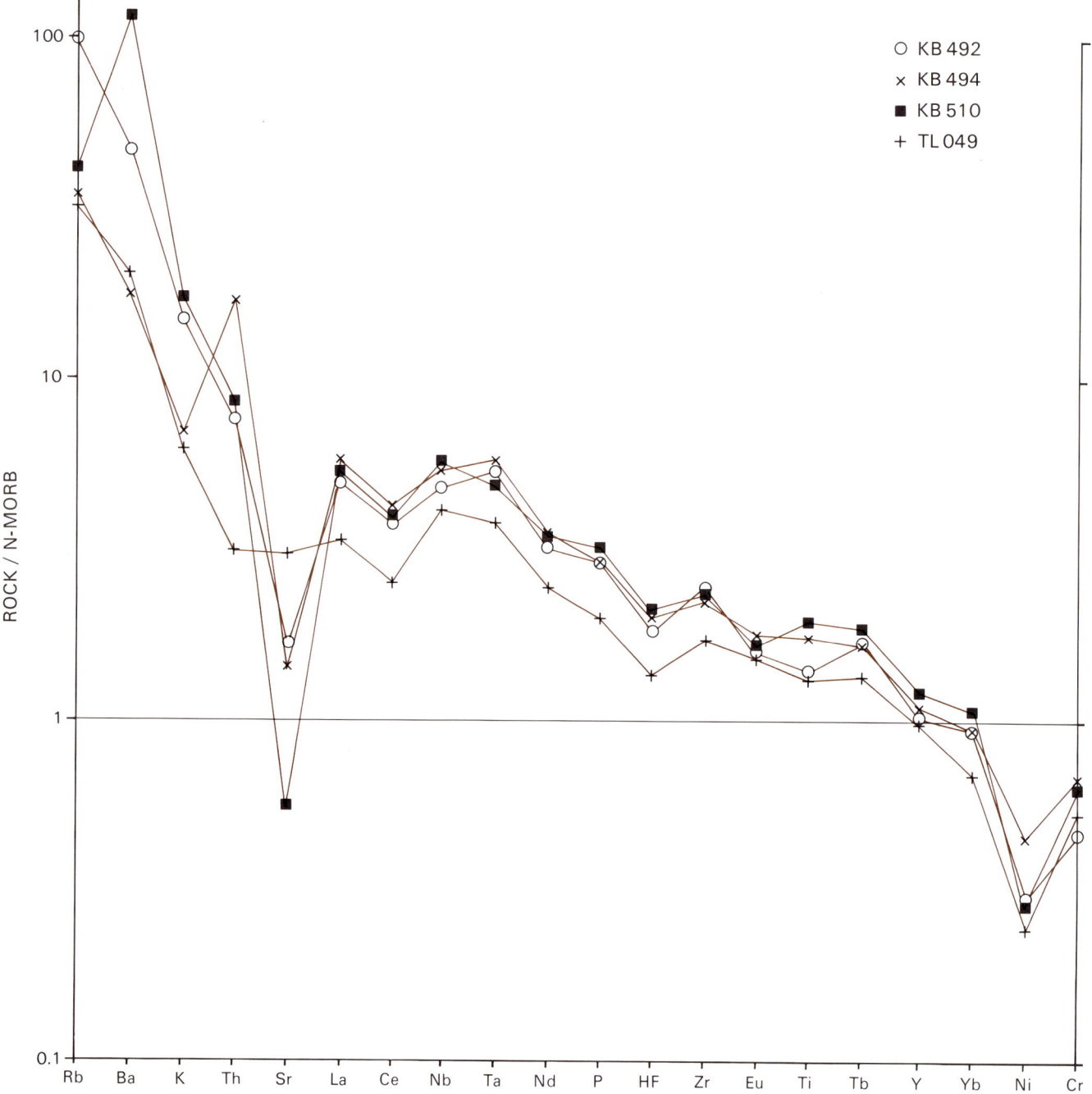

Figure 74   Bedded Pyroclastic Formation, N-MORB normalised multi-element diagram (normalising factors after Pearce, 1982a).

icular basaltic scoriae, and their frequency indicates that the source vent cut the subjacent Bedded Pyroclastic Formation. Towards its top, the ash-flow tuff grades into a fine-grained and silicified tuff.

The locally high proportion of the fine-grained sericite and chlorite matrix, and the heterogeneous character of the tuff are distinctive. The matrix could represent silt or mud, incorporated with the pyroclastic debris, either by remobilisation of unconsolidated, previously deposited debris, or directly from the vent. However, the tuff mainly overlies the basic

tuffs and tuffites of the Bedded Pyroclastic Formation, which covered an extensive area in central Snowdonia. Consequently, the matrix was probably a primary pyroclastic component and represents devitrified and recrystallised vitric dust. The heterogeneity of the tuff suggests its transport was limited.

The ash-flow tuff is overlain by up to 40 m of bedded acidic tuffs and tuffaceous siltstones, with plane-parallel and cross-laminated bands. These tuffs are generally fine to medium grained and silicified. The bed forms indicate subaqueous ac-cumulations of fine vitric dust, possibly as tuff-turbidites

Plate 40 The ridge of Clogwyn y Person [SH 615 554] viewed from Cwm Glas. The ridge comprises well-jointed, acidic ash-flow tuff (T) at the base of the Upper Rhyolitic Tuff Formation. Below, lie bedded basaltic sediments, basalt and hyaloclastite of the Bedded Pyroclastic Formation (BPF), and above an intrusive rhyolite (R).

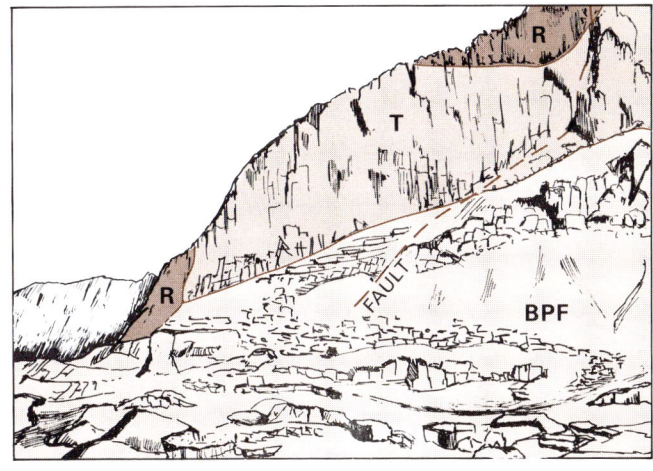

from redistributed, water-settled ash fall. The ash may have been derived either by elutriation from the underlying ash-flow during its transport or directly from the same vent.

On Clogwyn y Person (Plate 40), the ash-flow tuff and bedded acid tuffs are intruded by a rhyolite dome which probably had an extrusive expression. The dome (Group B3) is connected to a dyke-like feeder below. The dyke intrudes the Bedded Pyroclastic Formation bedded basic tuffs beneath the Upper Rhyolitic Tuff Formation ash-flow tuff at the north end of the ridge. The intrusion was facilitated by fluidisation at its contacts with the water-saturated, unlithified tuffs (Plate 31B). From here, the dyke clearly transects the overlying ash-flow tuff and extends into the dome. The sparsely porphyritic rhyolite is flow-folded and autobrecciated, with well-developed, columnar joints.

Dolwyddelan syncline

The Upper Rhyolitic Tuff Formation, up to 75 m thick, crops out in the core of the Dolwyddelan syncline to the east of the Snowdon massif (Figure 69). The sequence comprises poorly bedded and massive acid tuffs, tuffites and tuffaceous mudstones. Locally, beds up to 3 m thick, of acid tuff can be distinguished, but generally they are laterally impersistent. Most of the sequence consists of heterogeneous admixtures of pyroclastic and epiclastic debris.

The pyroclastic component includes devitrified fine and elongate shards and blocky, angular, cusp-edged bubble walls up to 2.5 mm, altered albite-oligoclase crystals and fragments and, most distinctively, tubular pumice clasts with splayed and ragged terminations. The epiclastic fraction is dominantly argillaceous, mud- and silt-grade material, which has subsequently been recrystallised. The formation becomes more dominated by tuffaceous mudstones towards the eastern closure of the syncline. Bedding is generally poorly defined and the massive units are interpreted as deposits of high-density turbidity flows generated by remobilisation of previously deposited pyroclastic debris and unlithified mud.

The lithologies of the Upper Rhyolitic Tuff Formation at Dolwyddelan are broadly similar to those of the Upper Crafnant Volcanic Formation of north-eastern Snowdonia. However, from the evidence of the facies variations related to an east-facing palaeoslope in the underlying Bedded Pyroclastic Formation, and the similar variations in the Upper Rhyolitic Tuff Formation, it is suggested that the provenance of the latter also lay in the west, in central Snowdonia. The prevailing mud facies reflects a continuance of the low-energy environments established here immediately prior to the emplacement of the Lower Rhyolitic Tuff Formation. The contrasting environments of central and north-eastern Snowdonia throughout the deposition of the Lower Rhyolitic Tuff Formation and the Bedded Pyroclastic Formation had been largely suppressed by Upper Rhyolitic Tuff Formation times.

East of Lliwedd

On the south side of the Snowdon massif, the Upper Rhyolitic Tuff Formation is exposed in small synclinal outliers on the ridge east of Lliwedd. The sequence preserved is only 25 m thick and comprises thin, flaggy-bedded, acid tuffs and tuffites, with intercalated tuffaceous siltstones, sandstones and conglomerates. The last consists of pebbles and cobbles of acid tuff and rhyolite, supported in a tuffaceous siltstone matrix. In places, the clasts are strongly deformed.

In the largest outliers, the bedded tuffs are associated with high-level rhyolite intrusions (Group B3) and in their vicinity the tuffs have been subjected to severe hydrothermal alteration.

GEOCHEMISTRY   On the Zr/TiO_2 vs. Nb/Y diagram (Figure 75) the acidic tuffs, representative analyses of which are presented in Appendix 1, Table 10, mainly straddle the boundary between the rhyolite and comendite/pantellerite fields. A single basalt plots in the andesite/basalt field. The associated rhyolite analyses plot in a similar area to the acid tuffs and are representative of the peralkaline Group B3 intrusions (Figure 63).

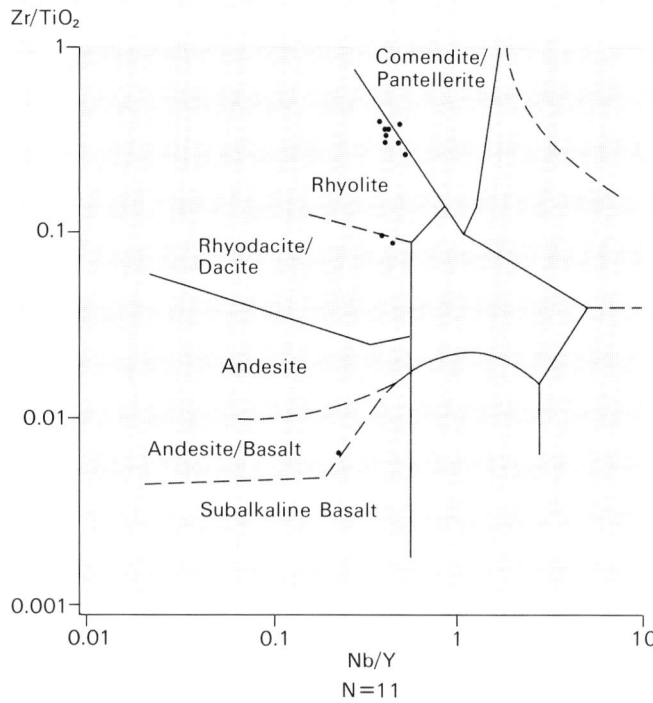

Figure 75 Upper Rhyolitic Tuff Formation, Zr/TiO_2 vs. Nb/Y diagram.

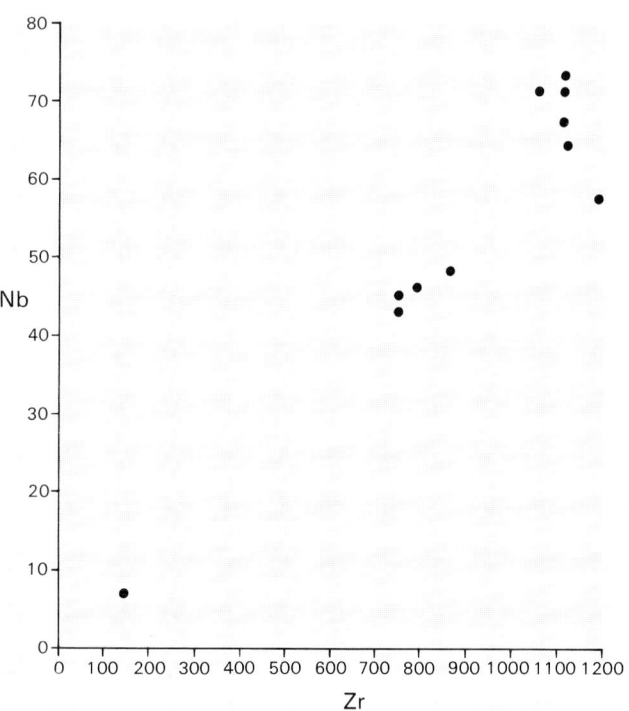

Figure 76 Upper Rhyolitic Tuff Formation, Nb vs. Zr diagram.

A plot of Nb vs. Zr (Figure 76) suggests that two distinct compositional clusters are present but this would only be corroborated by further sampling.

Evolution of the Snowdon Centre: a model

From consideration of all the available evidence, a new model of the evolution of the Snowdon Centre is proposed that represents a refinement of previously published accounts (e.g. Williams, 1927; Beavon, 1980; Howells et al., 1986). The stages of evolution, summarised in cartoon form in Figure 77, are described below.

Stage 1   Near-surface, upward migration of magma (? A2 rhyolite) occurred along the Beddgelert Fault Zone with the consequent formation of the Beddgelert pericline. A shoreline was established about the north, north-east and east margins of the pericline.

Stage 2   Basalt (arc-tholeiite) was extruded, both subaqueously, as pillowed lavas and hyaloclastites, and subaerially, to the north-west and north of the Beddgelert Fault Zone. High-level dolerite sills (arc-tholeiite affinities) were emplaced beyond the periphery of the Beddgelert pericline. Movement of subalkaline rhyolite magma (A1) along the Yr Arddu Fracture led to the subaqueous, and possibly later subaerial eruption and ponding of the Yr Arddu ash-flow tuffs in a small downsag caldera. Intrusion of A1 rhyolite domes followed at the Yr Arddu centre and elsewhere to the north along the Yr Arddu Fracture.

Stage 3   Uplift and faulting, accompanying near-surface migration of magma at the crest of the Beddgelert pericline, caused downslope collapse of the eastern flank of the pericline to form an extensive apron of mudflow megabreccia. Fissure-vents developed near the axis of the pericline resulting in the explosive subaerial eruption of subalkaline rhyolitic (A2) magma, which generated pyroclastic flows from which the basal welded tuff of the Lower Rhyolitic Tuff Formation was emplaced. Caldera collapse was initiated.

Stages 4 and 5   The main locus of eruption migrated northwards, beyond the shoreline, to the northern bounding fracture of the caldera. The main period of most voluminous ash-flow tuff eruption and caldera collapse ensued with maximum subsidence in the northern sector of the caldera. A single outflow tuff was emplaced subaqueously to the north and east of the caldera.

Stage 6   Shallow-marine reworking of the intracaldera tuffs accompanied regional subsidence. This was interrupted locally by resurgence along the caldera margin and Beddgelert Fault Zone, allowing the accumulation in subsiding areas of thick epiclastic sequences derived from the resurgent zones. Tuffs were redistributed northwards, from the northern margin of the caldera, as subaqueous sediment gravity flows. The upward movement and eventual eruption of rhyolitic magma as exogenous domes (? B1, B2 and B3), lava flows (B3) and small volume pyroclastic flows accompanied resurgence.

Stage 7   Subaqueous and subaerial basalt eruptions (transitional calcalkaline tholeiite to ocean-island basalt) commenced from vents mainly at the periphery of the caldera. Dolerite sills were emplaced towards the margins of the

caldera. Further rhyolite domes (B1, B2) developed and rhyolitic cores (B2) were emplaced in dolerite sills to form the multiple intrusions, mainly in the south-east sector of the caldera. Regional subsidence continued and thick sequences of shallow-marine, reworked, basaltic debris accumulated. A final phase of peralkaline rhyolitic (B3) activity ensued, mainly controlled by arcuate fractures in the north and north-east sectors of the Lower Rhyolitic Tuff Formation caldera, the western caldera-bounding fracture and an arcuate fault to the south of the caldera. Ash-flow tuffs, rhyolite domes and flows were emplaced.

CRAFNANT CENTRE: CRAFNANT VOLCANIC GROUP

In eastern and north-eastern Snowdonia, a deep marine environment prevailed prior to, and during the eruptive activity at the Snowdon Centre. Silt and mud deposition predominated and the mud component increased progressively with time. It was into this environment that the outflow, ash-flow tuff from the Lower Rhyolitic Tuff Formation eruptive centre was emplaced. Subsequently, in this area, the influence of the activity from the Snowdon Centre was minimal and the volcanic sequence (Crafnant Volcanic Group) which overlies the Lower Rhyolitic Tuff Formation outflow tuff (the basal member of the Lower Crafnant Volcanic Formation) reflects activity from the local Crafnant Centre (Figure 78). The absence of coarse terrigenes and shallow-marine reworking throughout the deposition of this sequence reflects the deep subaqueous containment of the volcanism. In comparison with the Snowdon Centre the area is not as well exposed and as a result the phases of development of the Crafnant Centre are less clearly established.

However, by mapping to establish correlation and thickness variations, and by interpretation of the distribution of lithologies and facies, a centre was defined (Howells et al., 1973, 1981a). In the light of later work on the Snowdon Centre, the early work is reinterpreted in this account.

Early acidic volcanism

Volcanic activity at the Crafnant Centre was dominantly acidic. The deposits comprise primary ash-flow tuffs, block and ash-flow tuffs, disrupted and slumped beds, and sequences of admixed pyroclastic and epiclastic debris. From their distribution and thickness variations, an outflow and intracaldera/proximal facies can be distinguished.

OUTFLOW TUFFS: LOWER CRAFNANT VOLCANIC FORMATION

The earliest activity of the centre is represented in three primary, nonwelded, acidic ash-flow tuffs. These tuffs (the upper three members of the Lower Crafnant Volcanic Formation) were designated Nos 2A, 2 and 3 by Howells et al. (1973, 1978, 1981a). They are interbedded with marine siltstones and mudstones overlying the outflow tuff from the Snowdon Centre. The lowest of the three tuffs, No. 2A (Howells et al., 1981a), up to 40 m thick, is the most restricted in distribution (Figure 78). It is silty and well cleaved in its basal 2 m. Above this, the tuff is massive, less well cleaved and sparsely porphyritic, with isolated feldspar crystals. At its top the tuff is extremely fine grained and siliceous.

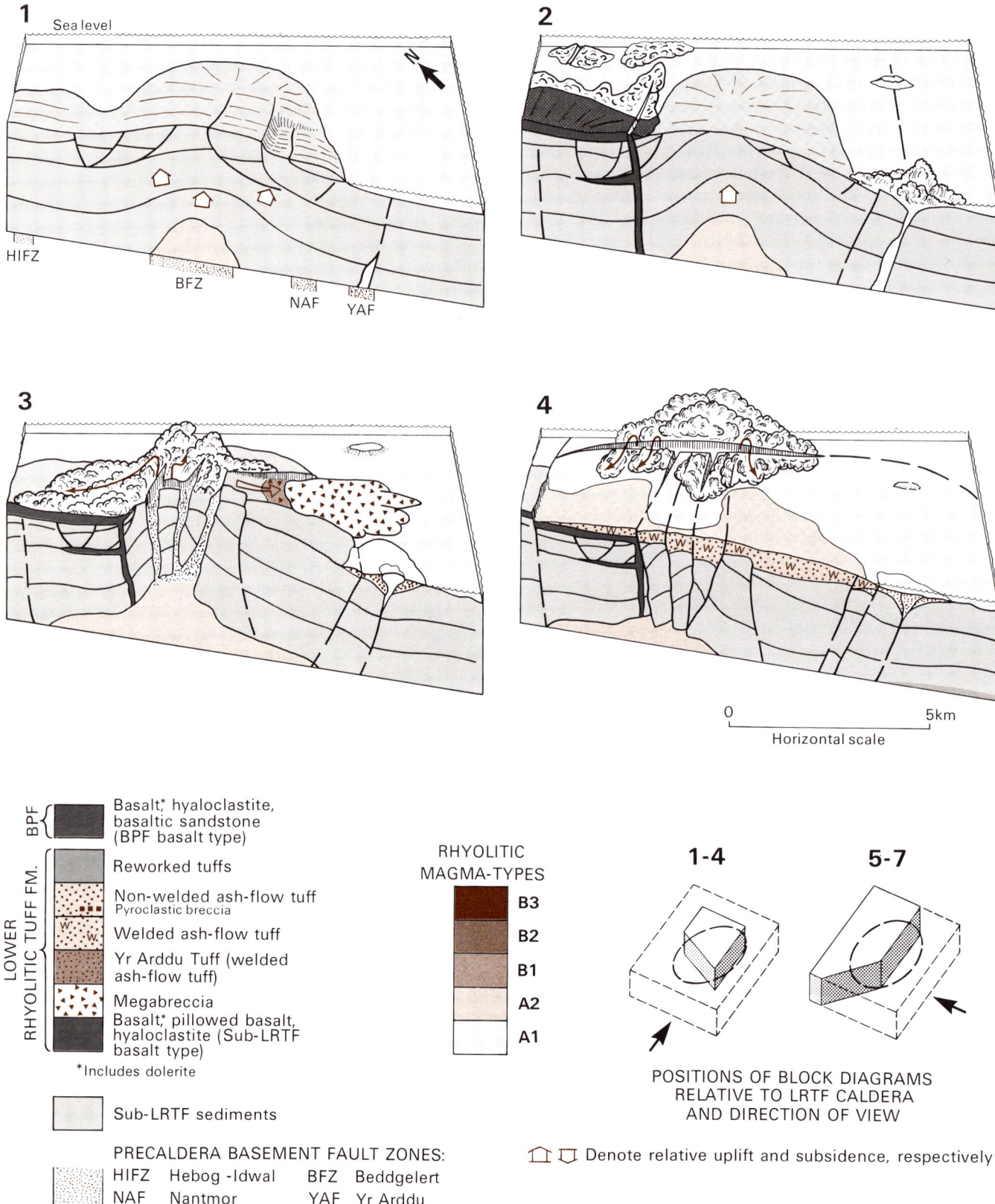

Figure 77   Model for the evolution of the Snowdon Centre. For explanation of Stages 1–7 see text.

5

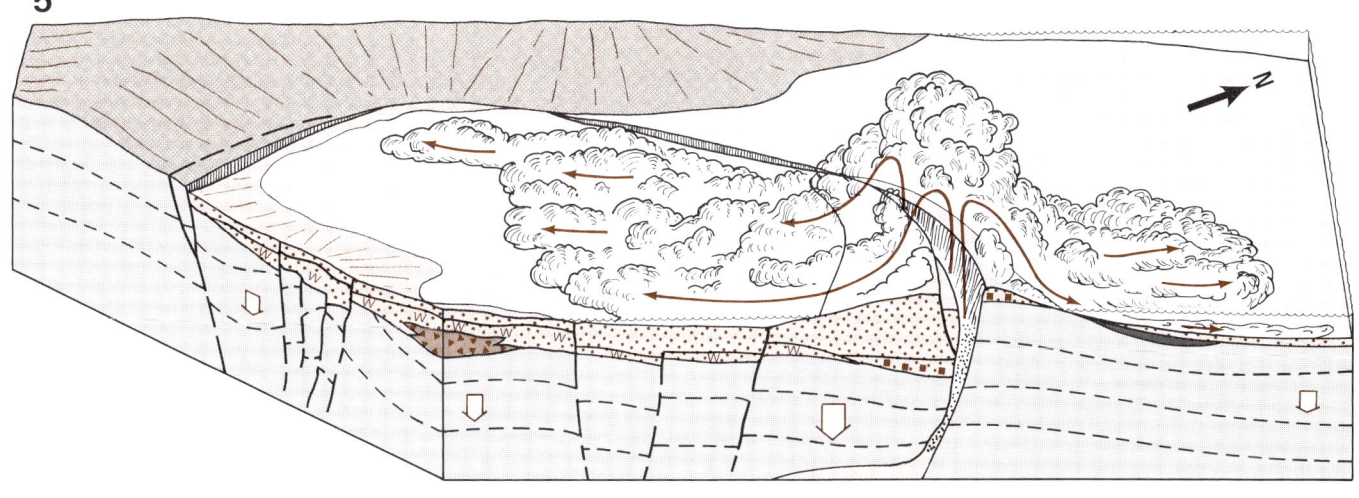

6

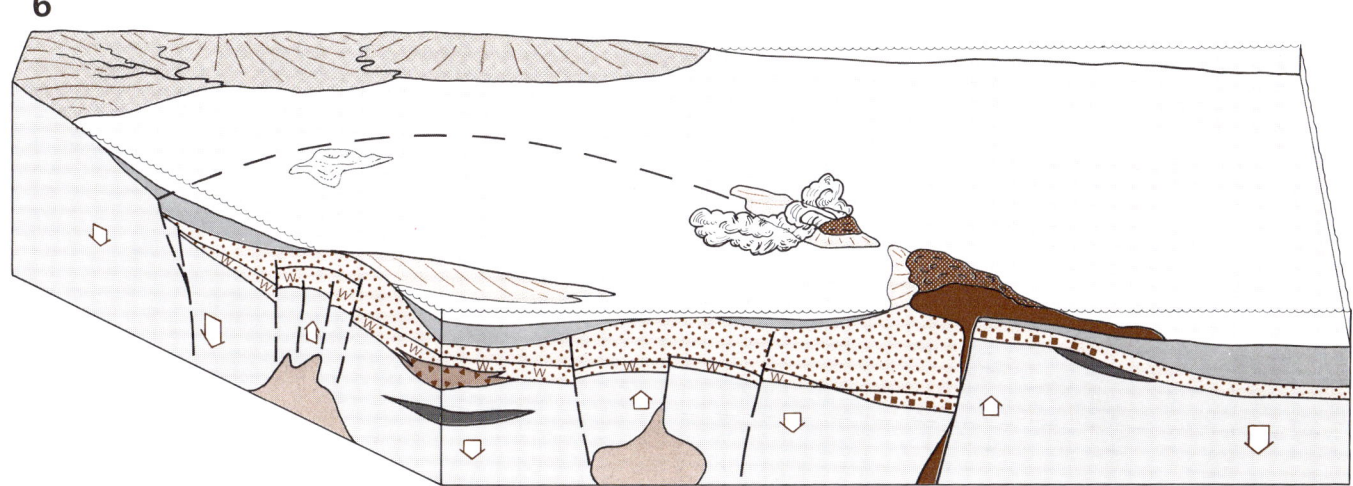

7

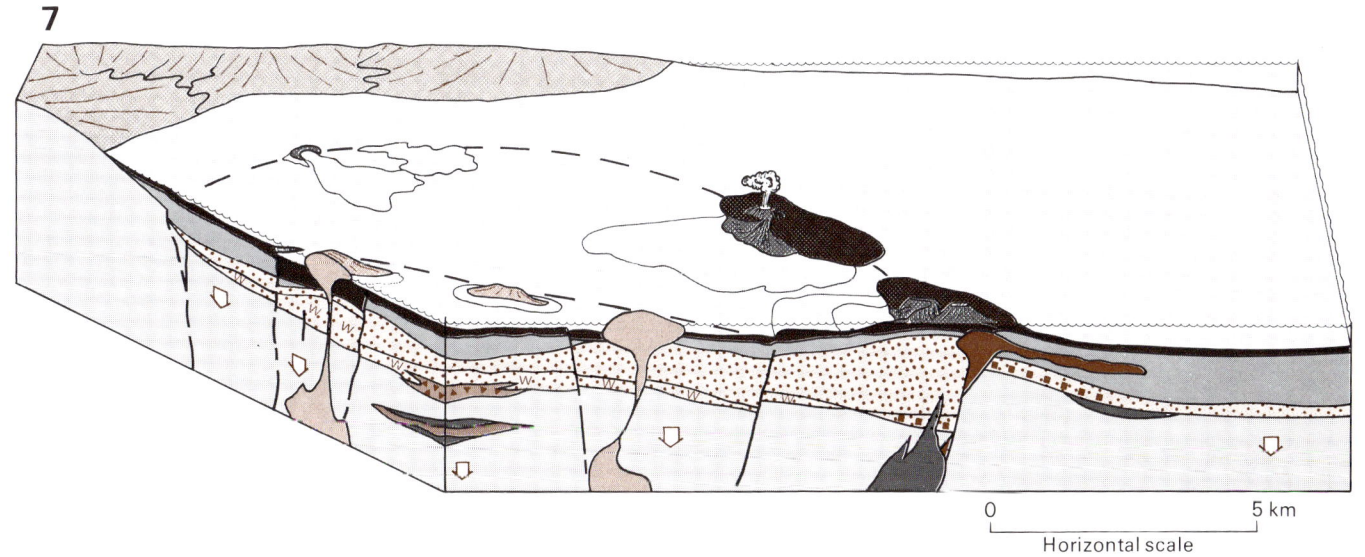

0   5 km

Horizontal scale

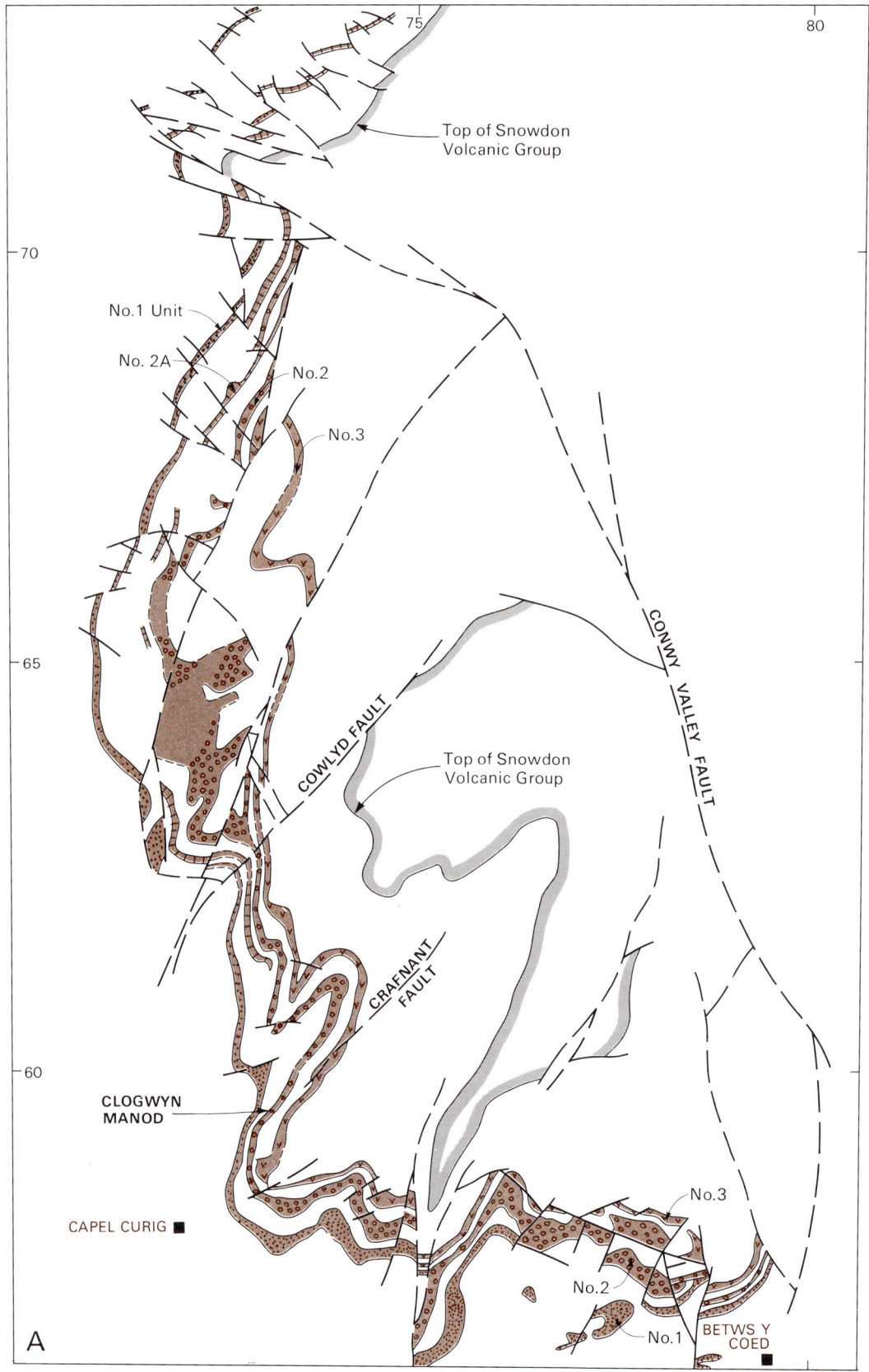

Figure 78   Lower Crafnant Volcanic Formation. Outcrop (A) and typical serial sections of outflow tuffs (B and C).

Siliceous nodules

Carbonate nodules

Clasts

Bedding, diffuse

Joints

Siliceous matrix

Micaceous matrix

Siltstones

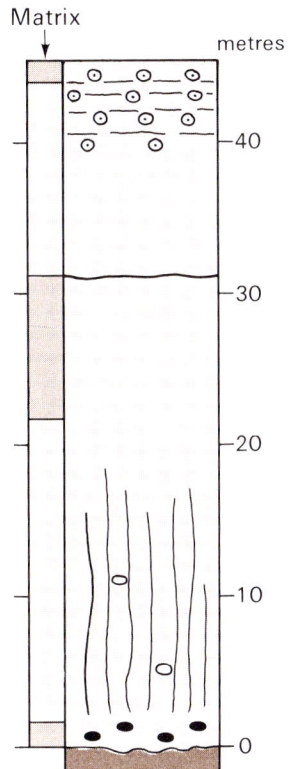

Matrix

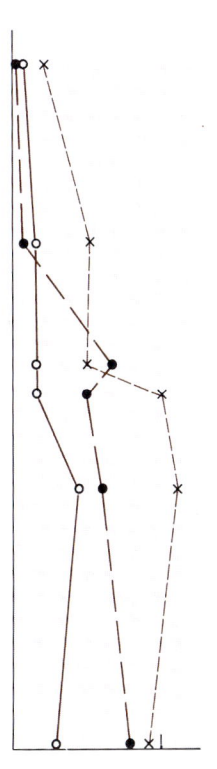

metres

40

30

20

10

0

○ Average shard size
× Maximum shard size
● Maximum crystal size

B. Typical section of No.2 tuff at Clogwyn Manod [SH 7326 5962]

Shard textures

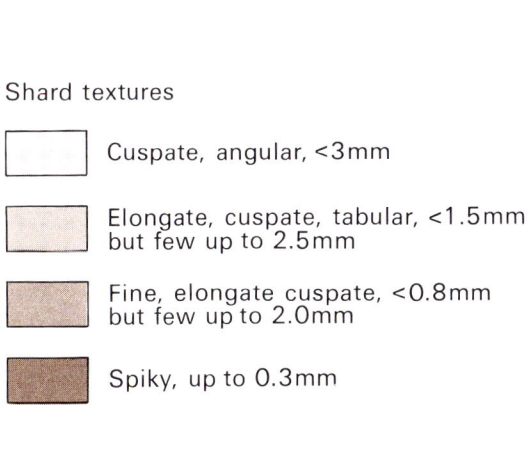

Cuspate, angular, <3mm

Elongate, cuspate, tabular, <1.5mm but few up to 2.5mm

Fine, elongate cuspate, <0.8mm but few up to 2.0mm

Spiky, up to 0.3mm

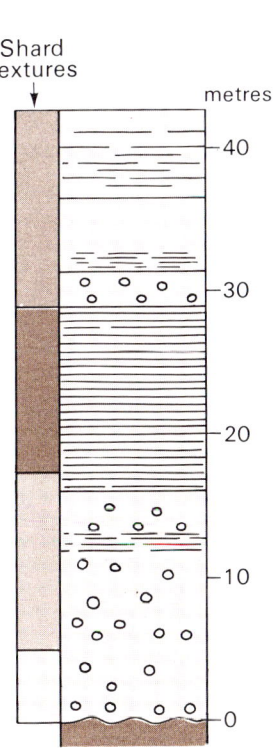

Shard textures

metres

40

30

20

10

0

C. Typical section of No.3 tuff at Clogwyn Manod [SH 7370 5976]

To the south, the tuff passes laterally into 12–15 m of cleaved, ochreous, weathered tuff overlain by about 8 m of grey, silty tuff, and further, into about 6 m of bedded tuff/tuffite with weathered-out carbonate nodules. This transition is interpreted as representing the reworking and incorporation into the background sediments of pyroclastic debris from the distal portions of the primary flow.

The No. 2 tuff, 18–75 m thick, crops out throughout east and north-east Snowdonia (Howells et al., 1973, 1978, 1981a). Typically, the nonwelded tuff is massive, distinctly siliceous and fine grained, with impersistent internal laminae. It is best developed to the east and in places it contains carbonate and siliceous nodules. The basal zone, about 2 m thick, is silty and in a few places contains brachiopod casts and oolites.

The No. 3 tuff, 20–100 m thick, is typically a massive, heterogeneous, block and ash-flow tuff although locally it contains few blocks and is well bedded (Figure 78C). Clasts are mainly of acidic tuff, but they include perlitic rhyolite, basalt, basaltic tuff and siltstone. In places, thin, crenulated, discontinuous, chlorite laminae, parallel to bedding, are well developed and particularly prominent on weathered surfaces. The local intense concentration of blocks of acid tuff composition, with little matrix, is often obscured by recrystallisation.

In their constituents and massive bedforms the three tuffs are typically the products of ash-flow eruptions. The preponderance of primary pyroclasts suggests that their emplacement was the direct result of eruption and that they are not secondary accumulations from the sloughing of previously accumulated pyroclastic debris (cf. Fiske and Matsuda, 1964). The association of the tuffs with marine sediments and the incorporation of marine fossils at their bases indicate that they were emplaced in a submarine environment. The local development of internal laminae reflects loss of energy within the flow and ingestion of water at the flow front. The fine-grained tops represent vitric dust elutriated from the water column after emplacement. The better developed bedforms, with contamination by epiclastic debris and siltstone intercalations in the thinner distal sequences, as at the southern edge of the No. 2A tuff outcrop, suggest periodic remobilisation of pyroclastic debris at the flow front subsequent to emplacement.

From the lateral variations across outcrop, the three flows are considered (Howells et al., 1973, 1978, 1981a,b) to be derived from a local source, in contrast to the underlying Lower Rhyolitic Tuff Formation outflow tuff (Howells et al., 1986).

PETROGRAPHY   The lowermost two ash-flow tuffs (Nos. 2A, 2) of the sequence are petrographically similar. They comprise devitrified and recrystallised shards, feldspar and quartz crystals and a few lithic clasts in a fine-grained matrix. Shards are locally well preserved as small rods and spikes, and good cuspate and rare bubble forms. They are generally less than 0.5 mm across and show slight fining upwards (Figure 78B). In the basal zone (<2 m) they are locally replaced by fine white mica and above, generally by a fine quartzose mosaic.

Feldspar crystals, mainly albite-oligoclase, are more common than small rounded quartz crystals. The feldspars occur as subhedral crystals, often resorbed, and as fragments. Crystals are most commonly concentrated in the basal zone and are sparsely distributed above. The matrix comprises a fine-grained aggregate of quartz, feldspar and sericite after devitrified (argillised) vitric dust. Locally, the basal zone includes much sericite and chlorite which probably represent a mud component derived from the substrate during transport.

The upper block and ash-flow tuff (No. 3) is distinctive in its shard component and the almost complete absence of crystals. The devitrified shards vary in shape from fine elongate and cuspate to equant bubble walls and in size from 0.1 mm to 4.2 mm across. Typically the shards are quartz aggregate pseudomorphs.

Accessory minerals in the three tuffs include opaque iron-oxides, sphene and apatite.

GEOCHEMISTRY   Selected analyses of ash-flow tuffs Nos. 2 and 3 of the Lower Crafnant Volcanic Formation are presented in Appendix 1, Table 12. On a plot (Figure 79) of Zr/TiO_2 vs. Nb/Y, both tuffs lie mainly within the rhyolite field. However, they are clearly discriminated in terms of their Zr/TiO_2 ratios, with that for No. 3 tuff being considerably higher (0.323–0.405, $N = 5$) than that for No. 2 (0.137–0.180, $N = 5$). No. 3 tuff therefore lies close to, and within (one analysis) the comendite/pantellerite field, which reflects its considerable enrichment in Zr (1001–1578 ppm) relative to No. 2 tuff (397–568 ppm). Similar enrichment in Nb, Y and Th is also present in No. 3 tuff (respectively 55–89 ppm, 118–269 ppm and 25–39 ppm) compared to No. 2 (respectively 30–41 ppm, 79–139 ppm and 13–24 ppm), making discrimination using various immobile element pairs, e.g. Nb vs. Zr (Figure 80) straightforward. A Nb vs. Zr plot also discriminates the No. 2 tuff, No. 3 tuff and the underlying No. 1 outflow tuff of the Lower Rhyolitic Tuff Formation. The last, although closer compositionally to No. 2 tuff than No. 3 tuff in terms of Zr, Nb, Y etc., differs from No. 2 in its general enrichment in Nb relative to Zr and Y (Appendix 1, Table 12).

Late acidic volcanism

The three outflow tuffs (Nos. 2A, 2 and 3) described above define the lower part of the outcrop of the Crafnant Centre volcanic accumulation in north-eastern Snowdonia, where they maintain their integrity as primary, nonwelded, ash-flow tuffs. The evidence of the subsequent volcanic activity indicates that the eruptive centre lay to the east of their outcrop; however, the outflows of this later activity are clearly recognised only in the outcrops to the south of the centre.

OUTFLOW TUFFS

Middle Crafnant Volcanic Formation

To the south of the centre the earlier ash-flow tuffs (Nos. 2A, 2 and 3) are overlain by a sequence, up to 90 m thick, comprising evenly flaggy-bedded, thin, primary ash-flow tuffs, remobilised tuffs and tuffites which are intercalated with black pyritic graptolitic *Diplograptus multidens* mudstones and siltstones (Middle Crafnant Volcanic Formation) (Howells et al., 1978) (Figure 81). Locally, the intercalated mudstones contain scattered feldspar fragments (Plate 43C) and, most distinctively, a few thin, normally graded, tur-

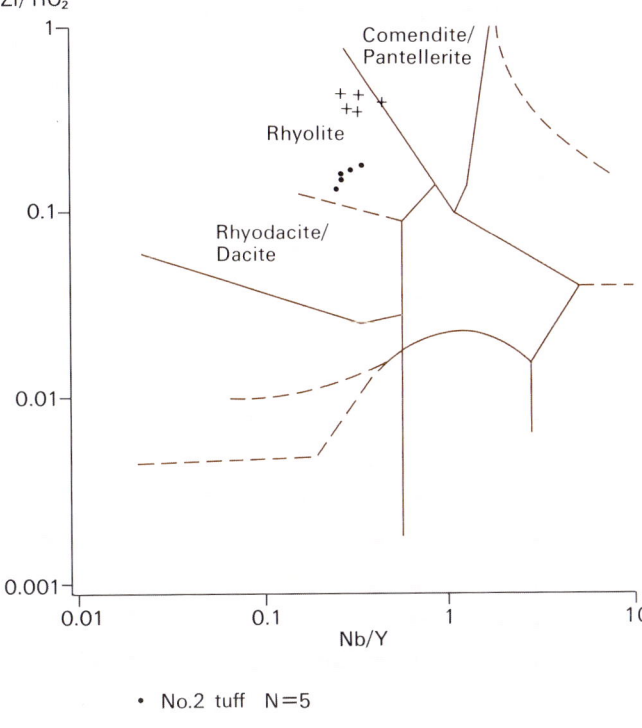

Figure 79 Lower Crafnant Volcanic Formation, Nos. 2 and 3 tuffs, Zr/TiO$_2$ vs. Nb/Y diagram.

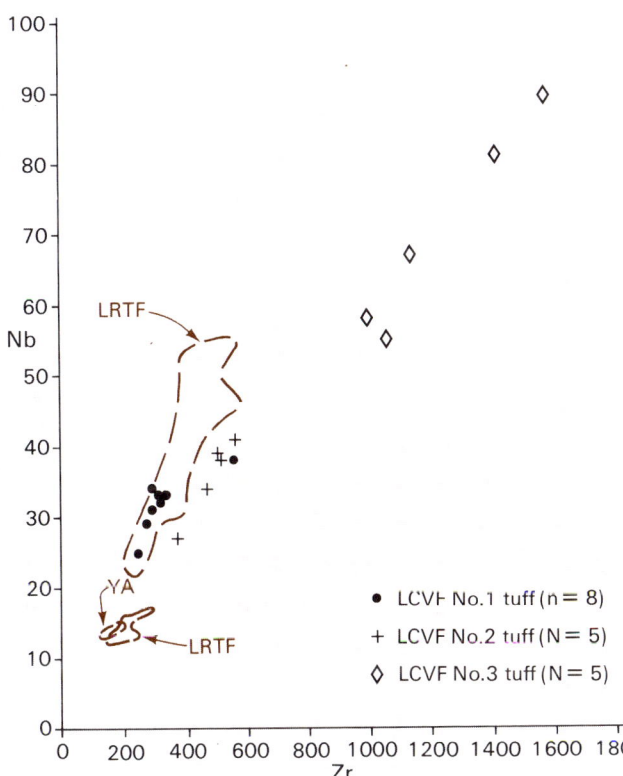

Figure 80 Lower Crafnant Volcanic Formation, Nb vs. Zr diagram with comparative fields for the Lower Rhyolitic Tuff Formation (LRTF) and Yr Arddu tuffs (YA).

biditic sandstones with pyrite traction carpets in loaded scour casts at the base.

Within these background sediments, the pyroclastic units are common and tuffs grade through tuffites into tuffaceous sediments. The tuffs range from ashes to breccias in grade and from well sorted to poorly sorted. The fine-grained, flaggy-bedded tuffs comprise recrystallised, delicate, cuspate shards in a fine micaceous matrix. Small vesicle-like cavities on exposed surfaces reflect weathered-out carbonate plates. Locally, scoured and loaded bases and internal cross-lamination (Plate 43B) are common and indicate turbiditic emplacement, probably from ash temporarily stored (Carey and Sigurdsson, 1984) in shallower parts of the environment.

The coarsest tuff breccias form massive beds, up to 1.3 m thick, with isolated blocks, up to 0.5 m, of grey, indurated mudstone and tuffite which show no preferred orientation or position in the beds. Some of the included blocks show a faint internal planar-lamination parallel to their indented peripheries which indicates the blocks were unlithified when incorporated (Plate 42A).

The tuffites are typically fine grained, banded and pale grey, grading into dark blue-grey mudstones and siltstones forming distinctively striped sequences. Locally, the admixture of pyroclastic and epiclastic components is more heterogeneous with irregular patches of shards and crystals in a mudstone/siltstone matrix (Plate 43D). In the striped sequences, convolute lamination (Plate 43A) and penecontemporaneous microfaults are well displayed. The striped tuffites occur as thick beds, up to 1.5 m, within the mudstones and possibly represent both tuff-turbidites and ash settled from the water column. Thinner beds, generally 0.5 m, of striped tuffites also directly overlie coarser, massive, ash-flow tuffs (Plate 42B) and in these instances they represent fine ash elutriated into the water column during ash-flow transport, which subsequently settled.

This well-bedded and ordered sequence is restricted to the outcrops east and north-east of Capel Curig (Figure 81). To the north, the sequence becomes progressively less well ordered because of slumping, and lateral correlation between outcrops is difficult to establish. The lithologies are similar with thick and thin flaggy and massive beds of fine-grained, vitric dust tuffs and tuffites and tuff breccias, with intercalated graptolitic mudstones and siltstones. Locally, large carbonate nodules, up to 1.5 m in diameter, occur.

In this area, the top of the sequence is marked by cleaved black graptolitic mudstone, up to 20 m thick.

Upper Crafnant Volcanic Formation

The final expression of acidic volcanism at the Crafnant Centre is a thick, up to 70 m, massive, heterogeneous tuffite (Upper Crafnant Volcanic Formation) (Figure 81) (Howells et al., 1978, 1981a,b), which overlies the upper black mudstone of the underlying sequence along the southern edge of the outcrop. The tuffite, equivalent to the 'Upper Tuff Bed' of Davies (1936), is almost entirely without internal bedding structures. The pyroclastic element consists of cuspate shards, from coarse bubble walls to small fragmented rods and spikes, together with fragmented, rounded and resorbed, albite feldspar crystals and pumice fragments. The epiclastic material is predominantly muddy in character and consists of a fine aggregate of chlorite, sericite, carbonaceous material

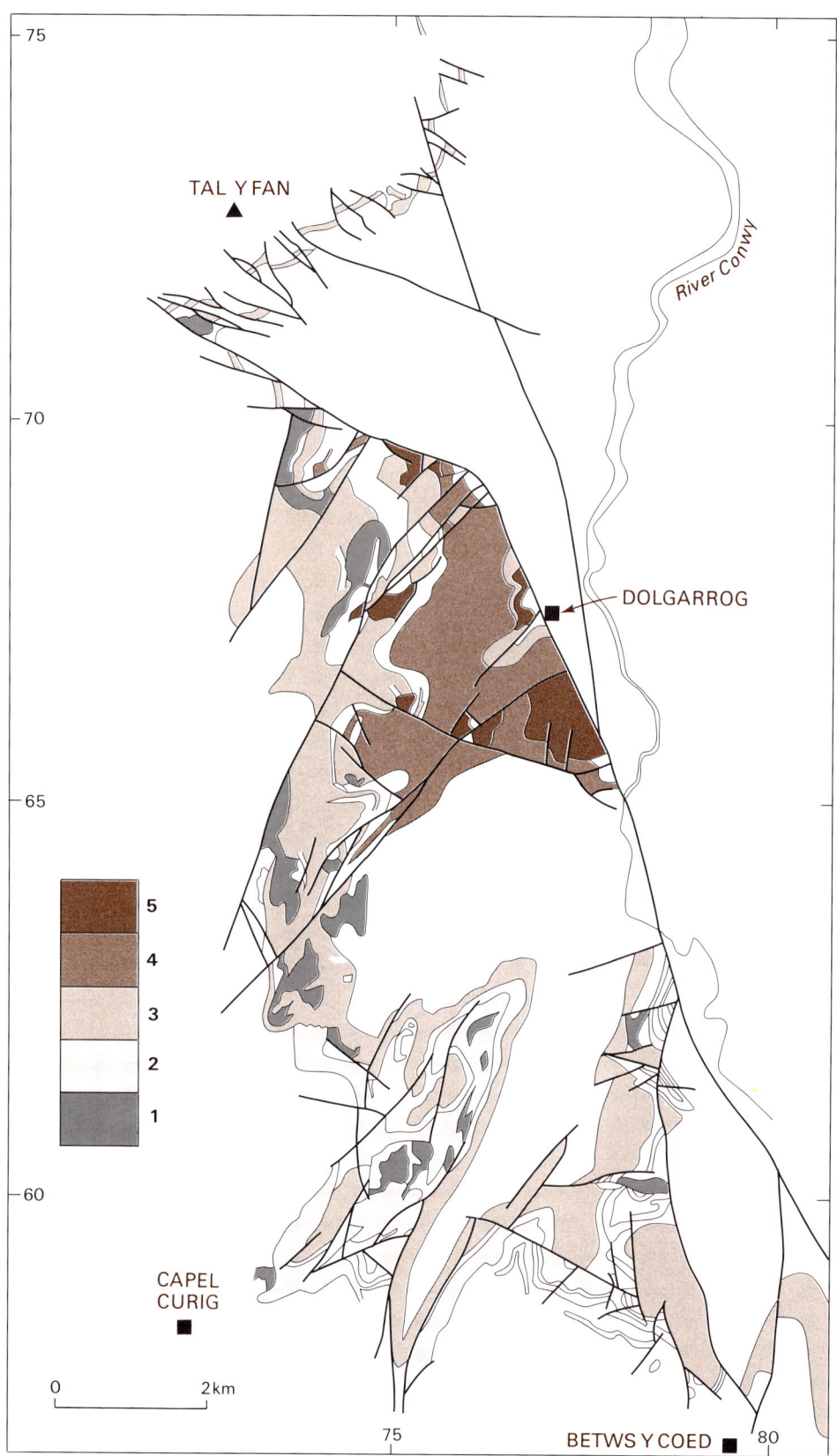

Figure 81 Main outcrop of the Middle and Upper Crafnant Volcanic Formations and the Dolgarrog Volcanic Formation.

1: Dolerite; 2: Middle Crafnant Volcanic Formation; 3: Upper Crafnant Volcanic Formation and Middle/Upper Crafnant Volcanic Formation undivided; 4: Dolgarrog Volcanic Formation.

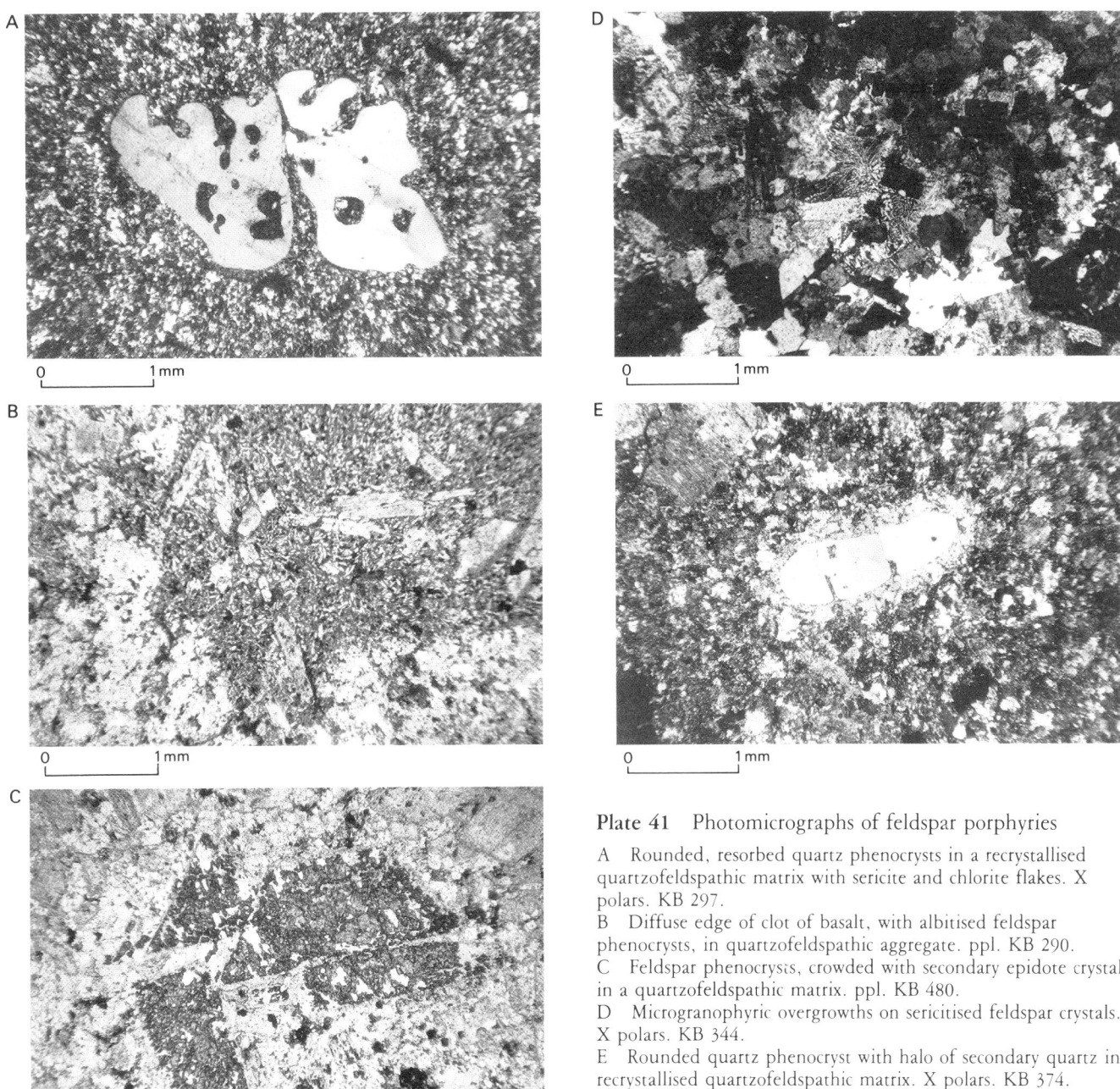

Plate 41 Photomicrographs of feldspar porphyries

A   Rounded, resorbed quartz phenocrysts in a recrystallised quartzofeldspathic matrix with sericite and chlorite flakes. X polars. KB 297.
B   Diffuse edge of clot of basalt, with albitised feldspar phenocrysts, in quartzofeldspathic aggregate. ppl. KB 290.
C   Feldspar phenocrysts, crowded with secondary epidote crystals, in a quartzofeldspathic matrix. ppl. KB 480.
D   Microgranophyric overgrowths on sericitised feldspar crystals. X polars. KB 344.
E   Rounded quartz phenocryst with halo of secondary quartz in a recrystallised quartzofeldspathic matrix. X polars. KB 374.

and opaque iron-oxide. In places, lithic clasts of siltstone, acid tuffs, dolerite and basalt are common, albeit patchy in distribution.

The proportion of the various constituents, from tuffaceous mudstones to tuffs, is highly variable, but there is seemingly no consistent pattern to the variations. Generally, the only bedding features are impersistent ribs of mudstone within the heterogeneous tuffite. Carbonate nodules, up to 0.7 m, are restricted in distribution.

The heterogeneous admixture of pyroclastic and epiclastic debris, the almost complete lack of sorting and minimal internal bedding suggest that the deposit was emplaced from high density gravity flows. Such flows could result from the remobilisation of pyroclastic material that had been previously rapidly emplaced on unlithified muds. Such a process, however, would not be expected to result in such comprehensive mixing of both elements during transport. Alternatively, the deposit may represent an explosive eruption, from a reactivated volcanic centre, through and incorporating unlithified mud which had accumulated during a period of volcanic quiescence.

GEOCHEMISTRY   Only two samples of the Upper Crafnant Volcanic Formation have been analysed (Appendix 1, Table

Plate 42   Middle Crafnant Volcanic Formation [SH 775 588]

A   Black mudstone incorporated into block and ash-flow tuff.
B   Laminated, water-settled, fall-out tuffs separating thin ash-flow tuff units.

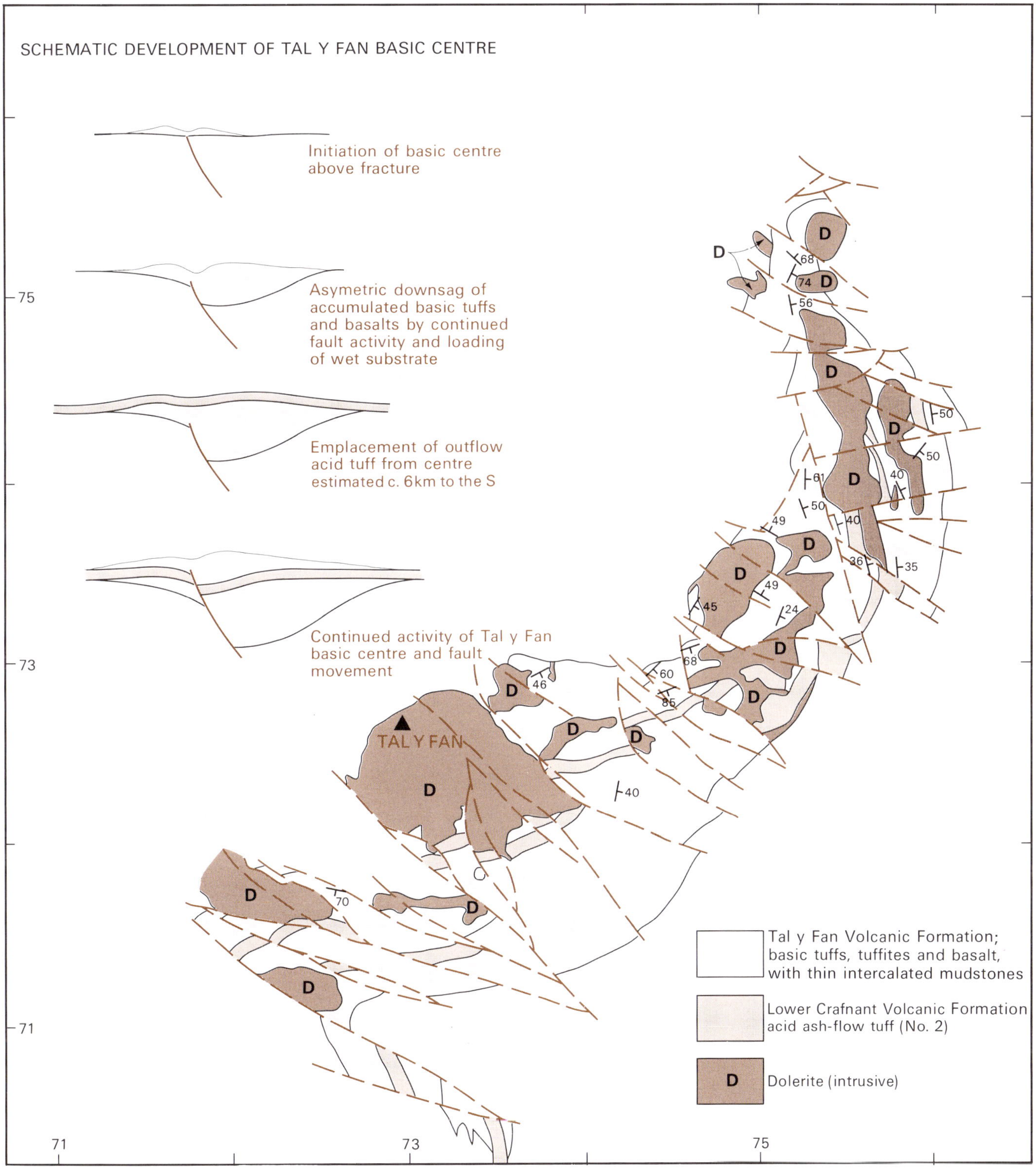

SCHEMATIC DEVELOPMENT OF TAL Y FAN BASIC CENTRE

Initiation of basic centre above fracture

Asymetric downsag of accumulated basic tuffs and basalts by continued fault activity and loading of wet substrate

Emplacement of outflow acid tuff from centre estimated c. 6km to the S

Continued activity of Tal y Fan basic centre and fault movement

TAL Y FAN

	Tal y Fan Volcanic Formation; basic tuffs, tuffites and basalt, with thin intercalated mudstones
	Lower Crafnant Volcanic Formation acid ash-flow tuff (No. 2)
D	Dolerite (intrusive)

Figure 82   Outcrop of the Tal y Fan Volcanic Formation and a model for its evolution in a fault-controlled downsag.

12). On a plot of Zr/TiO_2 vs. Nb/Y, one analysis plots on the boundary between the rhyodacite and rhyolite fields while the other plots within the rhyolite field. Both are relatively enriched in HFS elements (e.g. Zr = 394 and 340 ppm, Nb = 52 and 48 ppm) and the LREE, but these concentrations are lower than in the Nos. 1 and 2 tuffs of the Lower Crafnant Volcanic Formation. Th contents are also high (27 and 28 ppm) and are similar to those of the No. 3 tuff of the Lower Crafnant Volcanic Formation. Ba contents are extremely high (3501 and 3074 ppm) though these may not represent primary concentrations.

INTRACALDERA AND PROXIMAL TUFFS: MIDDLE/UPPER CRAFNANT VOLCANIC FORMATIONS

The relatively well-ordered sequence of the outflow facies from the activity of the Crafnant Centre is restricted to the southern end of the outcrop (Figure 81). To the north, the sequence is progressively more disturbed and the lithologies more admixed. Coincidentally, the sequence thickens to about 600 m. The constituent lithologies are identical to those of the outflow facies and their admixture occurs both as slumping of partly lithified beds, which largely retained their integrity, and grain-to-grain admixtures of pyroclastic and epiclastic components. The heterogeneity of the sequence and the restriction of outcrop prohibit recognition of a systematic pattern to the disruption. However, the distribution of the facies and the thickness variations are interpreted to represent a transition through a proximal facies into an eruptive centre.

The structure of the centre cannot be clearly distinguished and this is not purely because of the exposure. The disturbed sequence reflects unlithified accumulations of acidic pyroclastics and mudstone and in such an accumulation the volcanotectonic and contemporaneous tectonic structures were not preserved. The resultant effect is more extreme than the downsag caldera structure distinguished at the Snowdon Centre and probably reflects the deeper marine setting of the centre. The dominant north-east–south-west-trending faults in the vicinity of the thickest accumulation are later reactivations of faults that had earlier probably controlled the position of the volcanic activity.

Basic volcanism: Tal y Fan and Dolgarrog Volcanic formations

Two major expressions of basaltic volcanic activity (Tal y Fan Volcanic Formation, 500 m thick, Figure 82; Dolgarrog Volcanic Formation, 400 m thick, Figure 81), crop out in north-eastern Snowdonia. The Dolgarrog Volcanic Formation is closely associated with the Crafnant Centre and the Tal y Fan Volcanic Formation is situated to the north-west. They comprise basaltic lavas, pillow breccias, hyaloclastites and basaltic tuffs with a few intercalations of black mudstone. The locally thick accumulations wedge out along strike over a total distance of about 4–5 km. In places, the sequences were disrupted by slumping although there is no evidence of shallow-marine reworking.

The Tal y Fan Volcanic Formation developed in the interval between the emplacement of the lower two outflow ash-flow tuffs of the main Crafnant Centre. The upper (No. 2) tuff can be traced through the thick local accumulation of basaltic rocks, which indicates that the latter developed in a depression and did not form a significant topographic high on the sea floor which would have obstructed the emplacement of the ash-flow. Subsequent to the emplacement of the ash-

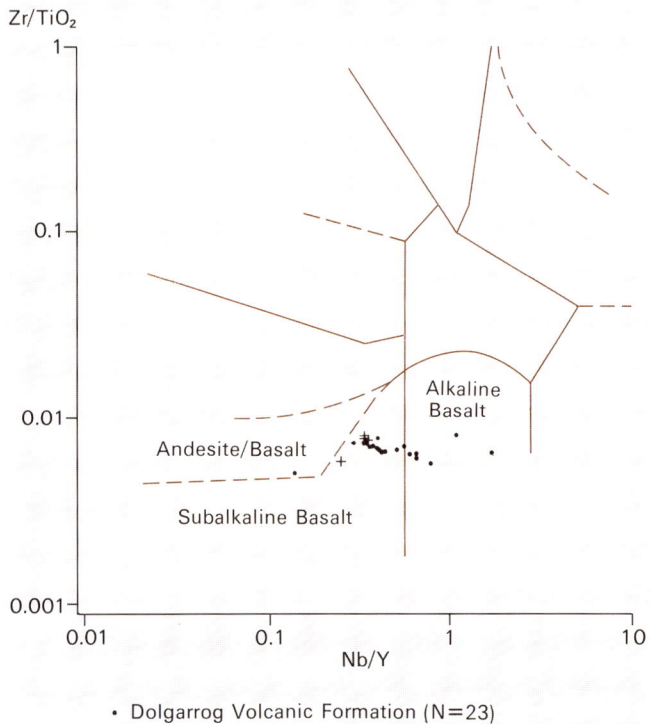

Figure 83   Dolgarrog and Tal y Fan Volcanic formations, $Zr/TiO2$ vs. Nb/Y diagram.

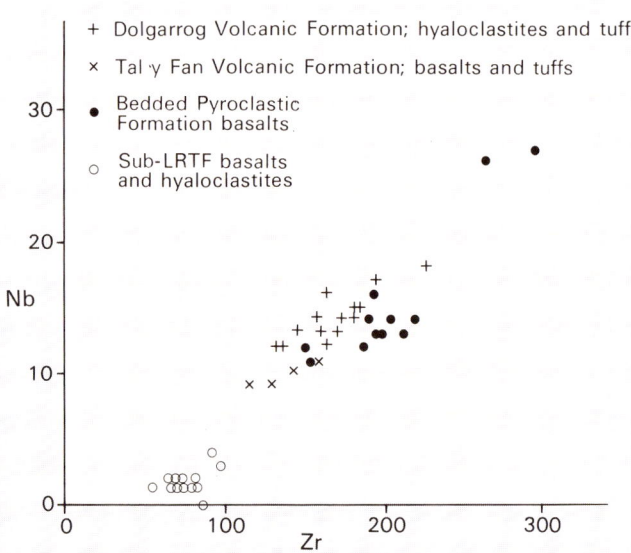

Figure 84   Dolgarrog and Tal y Fan Volcanic formations, Bedded Pyroclastic Formation and sub-LRTF basalts, Nb vs. Zr diagram.

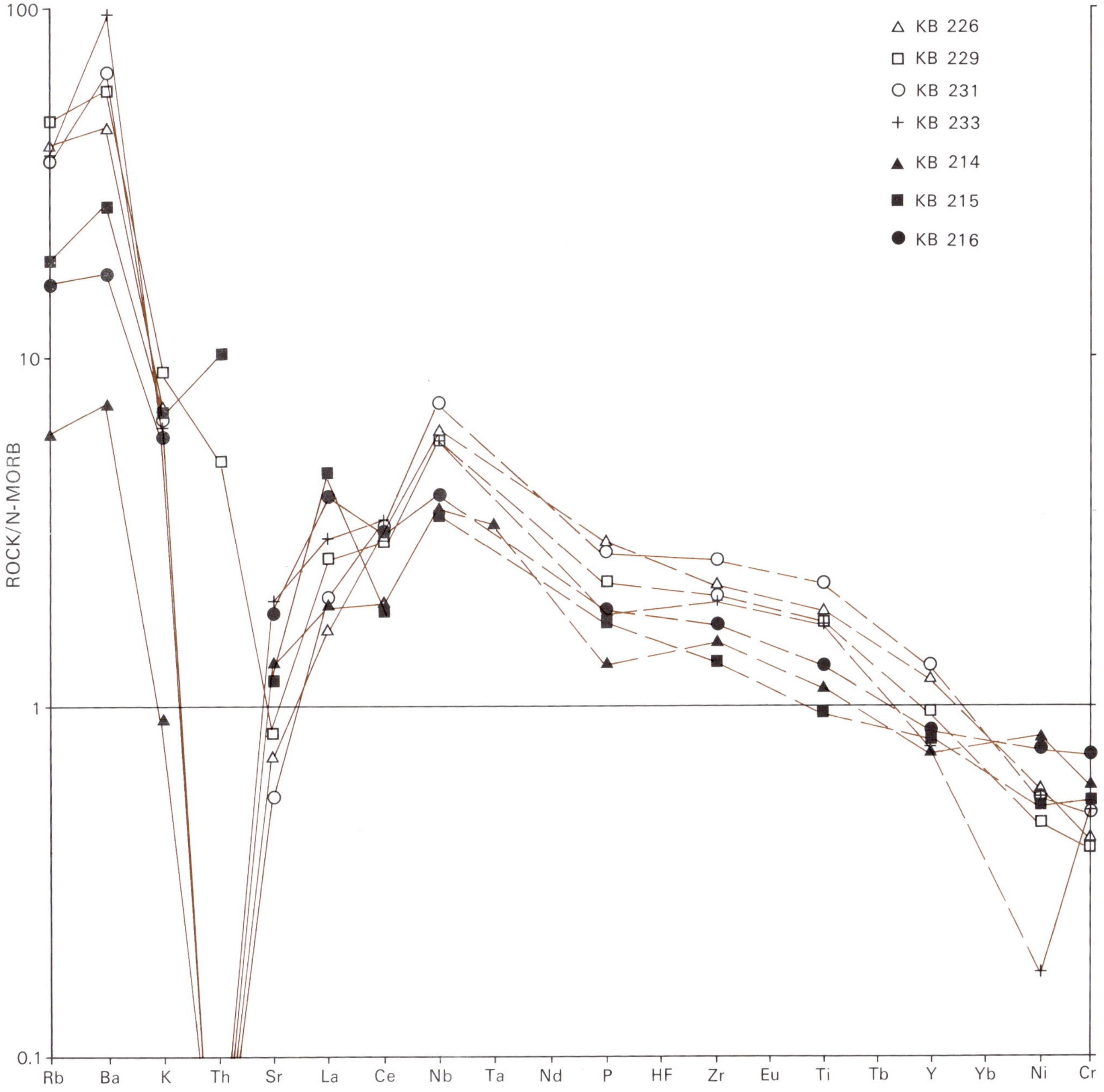

Figure 85   Dolgarrog and Tal y Fan Volcanic formations, N-MORB normalised multi-element diagram (normalising factors after Pearce, 1982a).

A

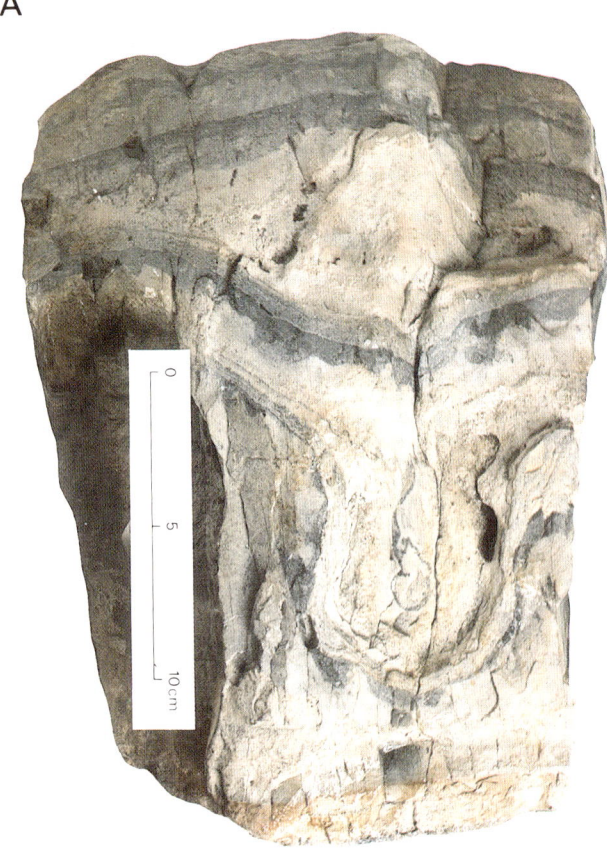

B

Plate 43   Specimens from the Middle Crafnant Volcanic Formation, Sarnau [7755 5884], near Betws y Coed. Bedding structures and lithologies.

A   Soft sediment deformation of fine-grained acid tuffs with thin interbeds of mudstone. Adjacent beds undeformed.

B   Fine-grained acid tuffs with interbeds of dark grey mudstone. Lower acid tuff bed shows cross-lamination. Bed of acid tuff near top of specimen shows extreme deformation at its base with flames of mudstone.

flow, activity at the basaltic centre continued and further subsidence preserved the ash-flow tuff in the sequence.

The Dolgarrog Volcanic Formation developed in the vicinity of the Crafnant Centre soon after cessation of its acidic activity. Its deposits are overlain by black mudstones, which elsewhere overlie the acidic deposits of the Crafnant Centre. As in the northerly accumulation, it suggests that the deposits were entrapped in a depression.

Massive and pillowed basalt flows are most common in the lower parts of the sequences. Typically, the basalts are highly vesicular, with irregularly shaped vesicles filled with quartz and chlorite, commonly concentrically arranged. The groundmass contains flow-aligned and stellate-arranged albitised feldspar laths with much chlorite, ilmenite, sphene and carbonate. Tabular albitised feldspar phenocrysts, up to 6 mm, are locally common.

The hyaloclastites consist of fragments of chloritised, vesiculated, basaltic glass and feldspar crystals in a groundmass of comminuted, chloritised, basaltic glass with much ilmenite. The feldspars are replaced by sericite, carbonate and quartz, and the vesicles are filled with quartz and carbonate.

Locally, devitrification and recrystallisation obscured the fragmentary character of the matrix though elsewhere it accentuated the clastic character. Common rounded basaltic clasts, up to 8 mm in diameter, bear feldspar microlites set in a devitrified glassy groundmass crowded with ilmenite grains. In places the hyaloclastite contains whole and fragmented pillows (Plate 44) and grades into pillow breccia.

Intercalations of thinly bedded, up to 20 cm, basaltic tuffs are composed of degraded, altered, basaltic fragments. Sparse evidence of normal grading and internal cross-laminations suggests that the tuffs were deposited by small debris flows remobilised from unstable accumulations of hyaloclastite. Impersistent, thin intercalations of black mudstone occur throughout the sequences.

The components of the accumulations and the associated black graptolitic mudstones indicate a submarine environment. The absence of shallow-marine reworking reflects an outer shelf setting below the level of storm wave-base. However, the water depth was not sufficient to inhibit vesiculation or explosivity.

The dominance of massive, hyaloclastite lithologies and

C

5cm

0

D

0   5cm

Plate 43 *continued*

C   Dark grey mudstone with scattered isolated feldspar crystals and clots of acid tuff.
D   Specimen from near the top of acid tuff bed showing gradation into overlying mudstone and included mudstone clast.

the restricted occurrence of included pillows, indicate that the major process of fragmentation was the expansion of entrapped volatiles in the erupted fragments and not the decrepitation of basaltic lava.

The close spatial relationship of an extensive transgressive dolerite sill with the extrusive basaltic rocks in the Tal y Fan Volcanic Formation suggests a genetic association (Campbell et al., 1988). Locally, the contact of the sill is difficult to position accurately. It is considered that the intrusion was emplaced into unconsolidated, water-saturated, basic pyroclastics and, in places, may have reached the surface. The base of the Dolgarrog Volcanic Formation is intruded by dolerite dykes, which probably lie along fissures related to the local basaltic eruptive centre.

The containment of the deposits in depressions would have been facilitated both by their accumulation on unlithified substrate and by extension along contemporary fractures, which probably controlled magma movement to the surface.

The process would have been incremental during accumulation and was gravity controlled.

GEOCHEMISTRY   Selected analyses of the Tal y Fan and Dolgarrog Volcanic formations are presented in Appendix 1, Table 12. On a plot (Figure 83) of Zr/TiO_2 vs. Nb/Y, analyses of the Tal y Fan Volcanic Formation plot within the field of subalkaline basalt. However, the Dolgarrog Volcanic Formation appears more variable, possibly because of the greater number of analyses, with most plots within the field of subalkaline basalt, although some overlap into the fields of alkaline basalt and andesite/basalt. In terms of Nb and Zr (Figure 84), the Tal y Fan Volcanic Formation is relatively depleted in Nb and Zr, although their respective Nb/Zr ratios are similar (about 0.125). Both formations are similar to basalts of the Bedded Pyroclastic Formation, but are slightly depleted in Zr relative to Nb. On a N-MORB normalised multi-element diagram (Figure 85), the Dolgarrog Volcanic

Plate 44   Basaltic breccia formed of fragmented pillows in hyaloclastite matrix. The Dolgarrog
Volcanic Formation [SH 759 673]. (BGS Photograph L1459).

Formation is significantly enriched in Ce and the HFS
elements relative to the Tal y Fan Volcanic Formation, which
is, in turn, slightly enriched in Sr, Ni and Cr. Both are
characterised by relative positive Nb and negative Sr
anomalies. In general terms, the N-MORB normalised HFS
element patterns of both formations are similar to those of
within-plate basalts typified for example by Hawaiian basalt
(cf. Pearce, 1982b), and they have the general characteristics
of E-type (enriched) MORB.

In comparison with the extrusive basaltic magmas of the
Snowdon Centre, both formations are considerably enriched
in LREE and HFS elements relative to the sub-Lower Rhyolitic
Tuff Formation basalts and neither shows negative N-MORB
normalised Nb anomalies. Therefore, both are more closely
comparable with the Bedded Pyroclastic Formation basalts,
although the Tal y Fan Volcanic Formation is relatively
depleted in Ce and the HFS elements, and slightly enriched

in Ni, while the Dolgarrog Volcanic Formation is depleted in
La and slightly enriched in Nb.

SUBVOLCANIC INTRUSIONS RELATED TO THE SNOWDON AND CRAFNANT CENTRES

These intrusions can be broadly subdivided into acidic and
basic. The acidic intrusions include rhyolites, feldspar porphyries and microgranites and the basic intrusions are
dolerites and basalts. Only a single intermediate, andesitic intrusion (Llyn Teyrn) is exposed.

Amongst the acidic intrusions, the rhyolites have been
discussed above because the geochemically determined
groups could be distinguished in both intrusions and extrusions, and both could be related in detail to the evolution of
the volcanic centres. However, in such a complex pattern of

intrusions, there are a few anomalies which cannot be so satisfactorily categorised. These include the multiple intrusions of the south-east sector of the Snowdon Centre and three major acidic plutons.

Multiple intrusions in the south-east sector (Snowdon Centre)

These intrusions were recognised by Beavon (1963) and referred to as composite sills. However, the contrasting acidic and basic components represent two separate phases of intrusion, so here they are termed multiple intrusions. The intrusions are predominantly sill-like, between 2 m and about 100 m thick, comprising thin, precursor, basic margins and a younger core of either feldspar porphyry or, in places, rhyolite. The latter are predominantly of rhyolite and rhyodacitic composition and most straddle the compositional fields of the group B1 and B2 rhyolites (Figure 66). In some instances the basic component only occurs on one edge of the intrusion. Only locally is the acidic component recognisably chilled against the basic selvage. In the south-east sector of the Snowdon Centre, several feldspar porphyry and microgranite sills occur with no basic component.

The feldspar porphyries comprise subhedral albite-oligoclase and alkali feldspar phenocrysts as individual crystals, up to 3.8 mm, and as clusters, up to 8 mm, in variable concentrations up to 50 per cent of the total rock. Quartz phenocrysts, up to 1.8 mm, are rounded, resorbed and sparsely distributed; they are commonly overgrown by secondary quartz haloes (Plates 41A and E). The groundmass is variably recrystallised to a fine-grained, quartzofeldspathic aggregate with sericite and chlorite flakes and segregations. In the groundmass, spherulitic, granophyric and platy feldspathic textures are locally well developed (Plate 41D). In places, secondary epidote grains are particularly distinctive and are commonly concentrated as inclusions in feldspar phenocrysts (Plate 41C). Accessories include apatite, carbonate, iron oxide, biotite, sphene and zircon. Clots of basalt with diffuse margins (Plate 41B) and clasts of dolerite are common, particularly in the vicinity of the basic margins. The former were incorporated in a magmatic state and represent local magma mixing.

The typical feldspar porphyries grade into microgranite which is more common in the thicker intrusions. The microgranite comprises an interlocking mosaic of subhedral, equant and lath-shaped crystals of plagioclase and alkali feldspar in a granophyric matrix.

The basic components of the multiple intrusions vary from fine-grained basalt to coarse-grained dolerite. The basalts are sparsely vesicular and invariably highly altered, although locally a flow-oriented felt of feldspar microlites and microphenocrysts of albitised plagioclase feldspar and augite can be distinguished. The dolerites are petrographically similar to those occurring throughout the district (see below).

Major acidic plutons

Within the district, three major acidic plutons (Figure 36) occur which physically cannot be easily related to either the 1st or 2nd Eruptive cycles. These are the Tan y Grisiau granite, and the Mynydd Mawr and Ogwen microgranites.

TAN Y GRISIAU GRANITE

The Tan y Grisiau granite crops out in an area of less than 4 km² in southern Snowdonia, on the south side of the Lower Rhyolitic Tuff Formation caldera (Figure 36). The extent of its hornfels zone (Bromley, 1965, 1969) and associated gravity (Institute of Geological Sciences, 1978b) and magnetic anomalies (Geological Survey of Great Britain, 1965) indicate a much larger body at depth. The gravity anomaly has been interpreted (Cornwell et al., 1980; Campbell et al., 1985) to reflect a steep-sided, subvertical body, with a north-north-west-dipping roof, which extends some 10 km to the north-east and some 5 km to the south-west of its outcrop.

At outcrop the granite intrudes Tremadoc strata but causes hornfelsing (thermal spotting) of strata as young as early Caradoc in age. It comprises a uniformly medium-grained, equigranular aggregate of albite-oligoclase and microperthite alkali feldspar (60 per cent) and quartz (35 per cent), with intersertal chloritised biotite (5 per cent) (Plate 39C). Granophyric intergrowth of quartz and alkali feldspar is common and sericitic alteration of the feldspars is locally intense. Accessories include iron oxide, apatite, zircon, anatase and carbonate. Bromley (1969) distinguished both a micrographic and a graphic, pegmatite facies, marginal to the main intrusion. In addition, thin granophyric sills, with brecciated and tourmalinised outer zones, cross-cut the adjacent strata.

K/Ar isotopic age determinations by Thomas et al. (1966) established a date of 408 Ma for the thermal metamorphism and consequently for the intrusions. More recently Evans (1989a,b) has determined an Rb/Sr isotopic age of 384 ± 10 Ma but argued that this represents a reset age.

MYNYDD MAWR MICROGRANITE

The Mynydd Mawr microgranite forms an upstanding, steep-sided, boss-like intrusion on the west side of the Lower Rhyolitic Tuff Formation caldera (Figure 36). At outcrop, it covers an area of about 3 km² and intrudes Upper Cambrian to Lower Ordovician (sub-Caradoc) sediments. Its aureole is generally well defined with well-indurated, hornfelsed sediments in immediate contact with the microgranite, and an outer zone of spotting up to 0.5 km wide. Immediately adjacent sediments are anomalously steeply dipping, locally overturned, and broadly concordant with the contact, but bedding becomes consistent with the regional dip away from the contact.

The microgranite is typically fine grained, comprising a granular quartz-feldspar aggregate with tabular phenocrysts, up to 2 mm, of alkali feldspar containing distinctive perthitic intergrowths. Quartz phenocrysts are sparsely distributed and typically occur as rounded crystals, less than 0.6 mm, with corroded peripheries and myrmekitic overgrowths. Riebeckite occurs as fine, subhedral, prismatic crystals and acicular aggregates within a fine network in the matrix. Accessories include iron oxide, muscovite plates, chlorite and aenite (Nockolds, 1938).

Within the deeper gullies of the intrusion's outcrop, Cattermole and Jones (1970) distinguished a coarser, more porphyritic facies which had 'irregular' contacts with the outer, finer-grained facies. However, the geochemical variations distinguished in the current sampling cannot be related to variations in the mineralogical phases determined with a petrological microscope.

The intrusion has been dated (Rb/Sr) at 438 ± 4 Ma (Evans, 1989a,b) one of the few intrusions whose Rb/Sr systematics have not been reset during early Devonian metamorphism (Chapter 6).

OGWEN MICROGRANITE

The Ogwen microgranite intrudes strata immediately underlying the deposits of the 2nd Eruptive Cycle on the north side of the Lower Rhyolitic Tuff Formation caldera (Figure 36). Its outcrop covers an area of about 0.75 km² and narrows to the north and south from its widest development near the centre of its outcrop. The form of the body at depth is unknown, although it is probably quite extensive as locally it profoundly affects the orientation of folds in the adjacent strata. The thermal alteration of adjacent sediments, however, is extremely restricted.

The microgranite is lithologically uniform. It comprises an aggregate of equant to tabular, subhedral to euhedral crystals of albite-oligoclase and alkali feldspar. Intersertal, granular quartzofeldspar encloses flakes and thin segregations of sericite and chloritised biotite. Feldspar crystals are variably sericitised but rarely intensely so. Albite phenocrysts, in individual rounded subhedral crystals, less than 1.8 mm, and larger clusters, less than 3 mm, are sparsely distributed. Accessories include iron oxide, carbonate, sphene and epidote.

The intrusion has been dated (Rb/Sr) at 410 ± 16 Ma (Evans, 1989a,b), which is considered to reflect its later age of Devonian metamorphic resetting.

GEOCHEMISTRY OF THE SUBVOLCANIC ACID INTRUSIONS (RELATED TO THE SNOWDON CENTRE)

Selected analyses for the principal acidic intrusions asociated with the Snowdon Volcanic Group are presented in Appendix 1, Table 9. Their relatively wide range of compositions is indicated by a plot of Zr/TiO_2 vs. Nb/Y (Figure 86). The Ogwen microgranite, together with two microgranite sills from the south-east area of the Lower Rhyolitic Tuff Formation caldera, plot in the field of rhyolite, close to the boundary with the rhyodacite field. The Tan y Grisiau granite is compositionally similar, but some data points just overlap into the rhyodacite field. The Mynydd Mawr microgranite, however, is predominantly of peralkaline composition, plotting in the comendite/pantellerite field, although one analysis actually plots in the phonolite field. Two other microgranite sills from the south-east and south-west of the Lower Rhyolitic Tuff Formation caldera straddle the boundary of the comendite/pantellerite and rhyolite fields. The feldspar porphyries from the south-east of the Lower Rhyolitic Tuff Formation caldera plot mainly across the boundary between the rhyolite and rhyodacite/dacite fields. There are, however, some feldspar porphyries which have significantly lower Zr/TiO_2 ratios, principally reflecting a higher TiO_2 content, and these overlap the rhyodacite/dacite and andesite fields.

On a plot of Nb vs. Zr, analysed samples (Appendix 1, Table 9) from the Mynydd Mawr microgranite can be resolved into two compositional groups, one with a Nb/Zr ratio of about 0.1 (Nb = 67–84 ppm, $N = 4$) and the other considerably enriched in Nb (122–128 ppm, $N = 4$) with higher Nb/Zr ratios (about 0.175). The latter group is also enriched

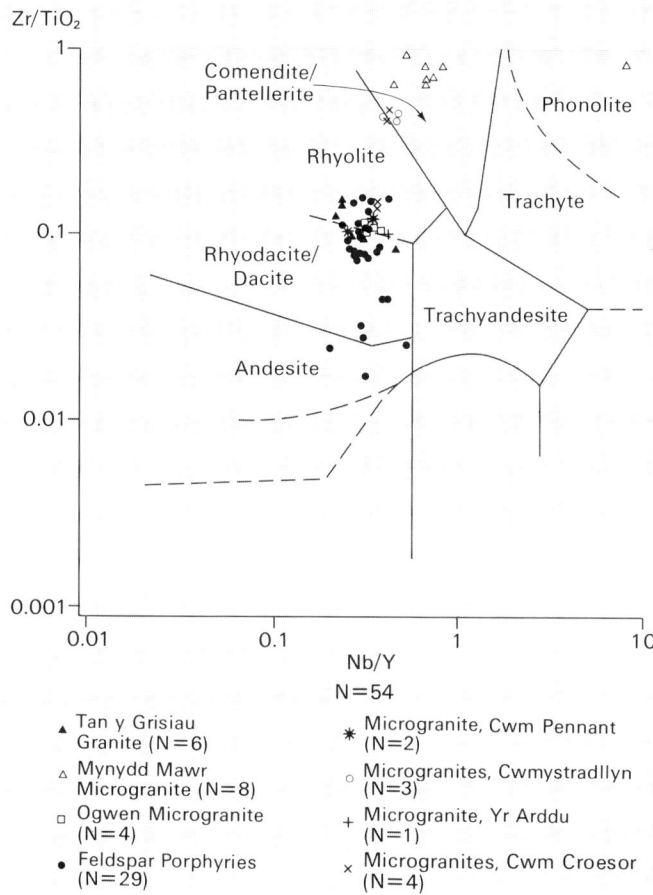

Figure 86 Intrusions related to the Snowdon Volcanic Group, Zr/TiO_2 vs. Nb/Y diagram.

in Y (up to 234 ppm) in all but one sample (16 ppm), and in Th, but is slightly depleted in TiO_2. The significance of the two groups is not yet apparent and requires further investigation; the only petrographic difference observed is that riebeckite is characteristic of the former but is infrequently seen in the latter. The Nb/Zr ratio of the low Nb group is similar to that of the ash-flow tuffs of the Pitts Head Tuff Formation and the Lower Rhyolitic Tuff Formation, and to the A1 and A2 rhyolites of the Snowdon Centre. It is most closely comparable to the rhyolite intruded into the intracaldera facies of the Pitts Head Tuff Formation (Appendix 1, Table 6). The high Nb group is compositionally unique in north and central Snowdonia.

The Tan y Grisiau granite has a Nb/Zr ratio (about 0.07) intermediate between the A1/A2 rhyolites and the B1/B2 rhyolites of the Snowdon Centre. It is markedly depleted in Th (12–20 ppm) relative to any of the other acidic intrusions of the Snowdon Centre but, with the exception of its lower Nb content, is otherwise most closely comparable to the B2 rhyolites and the feldspar porphyries (see below).

The Ogwen microgranite is even more closely comparable to the B2 rhyolites and has, unlike the Tan y Grisiau granite, similar concentrations of Nb (20–25 ppm, $N = 5$) and Th (27–30 ppm), but is slightly enriched in TiO_2 (0.33–0.38 wt%, $N = 4$). In all these respects, two further microgranite sills, one from the south-east of the Lower Rhyolitic

Tuff Formation caldera, and one to its west, in the vicinity of the Cwm Pennant Fracture zone, are also similar to B2 rhyolite compositions.

Close comparisons between the majority of the feldspar porphyries and rhyolites of B2 composition have been highlighted above. However, a small number of seemingly similar feldspar porphyry intrusions have significantly different chemistry. All are associated with multiple intrusions that have basic marginal facies. They differ in having less than 70 wt% $SiO2$ (56.64–65.88 wt%, N = 7), high TiO_2 (0.76–1.57 wt%), P_2O_5 (0.19–0.28 wt%), Fe_2O_3, MgO, Na_2O, V, and Sr, and low K_2O, Rb and Ba. On a plot of P_2O_5 vs. Nb, for example, these feldspar porphyries are compositionally intermediate between the main group of feldspar porphyries and the associated dolerites (and basalts) which constitute the marginal facies of the multiple intrusions. The petrographic evidence also suggests that the anomalous chemistry of these intrusions reflects the presence of small basaltic clasts.

Two further microgranite sills, one at Cwmystradlyn, to the south-west, and the other at Cwm Croesor, to the south of the Lower Rhyolitic Tuff Formation caldera, are very similar to the Mynydd Mawr microgranite (Figure 86) and hence to the B3 rhyolites. They have low Nb/Zr ratios (about 0.057) and very high incompatible element contents (Zr = 1210 ppm max., Nb = 66 ppm max., Y = 166 ppm max.).

Dolerite and basalt intrusions

The dolerite and basalt intrusions of north-west Wales are dominantly restricted to the outcrop of Ordovician strata. They do not occur in strata younger than Caradoc in age and have long been thought to be directly related to the Ordovician volcanic activity (e.g. Rast, 1969 and references therein). However, Shackleton (1954) argued that their main emplacement coincided with end-Silurian, Caledonian deformation. The discussion whether the ubiquitous dolerites were affected by faults, influenced by folds and to what extent they are cleaved has continued (e.g. Davies and Cave, 1976; Howells et al., 1977).

DISTRIBUTION, FORM AND CONTROLS ON EMPLACEMENT

The dolerite intrusions of central and northern Snowdonia (Figure 87) are confined to the Caradoc outcrop, in a belt about 35 km long and 15 km wide (about 20 km wide when restored palinspastically), in the north-east–south-west-trending Snowdonia graben (Campbell et al., 1988). The intrusions are mainly transgressive sills (Plate 46), up to 600 m thick, and can be traced laterally for distances up to 6 km. Aspect ratios (maximum thickness: lateral extent) are in the range of about 1:15 to 1:1 and many form concave-up, saucer-like bodies. All the intrusions are affected to some extent by regional Caledonian cleavage and, consequently, the broadly arcuate patterns within the belt inevitably reflect the regional structure. However, their restricted distribution suggests specific primary controls on their emplacement, namely the deep-seated fractures which controlled the development of the 'rift' and the shallower volcanotectonic structures within it (Campbell et al., 1988; Kokelaar, 1988). The progressive increase in basaltic magma emplacement during the volcanic cycles reflects an increase in the extensional stress

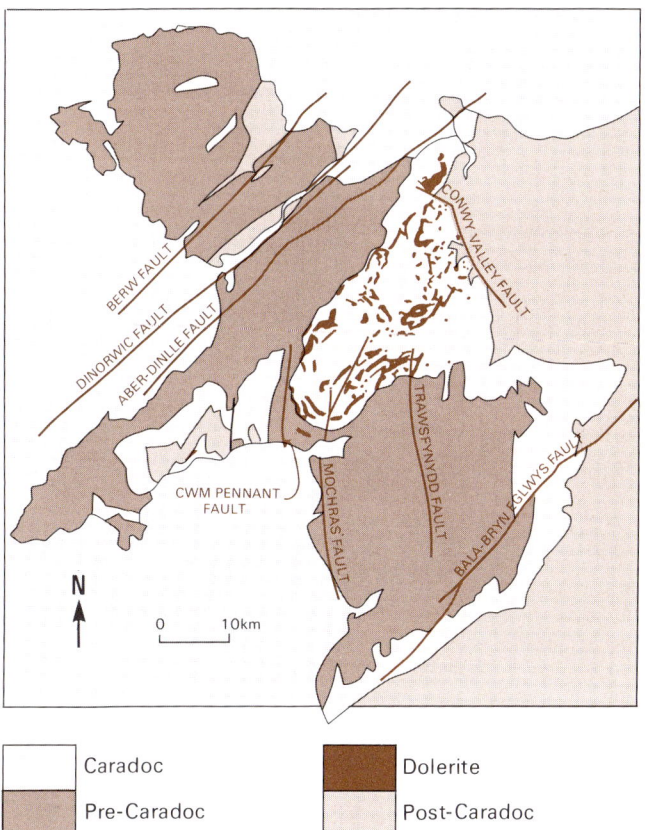

Figure 87   The distribution of dolerites in central and north-east Snowdonia and their relationship to the Caradoc outcrop and to major faults (modified after Campbell et al., 1988).

with time across the rift. However, by late Caradoc times the extension had failed to disrupt the lithospheric plate and create an ocean basin.

MESOSCOPIC AND MICROSCOPIC CHARACTERS

At outcrop, the dolerites are typically grey-green with a brown weathered surface. Columnar joints are commonly well developed. Variations in grain size bear little relationship to the thickness of the intrusion. Away from thin chilled margins they coarsen to a uniform, coarse to medium grade, although in places they are distinctively glomeroporphyritic. Well-defined internal layering and gabbroic textures are rarely present. Zones of alteration adjacent to the intrusions are generally narrow, less than 2 m, although rarely, substantially thicker zones both above and below the intrusions have been recorded (Smith, 1987). In the vicinity of basaltic eruptive centres (e.g. Moel Hebog, Snowdon and Tal y Fan) intrusions of basalt grade down into dolerite at deeper levels.

TAL Y FAN SILL

The subconcordant sill, up to 110 m thick, is typical of the intrusions within the graben. It was emplaced at the base of the Tal y Fan Volcanic Formation (Figure 82). It is considered to be coeval with the volcanic sequence. This intrusion was selected for detailed study (Merriman et al., 1986; Bevins and

TEXTURES   TEXTURAL ZONES

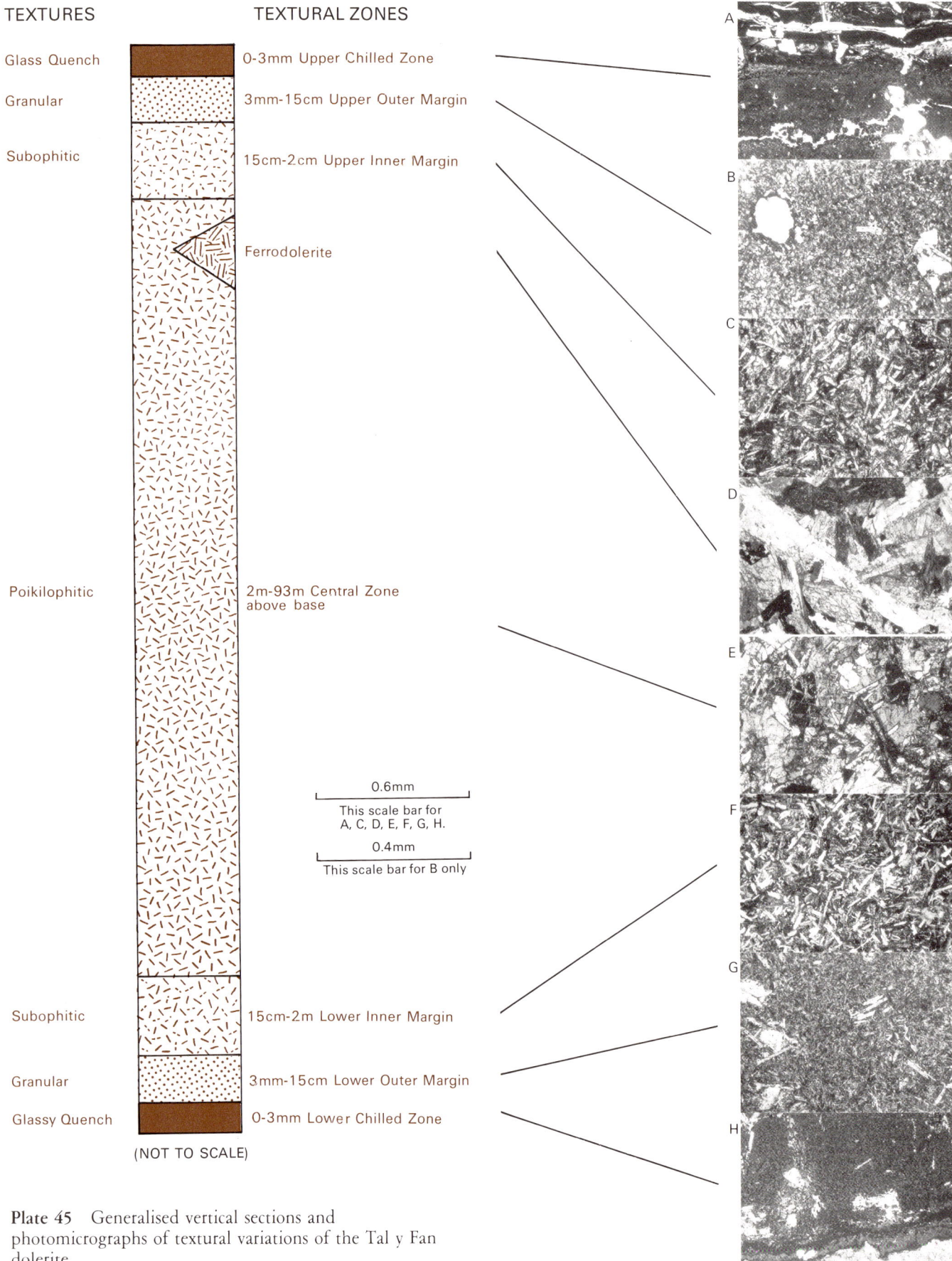

Glass Quench   0-3mm Upper Chilled Zone

Granular   3mm-15cm Upper Outer Margin

Subophitic   15cm-2cm Upper Inner Margin

                      Ferrodolerite

Poikilophitic   2m-93m Central Zone
                      above base

0.6mm
This scale bar for
A, C, D, E, F, G, H.
0.4mm
This scale bar for B only

Subophitic   15cm-2m Lower Inner Margin

Granular   3mm-15cm Lower Outer Margin

Glassy Quench   0-3mm Lower Chilled Zone

(NOT TO SCALE)

Plate 45   Generalised vertical sections and
photomicrographs of textural variations of the Tal y Fan
dolerite.

Plate 46 Prominent feature formed by a dolerite sill intruding the Lower Crafnant Volcanic Formation, Cwm Eigiau [SH 720 635]. (BGS Photograph L1497).

Merriman, 1988) to determine both the primary petrographic variability and element mobility during post-intrusive alteration (Chapter 6).

PETROGRAPHY   The primary mineralogy of the Tal y Fan sill comprises clinopyroxene + Ca plagioclase + ilmenite ± olivine. These primary phases are variably altered to secondary minerals including Na and K feldspars, amphiboles (predominantly actinolite), prehnite, pumpellyite, chlorite, epidote, sphene, white mica and stilpnomelane (Bevins and Merriman, 1988).

On textural characteristics, eight zones have been identified (Plate 45) (Merriman et al., 1986). The central zone comprises coarse dolerite with a poikilophitic texture of plagioclase and olivine (pseudomorphs) enclosed in large plates of clinopyroxene up to 2 cm in diameter. Ilmenite, apatite and rare kaersutite are also present. These primary phases have suffered variable alteration with the development of secondary, generally hydrous minerals. The grain size decreases in the outermost 3–4 m of the central zone and the texture becomes subophitic in the inner marginal zone. The outer marginal zone has a granular texture composed of

plagioclase laths (up to 1 mm) with granules of clinopyroxene and ilmenite (both as pseudomorphs). The thin, chilled contacts, formerly glassy basalt, are now principally altered to chlorite.

At the top of the central zone, coarse dolerite grades upwards into a medium-grained, feldsparphyric ferrodolerite (Plate 45) composed of albitised plagioclase feldspar laths with intergrowths enclosing optically continuous grains of clinopyroxene, blades of ilmenite and numerous acicular apatite crystals. Interstitial chlorite, pumpellyite and stilpnomelane probably represent an original iron-rich groundmass.

MINERAL CHEMISTRY   Variation in mineral chemistry, determined by electron microprobe analysis, can be accounted for by different cooling rates but possibly also by minor variations in magma composition. The most obvious effect of cooling rate is seen in the chemistry of the clinopyroxenes from the inner and outer marginal zones (Figure 88). Here rapid cooling rates have resulted in relatively high TiO_2 (2.94–3.52 wt%) and Al_2O_3 (2.94–5.28 wt%) contents. Away from the margins the contents are significantly lower, in the range of 0.82–2.31 wt% and 1.51–3.92 wt%

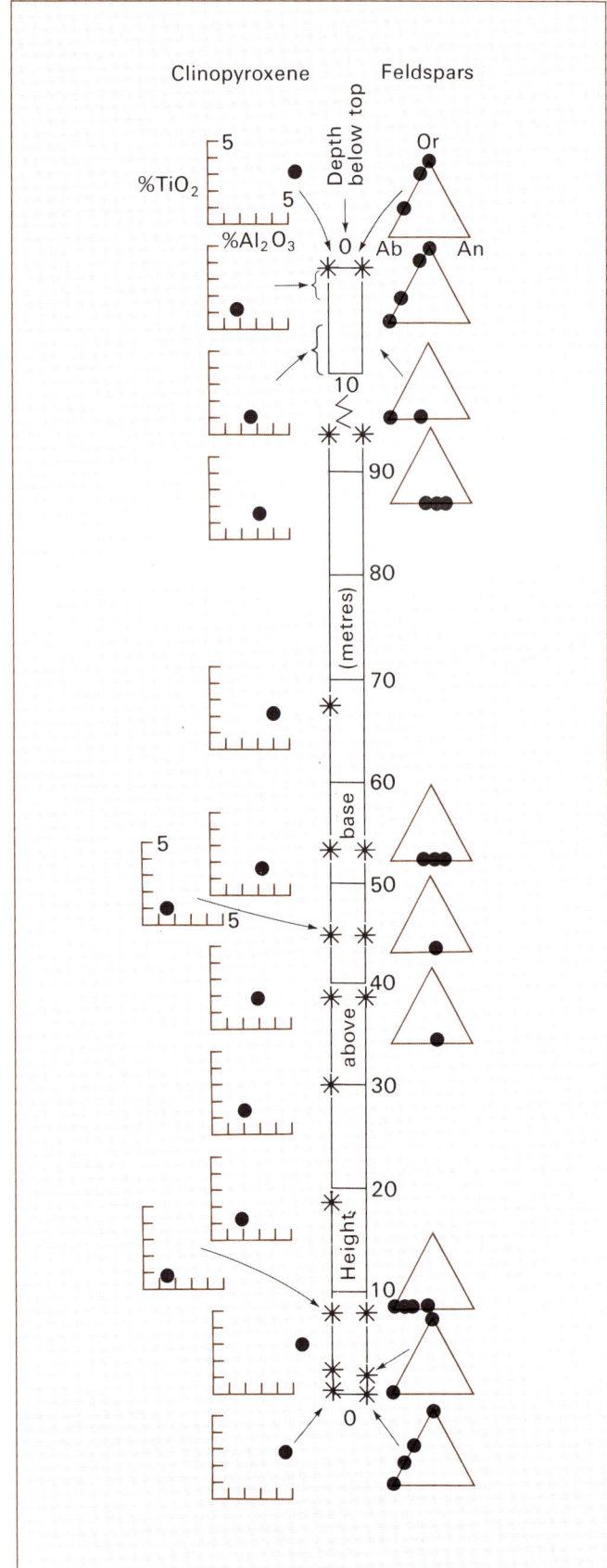

Figure 88 Tal y Fan dolerite, chemical variations in clinopyroxenes and feldspars (after Merriman et al., 1986).

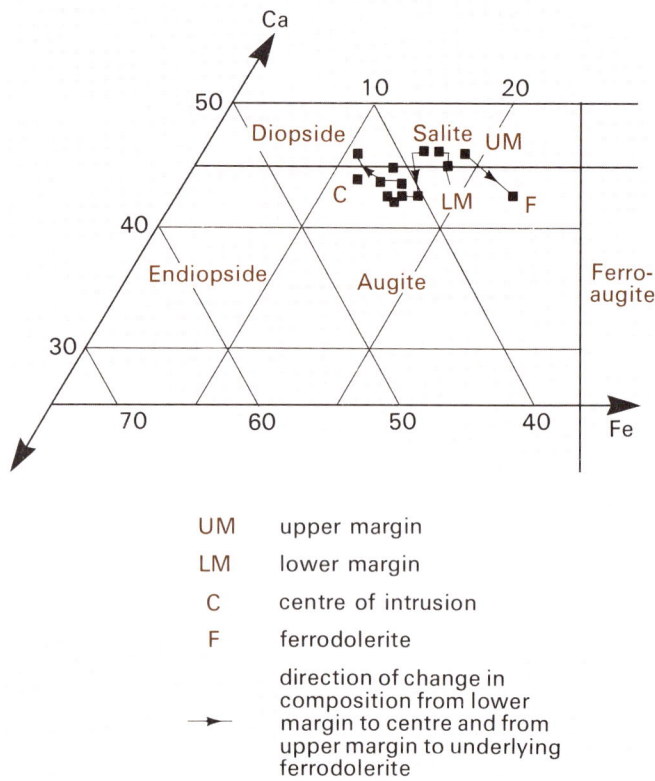

UM	upper margin
LM	lower margin
C	centre of intrusion
F	ferrodolerite
→	direction of change in composition from lower margin to centre and from upper margin to underlying ferrodolerite

Figure 89 Tal y Fan dolerite, analyses plotted on pyroxene quadrilateral (after Merriman et al., 1986).

respectively. The high Ti and Al contents at the margins reflect high crystal growth rates and low diffusion rates (Schiffman and Lofgren, 1982).

In the pyroxene quadrilateral (Figure 89), analyses from the marginal zones plot in the salite field—their relatively high Ti and Al contents reflecting high Ca Al_2SiO_2 and Ca Ti Al_2O_6 components. The majority of analyses from the central zone plot in the augite field, with those most central being most diopside-rich. This Ca, Mg-rich character is significant in establishing original compositional differences in the melt. Predictably, the ferrodolerite contains the most iron-rich augites.

Variations in feldspar composition through the intrusions are shown in Figure 88 and representative analyses in Appendix 1, Table 13. Plagioclase is extensively altered to albite plus various hydrous Ca-Al silicates. Complete replacement of the anorthite (An_{1-9}) has occurred in the outermost 4–5 m of the intrusion. Within 10 m of each contact, much of the plagioclase is partially albitised (An_{4-39}). Relict crystals of primary labradorite (An_{53-68}) are present in the central zone although commonly juxtaposition with albitised crystals. Late primary alkali feldspar (Or_{37-98}) occurs in the ferrodolerite where it overgrows and replaces plagioclase euhedra and infills interstices. The presence of microperthitic textures indicates that these are exsolved high temperature feldspars.

Chlorite pseudomorphs after olivine, up to 1 mm in diameter, form about 3 per cent of the mode in the inner marginal zone and about 15 per cent at 40 m above the base. Ilmenite, in the central zone, is either pristine or shows variable degrees of replacement. However, in the marginal zones, ilmenite is invariably replaced by sphene or by leucoxene. Unaltered primary ilmenite contains significant MnO, 3.28–4.70 wt%.

MECHANISM OF INTRUSION   Variations in primary mineral proportions and chemistry across the intrusion reflect processes of emplacement and solidification of the magma (Merriman et al., 1986). The variations could be produced either by flow differentiation (Hughes, 1982) or, possibly more likely, by crystallisation progressing from the margin inwards (Delaney and Pollard, 1982) as the magma was intruded. With the latter process, samples collected from the margins inwards would represent progressively later magmas. If these magmas had been drawn from a compositionally zoned, high-level chamber, then the earliest would be relatively felsic in composition, becoming more basic as lower zones in the magma chamber were tapped. Such a mechanism accounts for the most primitive Mg-rich magma occurring at the centre of the Tal y Fan intrusion.

The ferrodolerite represents in situ fractionation of a relatively felsic and iron-rich liquid containing significant concentrations of incompatible elements.

GEOCHEMISTRY   Major and trace element concentrations have been determined from 13 samples across the Tal y Fan dolerite intrusion (representative samples, Appendix 1, Table 13) (Merriman et al., 1986). The most obvious feature of the vertical element profiles (Figure 91) is that, with the exception of the ferrodolerite in the central zone, the variations are regular, whilst in the marginal and chilled contact zones they are more erratic.

Probably the most significant profile is that of MgO, together with the closely similar pattern of Ni. Samples show an increase in these elements towards the core, reaching a maximum of 10.93 wt% MgO and 135 ppm Ni at 45 m above the base. This distribution reflects the varying proportion of olivine pseudomorphs from 3 per cent at 1.6 m to about 15 per cent at 45 m above the base. This indicates that MgO and Ni concentrations correspond to original values in the central zone and that here the samples are relatively primitive. Compositions are more evolved towards the margins while the ferrodolerite represents fractionated liquid.

Silica shows a general decrease towards the core, coincident with the increase in olivine. Accordingly, the central part of the intrusion approaches an ultrabasic composition with MgO 10 per cent and SiO_2 44 per cent. The highest SiO_2, 51 per cent, occurs in the fractionated ferrodolerite.

FeO variation is regular in the central zone with concentration increasing steadily towards its base, but the highest concentration, recorded in the ferrodolerite, reflects its fractionated nature.

Variations in CaO, Al_2O_3 and Na_2O are irregular across the intrusion. Highest concentrations of CaO occur in the central zone, where Ca-rich plagioclase feldspar is largely preserved. Consequently, Na_2O is lowest in the central zone, but increases towards the margins where albitised feldspars are predominant, and reaches a maximum in the ferrodolerite. Similarly, Al_2O_3 is highest in the central zone and generally decreases towards the margins, although relatively high Al_2O_3 in the lower, outer margin, is probably related to Al-rich clinopyroxene.

The constancy of Nb/Zr ratios suggests that Zr enrichment in the ferrodolerite, and its relative depletion in the lower margin, are both primary features. This pattern is reflected in other HFS element ratios, e.g. Ti/Zr and Y/Zr. However, LIL elements Rb, Sr, K and Ba were highly mobile during alteration (Chapter 6).

GEOCHEMISTRY OF DOLERITES ASSOCIATED WITH THE SNOWDON AND CRAFNANT CENTRES

On a plot (Figure 90) of Zr/TiO_2 vs. Nb/Y, the dolerites intruded about the Snowdon and Crafnant volcanic centres lie predominantly within the field of subalkaline basalt, with a significant number of analyses overlapping into the andesite/basalt field. A few analyses overlap into the alkaline basalt field, though only samples of the Cae Coch intrusion, emplaced at the top of the Snowdon Volcanic Group near the middle of the Crafnant Centre, are clearly in that field. A cluster of three samples representing the Llyn Teyrn intrusion fall within the andesite field. Two samples, one from the south-east of the Lower Rhyolitic Tuff Formation caldera (KB 318), the other associated with the Crafnant Centre (KB 143), plot in the rhyodacite/dacite field. The former may owe its anomalous chemistry (Nb = 19 ppm, Zr = 395 ppm) to local mixing of magma in an area where multiple intrusions of acid and basic magma are common. The latter (Nb = 34 ppm, Zr = 590 ppm) may also reflect magma mixing, as it lies in close

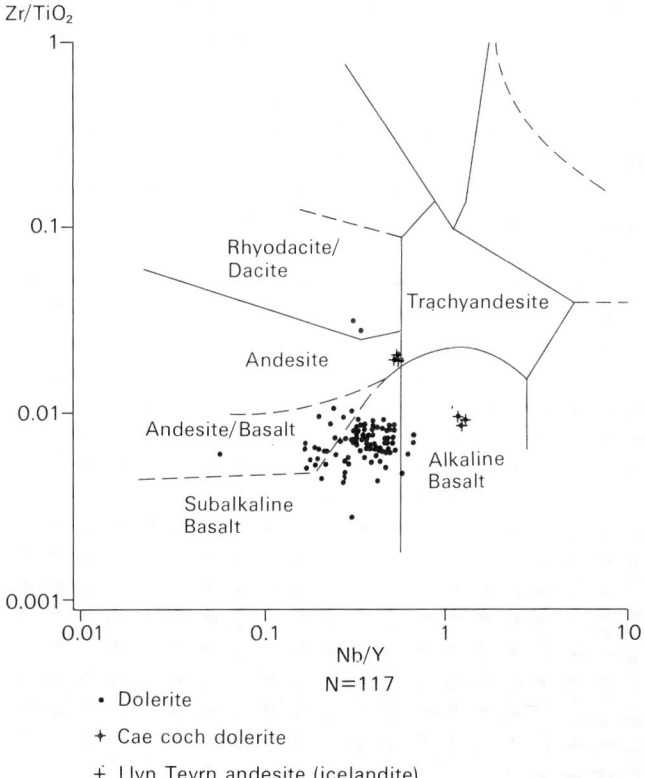

• Dolerite

+ Cae coch dolerite

+ Llyn Teyrn andesite (icelandite)

Figure 90   Dolerites of the Snowdon and Crafnant centres (including the Llyn Teyrn andesite), Zr/TiO_2 vs. Nb/Y diagram. The two analyses plotting in the Rhyodacite/Dacite field represent dolerites incorporating an acidic component (KB 143, 318).

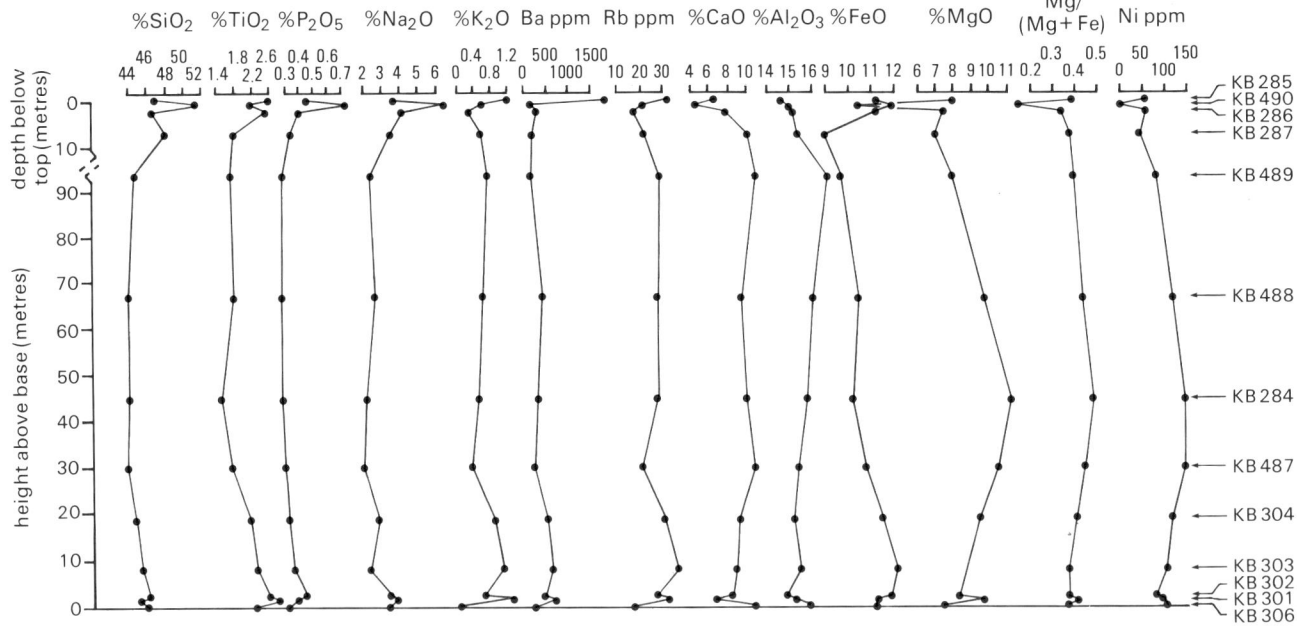

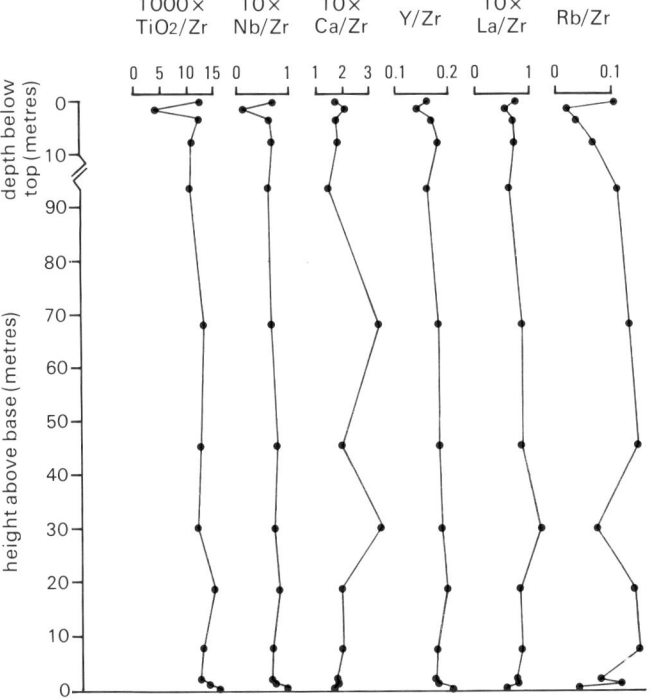

Figure 91 Tal y Fan dolerite, vertical profiles of major and trace element concentrations and trace element ratios (after Merriman et al., 1986).

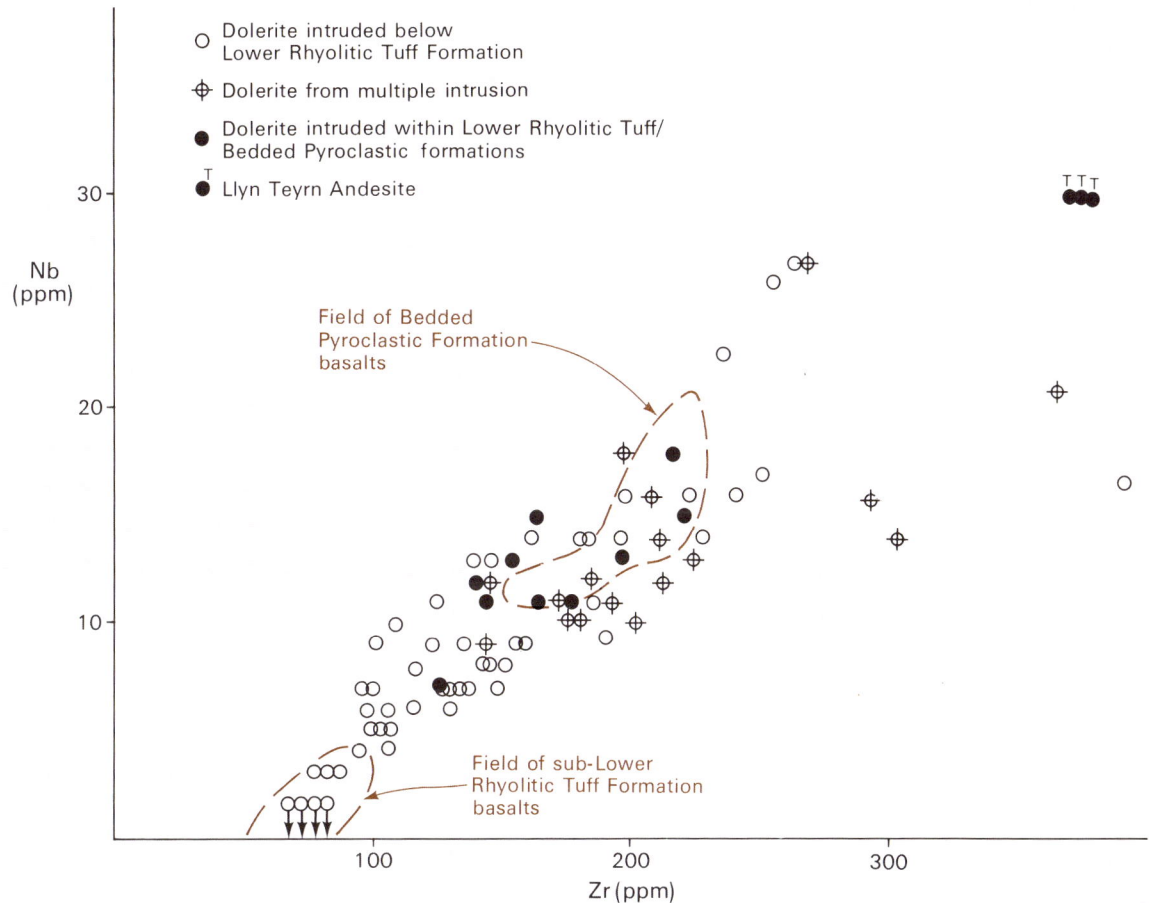

Figure 92   Nb vs. Zr diagram of dolerites; intrusions below the Lower Rhyolitic Tuff Formation, including multiple intrusions; intrusions within the Lower Rhyolitic Tuff Formation and Bedded Pyroclastic Formation; and the Llyn Teyrn icelandite intrusion.

proximity to a rhyolite intrusion. In thin section, areas of an originally glassy mesostasis can be discerned within the generally subophitic texture.

The dolerites are described below in relation to their geographical association with a) the Snowdon Centre and b) the Crafnant Centre.

Snowdon Centre

Dolerites in the vicinity of the Snowdon Centre, representative analyses of which are in Appendix 1, Table 11, are compositionally variable (e.g. Zr = 64–395 ppm, Nb = <2–27 ppm, Y = 19–60 ppm, N = 69), although their Nb/Zr ratios are broadly similar (about 0.075). Their compositional range spans that of the sub-LRTF basalts and the more HFS element enriched basalts, tuffs and basaltic sandstones of the Bedded Pyroclastic Formation (e.g. Figure 92). Accordingly, it is possible that the dolerites are related to more than one phase of basaltic volcanism. Furthermore, several of the dolerites are compositionally unrepresented by any recognised phase of extrusion. These generalisations can be further demonstrated by various N-MORB normalised multi-element plots (Figures 93, 94 and 95). The problem can be resolved to some extent by consideration of the stratigraphical level at which the intrusions occur.

Multi-element profiles of dolerites intruded within the Lower Rhyolitic Tuff Formation and Bedded Pyroclastic Formation (N = 10) are similar to those of basalts of the Bedded Pyroclastic Formation (Figure 93). The similarity is typified by their LREE and HFS element enrichment relative to N-MORB, and positive La and Nb and negative Ce, P and Ni N-MORB normalised anomalies. This is in marked contrast to profiles for the sub-LRTF basalts (Figure 94). The characteristics of the dolerites intruded within the Lower Rhyolitic Tuff Formation and Bedded Pyroclastic Formation are transitional between N-type MORB and E-type MORB. The dolerite intrusions below the base of the Lower Rhyolitic Tuff Formation are, however, more complex, with a wider compositional range (Figure 94) lying between and overlapping the sub-LRTF basalts and Bedded Pyroclastic Formation. The dolerites overlapping one or other of the two fields of N-MORB basalt compositions have broadly similar profiles to the respective basalts. Thus, the dolerites overlapping the sub-LRTF basalt, are also interpreted as being of broadly volcanic-arc tholeiite affinity. Those dolerites intermediate between the two basalts display both negative and positive N-MORB normalised Nb and Ce anomalies, and several dolerites have negative P anomalies as do the sub-LRTF

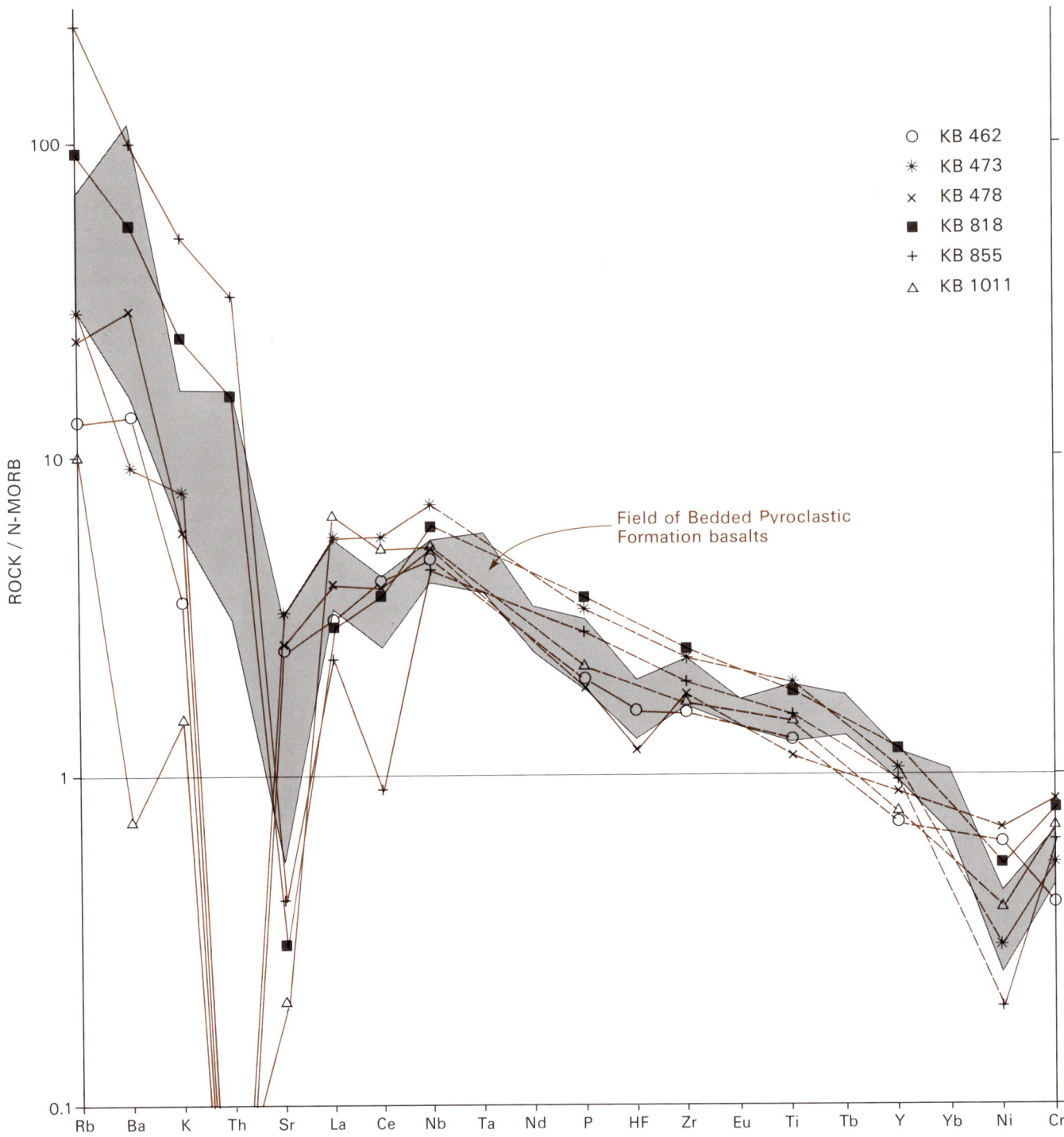

Figure 93   N-MORB normalised multi-element diagram of dolerites from the Snowdon Centre, in comparison with basalts of the Bedded Pyroclastic Formation (normalising factors after Pearce, 1982a). The compositional field of basalts of the Bedded Pyroclastic Formation, as defined in Figure 74, is shown for comparative purposes.

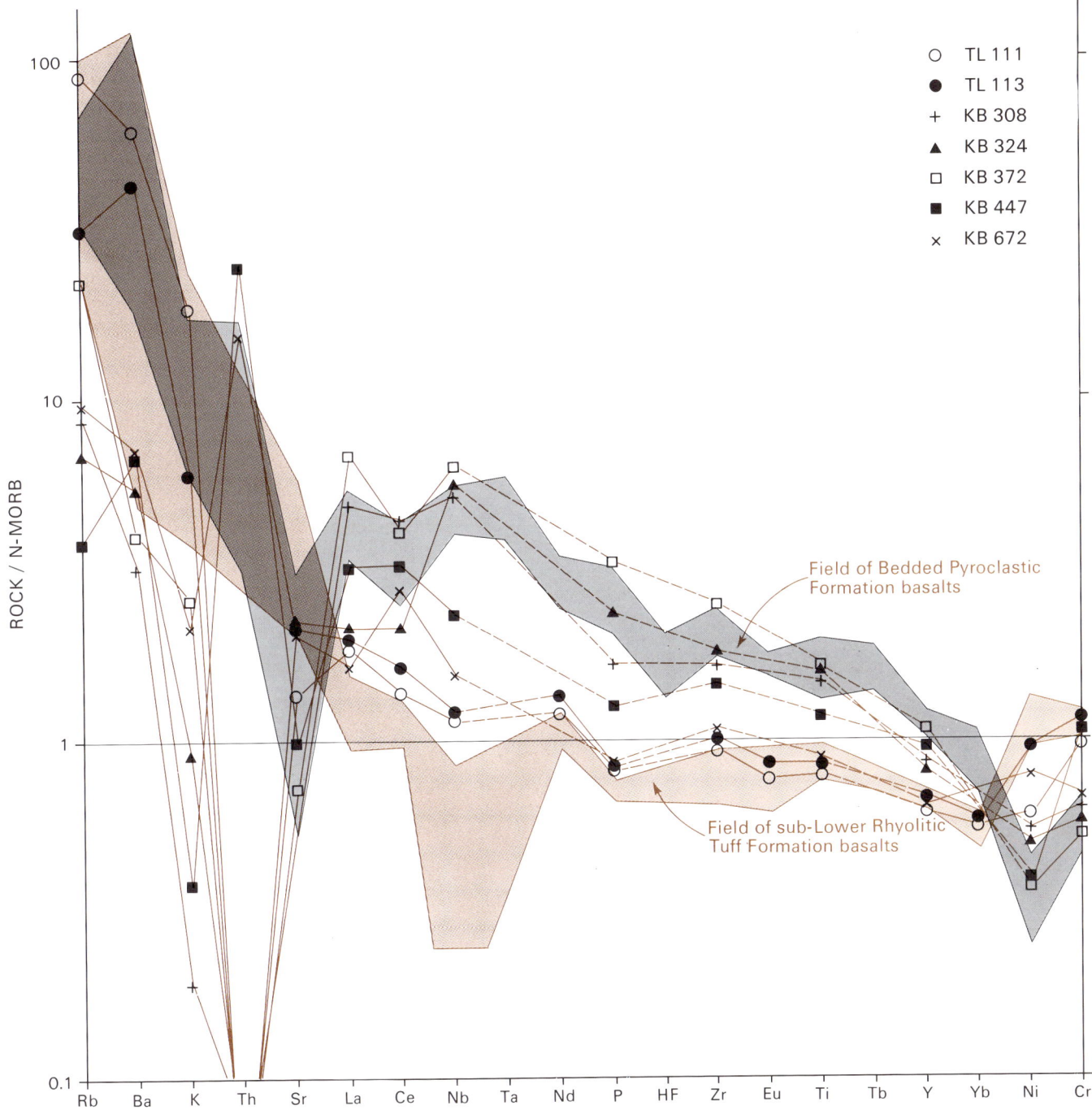

Figure 94 N-MORB normalised multi-element diagram of dolerites from the Snowdon Centre in comparison with sub-LRTF basalts. The compositional fields of the sub-LRTF basalts and basalts of the Bedded Pyroclastic Formation, as defined respectively in Figures 48 and 74, are shown for comparative purposes.

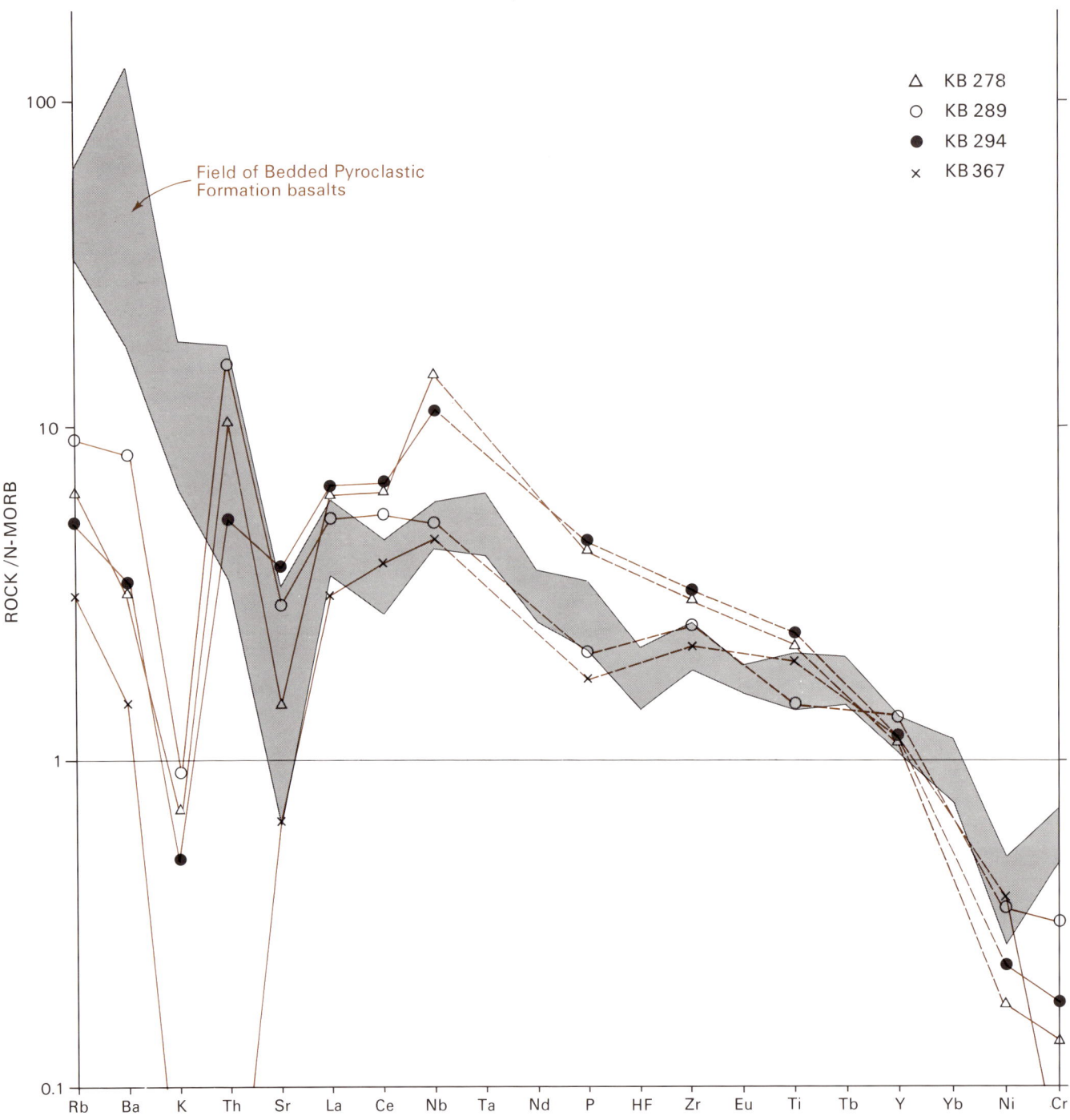

Figure 95   N-MORB normalised multi-element diagram of dolerites of the multiple intrusions, Snowdon Centre, in comparison with Bedded Pyroclastic Formation basalts. The compositional field of basalts of the Bedded Pyroclastic Formation, as defined in Figure 74, is shown for comparative purposes.

basalts, but none has as marked Ni and Cr positive anomalies.

The marginal dolerites of the multiple intrusions to the south-east of the Lower Rhyolitic Tuff Formation caldera (Figure 95), have generally similar profiles, apart from the mobile LIL elements, to the Bedded Pyroclastic Formation basalts, though in some instances they show negative N-MORB normalised P anomalies. Although all have the pronounced negative Ni anomalies, typical of the N-MORB normalised Bedded Pyroclastic Formation basalts, they have, in addition, an even greater negative Cr anomaly. It is notable that most of the dolerites with chemistry similar to the Bedded Pyroclastic Formation and intruded below the base of the Lower Rhyolitic Tuff Formation, including the multiple intrusions, are concentrated to the south-east of the Lower Rhyolitic Tuff Formation caldera in the sector extending from Moel Ddu to the Dolwyddelan syncline (Figure 51). Also, many of the thickest sills beneath the base of the Lower Rhyolitic Tuff Formation are those relatively depleted in the LREE and HFS elements. It must also be pointed out that some of the dolerite sills, for example those below the base of the Lower Rhyolitic Tuff Formation to the north-east of Moel Hebog, are probably multiple intrusions, since their chemistry reflects a compositional bimodality.

It is proposed therefore, that, as with the emplacement of rhyolites around the Snowdon Centre, dolerites were intruded during at least two main phases in the evolution of the centre, one before and the other after the development of the Lower Rhyolitic Tuff Formation caldera. Only dolerites with chemistry similar to the Bedded Pyroclastic Formation (Figure 93) have been recognised as having intruded the Lower Rhyolitic Tuff Formation and Bedded Pyroclastic Formation. No dolerites are known to have been intruded to higher stratigraphical levels than the Bedded Pyroclastic Formation. Although outcrops of these strata are very restricted, it seems unlikely that there was any major basaltic volcanism associated with the Snowdon Centre subsequent to the Bedded Pyroclastic Formation. Dolerites similar to the sub-LRTF basalts, or intermediate in composition between them and the Bedded Pyroclastic Formation, are restricted to levels below the base of the Lower Rhyolitic Tuff Formation.

Crafnant Centre

Selected analyses for the dolerites associated with the Crafnant Centre are presented in Appendix 1, Table 13. Unlike the Snowdon Centre, the majority of intrusions occur within the Snowdon Volcanic Group, and to some extent postdate the acidic volcanism. The compositional range (e.g. Nb, Zr, Y, Th) of the dolerites associated with the Crafnant Centre is similar to that of dolerites around the Snowdon Centre. Comparison with the geochemistry of the two phases of extrusive basaltic volcanism (Figure 96) associated with the Crafnant Centre demonstrates that the dolerites overlap the basaltic compositional fields but span a wider compositional range, most notably towards relatively depleted Nb and Zr compositions. Nb/P$_2$O$_5$ (Figure 96) and Nb/Zr ratios show little variation, with the exception of the alkaline basalt intruded into the Dolgarrog Volcanic Formation at Cae Coch (Figure

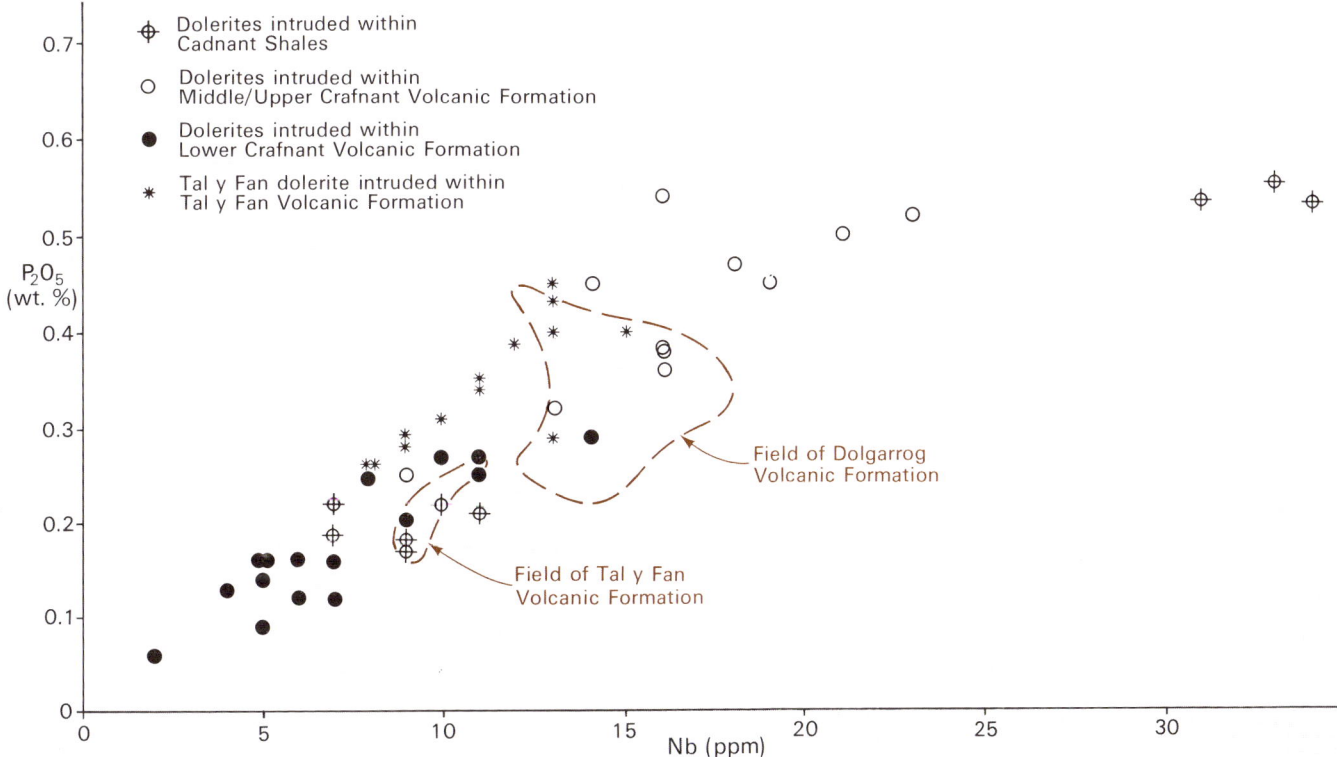

Figure 96 P$_2$O$_5$ vs. Nb diagram of dolerites of the Crafnant Centre and their relationship to the Tal y Fan and Dolgarrog Volcanic formations.

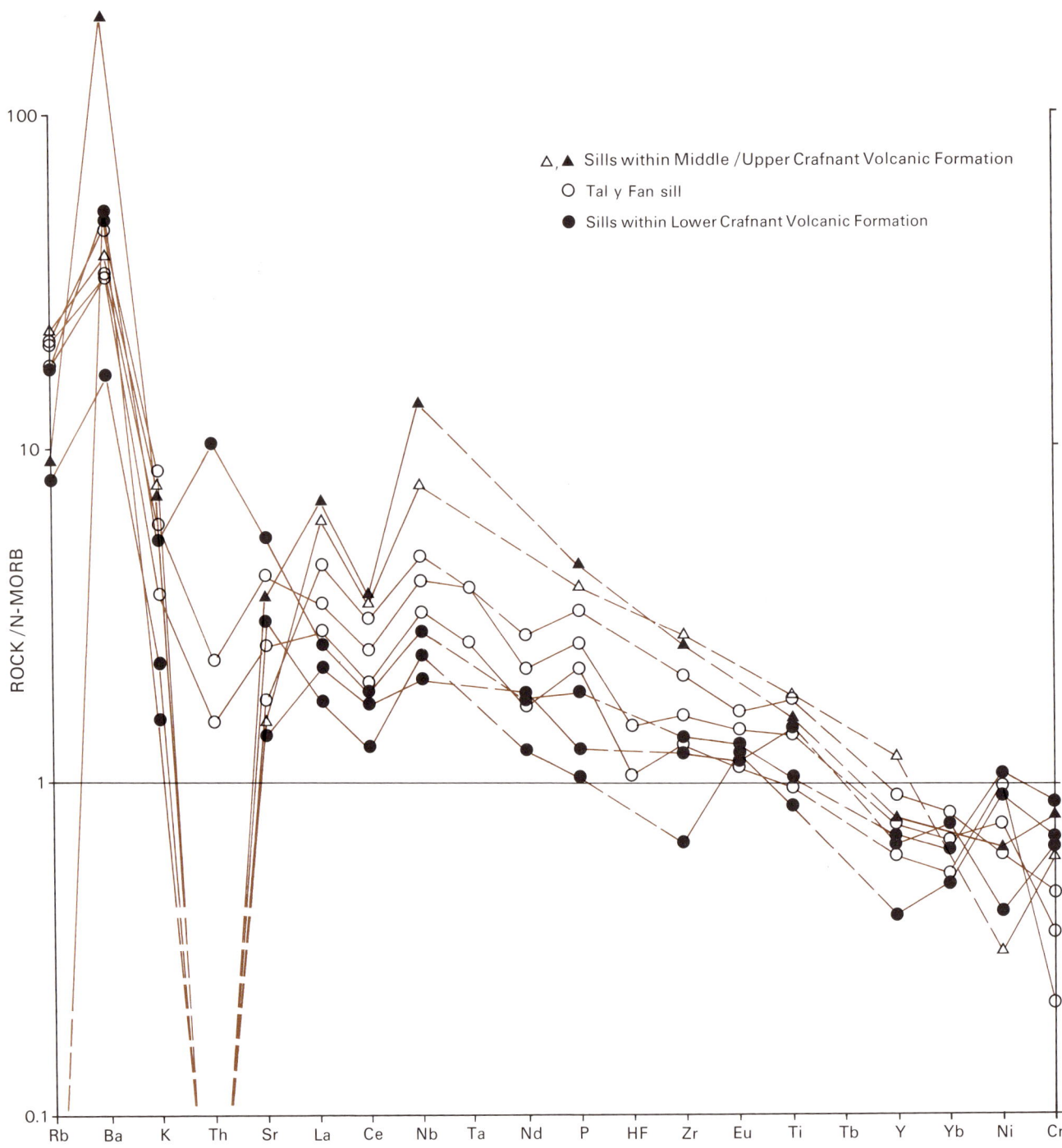

△,▲ Sills within Middle /Upper Crafnant Volcanic Formation
○ Tal y Fan sill
● Sills within Lower Crafnant Volcanic Formation

Figure 97   N-MORB normalised multi-element diagram of dolerites of the Crafnant Centre.

102) which is relatively Nb enriched. Some variation of dolerite chemistry with level of intrusion is indicated, in that all dolerites with Nb content greater than 15 ppm intruded the Middle and Upper Crafnant Volcanic formations and the overlying Cadnant Shales (Figure 96). Dolerites intruded within the Lower Crafnant Volcanic Formation tend to be relatively depleted in Nb, P_2O_5 (Figure 96) and Zr. However, some dolerites intruded at the highest level, in the Cadnant Shales, are similarly depleted (Figure 96). The range of compositions within any one intrusion may be considerable as is demonstrated by the Tal y Fan sill (Figure 91), and partially reflects in-situ fractionation (Merriman et al., 1986).

N-MORB normalised multi-element profiles of the dolerites at the Crafnant Centre are similar to those of the Tal y Fan and Dolgarrog Volcanic formations (Figures 85 and 97). In particular, the Tal y Fan sill closely matches the profile of the Tal y Fan Volcanic Formation with the notable exception of significant P depletion in the latter. Intrusions within the Middle and Upper Crafnant Volcanic formations and the Dolgarrog Volcanic Formation are the most closely comparable to the hyaloclastites of the Dolgarrog Volcanic Formation but differ in their relative depletion of Sr, La and P.

In comparing the N-MORB normalised profiles of dolerites associated with the Snowdon and Crafnant centres, the Tal y Fan sill is most closely comparable to the dolerites intruded within the Lower Rhyolitic Tuff Formation and Bedded Pyroclastic Formation (and also similar, therefore, to the Bedded Pyroclastic Formation basalts and volcaniclastics). It has relative LREE to HREE enrichment, positive La and Nb anomalies and negative Ce, Nd and Hf anomalies. It differs, however, in having less HFS element enrichment and a positive Ni and Cr anomaly. In general, intrusions within the Lower Crafnant Volcanic Formation have positive Nb anomalies; however, they typically have negative P values and their marked positive Ni and Cr anomalies are more typical of the sub-Lower Rhyolitic Tuff Formation basalts and related dolerites. The dolerites intruded within the Middle and Upper Crafnant Volcanic formations are enriched in Sr, Nb, P, and Zr compared to almost all dolerites associated with the Snowdon Centre.

Llyn Teyrn andesite intrusion

The substantial intrusion at Llyn Teyrn, 2.5 km east of Snowdon summit (Figure 36), although petrographically closely similar to dolerites associated with the Snowdon Centre, is andesitic in composition (Appendix 1, Table 11). It is significantly enriched in LREE and HFS elements relative to most other dolerites. Its N-MORB normalised multi-element profile is comparable to the Bedded Pyroclastic Formation basalt profile but at higher LREE and HFS element concentrations. The presence of areas of fine-grained mesostasis within an otherwise coarsely ophitic texture may indicate some degree of mixing of basalt/andesite compositions and this may account for its unusual intermediate whole-rock composition. Given its higher TiO_2 and Zr contents than orogenic andesites (cf. Pearce and Norry, 1979; Bailey, 1981; Pearce, 1982a) and its similarities to subalkaline, silica-oversaturated icelandites of ocean islands, the Llyn Teyrn intrusion is identified as icelandite. Icelandites have been identified amongst Caradoc lavas on the Lleyn Peninsula (Croudace, 1982; Leat

and Thorpe, 1986) and in the Silurian sequences of Skomer island in south-west Wales (Thorpe et al., 1989).

PETROGENESIS OF THE 2ND ERUPTIVE CYCLE: SNOWDON VOLCANIC GROUP AND RELATED INTRUSIONS

The 2nd Eruptive Cycle (Snowdon Volcanic Group) comprises extrusive volcanic rocks with a wide variety of compositions: the Bedded Pyroclastic Formation (Figure 72), the Tal y Fan Volcanic Formation and the Dolgarrog Volcanic Formation (Figure 83) of predominantly subalkaline basalt composition; sub-LRTF basalts of predominantly basalt/andesite composition (Figure 46); the Pitts Head Tuff Formation (Figure 41), the Lower Rhyolitic Tuff Formation (Figure 53), the Lower Crafnant Volcanic Formation (except No. 3 tuff) (Figure 79) and the Middle and Upper Crafnant Volcanic formations of rhyodacite to predominantly subalkaline rhyolite composition; and the Upper Rhyolitic Tuff Formation (Figure 75) and Lower Crafnant Volcanic Formation, No. 3 tuff (Figure 79) of peralkaline rhyolite composition.

Intrusions spatially associated with the Snowdon Volcanic Group include: basic intrusions of predominantly subalkaline basalt composition, with some of basalt and a few of alkaline basalt compositions (Figure 90); the Llyn Teyrn icelandite, the only substantial intrusion of intermediate composition; rhyolites (Figure 62), feldspar porphyries, and some microgranites and granites (Figure 86) of rhyodacite to subalkaline rhyolite composition; and rhyolites and a few microgranites of transitional subalkaline rhyolite to peralkaline (comenditic) rhyolite composition (Figures 62 and 86). However, considered as a whole, analyses of the Snowdon Volcanic Group and associated intrusions fall mainly in one of two groups, (Figures 37, 86 and 90), subalkaline basalt and subalkaline rhyolite.

Basalt petrogenesis

The fundamentally different end-member basaltic compositions are represented in the 2nd Eruptive Cycle. The sub-LRTF basalts represent one composition which, on a N-MORB normalised multi-element plot (after Pearce, 1982b; Saunders and Tarney, 1984), are characterised by slight overall HFS element depletion relative to N-MORB, with Nb depletion the most marked (Figure 48). This pattern is closely comparable with those of volcanic-arc basalts and specifically with island-arc tholeiites (cf. South Sandwich Islands; Pearce, 1982b). The LIL element enrichment must be interpreted with caution because of possible modification during alteration but, taken at face value, the pattern would imply a strong subduction zone component, as in suprasubduction-zone basalts (Pearce et al., 1984b), though rather more characteristic of back-arc systems (cf. Saunders and Tarney, 1984) than of volcanic-arc system basalts (cf. Pearce, 1982a,b). Their Ni and Cr enrichment relative to Y and Yb is an occasional feature of transitional volcanic-arc basalts and back-arc systems, such as the Mariana trough back-arc basin (Saunders and Tarney, 1984). The relatively high compatible element contents (e.g. Ni and Cr) of the sub-LRTF basalts suggest that they may be compositionally near to unmodified melts of mantle peridotite (Sato, 1977; Hart and Davis, 1978).

In contrast, the basalts of the Bedded Pyroclastic Formation (Figure 74) are, with respect to N-MORB, significantly enriched in the HFS elements Nb to Ti and markedly depleted in Ni and, to a lesser extent, Cr. The HFS element patterns are similar to those of within-plate basalts, typified for example by Hawaiian basalt (cf. Pearce, 1982b), and they have the general characteristics of E-type (enriched) MORB. The low Th/Ta ratios of some of the basalts (e.g. TL 049, TL 050, KB 863, Table 10) are particularly characteristic of ocean-island basalts (Leat and Thorpe,1989). The low Ni and Cr, and relatively high HFS element contents of the basalts of the Bedded Pyroclastic Formation suggest that they may represent the products of significant fractional crystallisation of a more primitive (mantle) melt. Associated intrusive dolerites and basalts (Figures 93 and 94) display compositions which either overlap or are intermediate between the two end-member basalt compositions, although dolerites similar to the Bedded Pyroclastic Formation basalts are numerically, and possibly volumetrically more abundant.

The variations in ratios of Nb and Ta to Th in the basalt groups (Figures 48 and 74; Appendix 1, Tables 7 and 10) cannot be accounted for by fractional crystallisation since the partition coefficients D_{Th}, D_{Ta} and D_{Nb} are all similarly low (< 0.1) in basaltic magmas. It would therefore require very large degrees of fractionation (> 90 per cent?), thus leading away from basaltic compositions, to achieve the observed variations in ratios. Although assimilation of continental crust with higher Th/Ta and Th/Nb ratios could account for these variations, mass balance calculations (Thorpe et al., in press) indicate that such assimilation would modify the basalt towards andesite (icelandite). The variations therefore probably reflect derivation of the basalts from a compositionally heterogeneous source or sources, of which ocean-island basalt represents one end of the spectrum, and transitional calcalkalic arc basalts to arc-tholeiitic basalts the other.

Variations, particularly in basalts of the Bedded Pyroclastic Formation, in the ranges of incompatible element abundances, (e.g. Nb, Zr, Y, Ta, Th, and Hf which vary by factors up to about 2.5) and of LREE abundances indicate, however, the importance of fractional crystallisation. Comparable variations are observed in the compatible elements, with the highest concentrations of Ni, Cr and V present in those basalts with relatively lower HFS element and REE contents; this is consistent with them being compositionally more primitive. The most REE enriched basalts (Figure 47) have small Eu anomalies and, together with the trace element variations, are consistent with fractionation involving removal of plagioclase, olivine, pyroxene and Fe-Ti-oxide. The two extrusive basalt formations, the Dolgarrog Volcanic Formation and Tal y Fan Volcanic Formation, have N-MORB normalised profiles (Figure 85) that are similar to those of the Bedded Pyroclastic Formation, although the Tal y Fan Volcanic Formation is slightly depleted in HFS elements. Their petrogenetic interpretation is, therefore, regarded as being essentially the same as that for the Bedded Pyroclastic Formation.

The parental basaltic magmas are assumed to be derived from a garnet-free spinel- (or plagioclase-) lherzolite mantle source, with up to 60 per cent fractional crystallisation of the olivine gabbro assemblage (Thorpe et al., in press). The

heterogeneity of the basalt magmas is considered to reflect mantle heterogeneity due to the varying influence of subduction related processes. Evidence for such mantle heterogeneity has been indicated in a study of earlier basaltic volcanism (Tremadoc–Llandeilo) in the Welsh Marginal Basin (Kokelaar et al., 1986).

Petrogenesis of intermediate to rhyolite magmas

Unlike the 1st Eruptive Cycle (Llewelyn Volcanic Group), there is no clear compositional continuum from intermediate to acidic compositions within the 2nd Eruptive Cycle (Snowdon Volcanic Group) and its associated intrusions. The only significant intermediate intrusion is of icelandite composition. Several rhyolitic (and rhyodacitic) compositions have been identified. All can be referred to one of the five rhyolite groups (A1, A2, B1, B2 and B3) (Campbell et al., 1987) associated with the Snowdon Centre. The exceptions are some feldspar porphyries (B2a) and the No. 1 tuff of the Lower Crafnant Volcanic Formation, which compositionally lies between A2 and B2 rhyolites. In the following discussion, both crustal fusion and fractional crystallisation are considered as possible mechanisms for genesis of the intermediate and acidic magmas.

Crustal fusion

Any assessment of the possibility of generating significant volumes of acidic magma by crustal fusion (either by bulk or partial melting) depends on adequate knowledge of the composition of available pre-Ordovician crust. No strata older than Tremadoc in age are exposed within the Caradoc Snowdon graben (Kokelaar, 1988) or 'failed rift' (Campbell et al., 1988). Accordingly, any estimates of the composition of the pre-Palaeozoic crust in this region must depend on analogy with adjacent areas and on theoretical considerations.

Considering these factors, Thorpe et al. (in press) concluded that the likely composition of Proterozoic upper crust in North Wales would, for example, have contained Th in the range 4.6–7.5 ppm and Th/Ta ratios varying between 1.8 and more than 11. Furthermore, they considered that the lower half of the crust must have had an even lower Th content. As none of the rhyolite groups considered here has Th/Ta and Th compositions within these ranges, they are not considered to have been generated by either bulk or partial melting of local Precambrian crust.

Evidence for some degree of crustal contamination, however, is indicated by the isotopic data of Thorpe et al. (in press). ϵ_{Nd} values (calculated at 445 Ma) for basalts from the Bedded Pyroclastic Formation are in the range $+ 3.6$ to $+ 4.5$. No isotopic data for the basalts at the base of the Lower Rhyolitic Tuff Formation are as yet available, though comparative data for a similar but earlier arc-tholeiitic basalt from the Rhobell Volcanic Complex to the south-south-west gives a slightly lower ϵ_{Nd} value of $+ 1.9$ (calculated at 508 Ma, Kokelaar et al., in press). The icelandite, however, has significantly lower values ($\epsilon_{Nd} = - 0.5$ and $+ 0.1$). As fractionation does not affect ϵ_{Nd} values, the icelandite could not be generated solely by fractionation of a Bedded Pyroclastic Formation basalt parent without contamination (?crustal) by a source enriched in [143]Nd. If the ϵ_{Nd} values of the sub-LRTF basalts are indeed lower than those for the Bedded Pyroclastic Formation basalts, the need for contamination of magmas descended from sub-LRTF basalts would be reduced. ϵ_{Nd} data for the rhyolites varies between -1 and $+2.6$. A1 rhyolites have the lowest values (-1.0 and -0.7) and B3 rhyolites the highest ($+1.7-+2.6$), with other rhyolites having intermediate values.

Available data for potential crustal contaminants are derived from rocks of the Sarn and Mona complexes on the Lleyn Peninsula and Anglesey ($\epsilon_{Nd} = -4.2$ to -6.2). These rocks could not have generated the rhyolites by bulk or partial melting, as such products would be too enriched in radiogenic ^{143}Nd. Even so, they could represent a crustal contaminant which, in conjunction with fractional crystallisation of basaltic magmas, could generate the suite of Snowdon Volcanic Group magmas as described below.

Fractional crystallisation

Due to the fundamental bimodality of the Snowdon Volcanic Group, variation trends which clearly relate the basaltic and acidic compositions are less readily defined than in the Llewelyn Volcanic Group. The choice of a suitable fractionation index to test relationships is again restricted by problems of element mobility. Thus, Rb and U must be considered unreliable. Th is suitable, but its concentration in most of the basalts (and dolerites) is at the limit of precision of the XRF analytical technique employed, and only limited precise Th data, analysed by INAA, are available. Consequently, Zr has also been used, albeit with caution, due to its increasing compatibility in more acidic melts.

The incompatible elements, Y, Nb, La, Ce, Ga and Th all display strong positive correlations with Zr, and Nb and Y in particular show little scatter from basic to intermediate (icelandite) and acidic compositions. On a plot of Nb vs. Zr, the A1 and A2 rhyolite groups lie on the trend defined by the basic–intermediate compositions and have similar Nb/Zr ratios (about 0.011) but enriched Nb and Zr contents. The B1, B2 and B3 rhyolites have significantly lower Nb/Zr ratios (about 0.05). Negative correlations of the compatible elements (e.g. Ni and Cr) extending from basic to acidic compositions are also observed. Marked inflection points at intermediate compositions are present for several elements (vs. Zr), including TiO_2, Fe_2O_3, P_2O_5, MgO, CaO, V and Co. Thus Fe-Ti-oxide and apatite fractionation is indicated and in the A1 and B1 rhyolites and possibly the A2 and B2 rhyolites, there is the possibility of some zircon fractionation.

The Bedded Pyroclastic Formation basalts have moderately LREE-enriched, chondrite-normalised REE patterns with slight Eu anomalies, reflecting plagioclase fractionation in the most REE-enriched samples. Chondrite-normalised plots for all the rhyolites are similar, indicating that they are genetically related (Campbell et al., 1987). The patterns are also similar to those for the basalts in their LREE-enrichment but they have considerably greater Eu anomalies, particularly in the A1, A2 and B3 rhyolites; this implies extensive plagioclase fractionation in their genesis. Overall REE-enrichment compared with the basalts is about 2.5–10. Relative LREE-enrichment is rather less in the B3 rhyolites than in the other groups.

However, the rhyolite groups have markedly different

Th/Nb (Campbell et al., 1987) and Th/Ta ratios (Figure 98) at roughly similar Th abundances, with the exception of groups A1 and B1 which are broadly comparable. This suggests that the rhyolites represent distinct batches of magma, not related directly to each other by fractional crystallisation of a common parent.

An attempt has been made to model fractional crystallisation of the mafic magmas using the Rayleigh Crystal Fractionation law (Thorpe et al., in press). A two-step process was assumed, involving (A) a basalt-intermediate (icelandite) stage and (B) an intermediate- rhyolite stage, with different phenocryst mass fractions representing the basalt-intermediate stage, and two intermediate-rhyolite intervals representing no zircon fractionation (B) and weak zircon fractionation (C).

Rhyolites of B3 composition can be generated by crystal fractionation via stages A and B (zircon-free), with icelandite and B3 compositions representing 50 per cent and 80–98 per cent crystallisation respectively (Figures 98, 99 and 100). In order to reproduce higher Th/Ta ratios (Figure 98) than

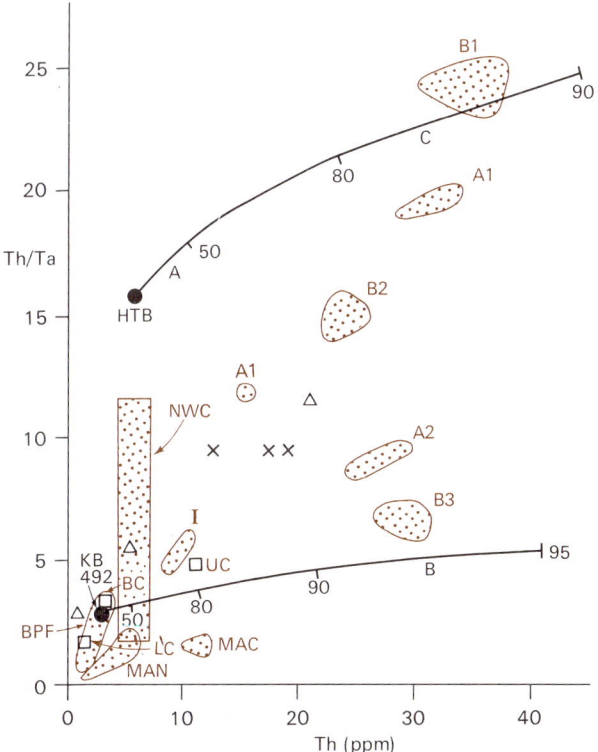

Figure 98 Th/Ta vs. Th diagram to model the generation of Snowdon Centre rhyolites (A1, A2, B1, B2, B3).

The model assumes crystal fractionation according to the Rayleigh Fractionation Law. The three fractionation pathways, A, B and C, reflect different modal assemblages (see text), and the extent of fractionation is expressed in percentages (50, 80, 90, 95). HTB: High Th Basalt; NWC: North Wales Crust; LC, BC, UC: lower, bulk and upper crust respectively, calculated by Taylor and Maclennan (1985); MAN: Manaslu Granite (Vidal et al., 1982); MAC: Macusani Tuffs (Noble et al., 1984); I: Icelandite, with Ta abundances calculated from Nb, assuming chondrite Nb/Ta ratio of 17.5; BPF: Bedded Pyroclastic Formation.

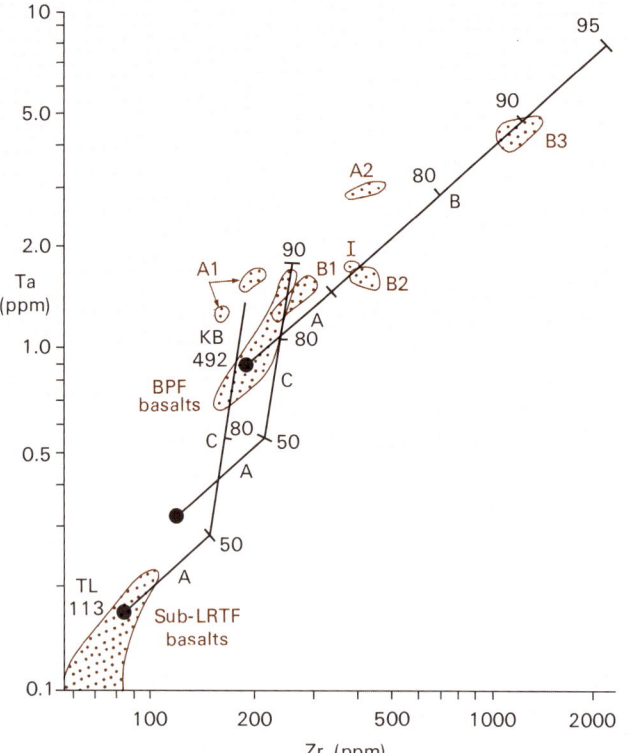

Figure 99   Ta vs. Zr diagram to model the generation of Snowdon Centre rhyolites.

Explanation of symbols as on Figure 98 with, in addition, the field of the sub-LRTF basalts also shown (Ta content predicted from Nb content using Nb/Ta = 17.5). I: Icelandite.

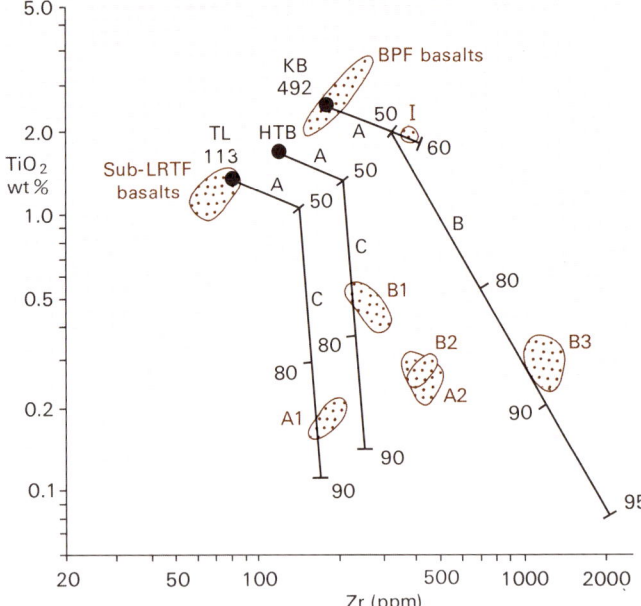

Figure 100 TiO_2 vs. Zr diagram to model the generation of Snowdon Centre rhyolites. Explanation of symbols as on Figure 98.

would be generated by direct fractionation and to lower the ϵ_{Nd} value sufficiently, it is necessary to invoke some assimilation of crust during the interval basalt-icelandite. The somewhat lower ϵ_{Nd} values of the identified icelandite imply, specifically, a rather greater degree of crustal assimilation in the genesis of the Llyn Teyrn icelandite intrusion.

In order to generate the A1 and B1 rhyolites, a modelled trend (Figures 99 and 100) is produced from basalts of sub-LRTF composition via stages A and B. This gives a reasonable fit for A1 rhyolites at very high degrees of crystallisation (>95 per cent), but it generates rhyolites with lower Ti content than those of B1 composition. The possibility of some degree of mixing with basic magma in B1 rhyolites, for which there is some field and petrographic evidence, as well as limited crustal assimilation, may explain the variation.

As A2 and B2 rhyolites occupy positions essentially between groups A1 and B1, and B3 in Figures 98, 99 and 100, they possibly represent products of fractional crystallisation of basalts with compositions between those of the parent end-members. Such compositions are represented in associated dolerite intrusions. Alternatively, but less likely, they may represent mixing between the other rhyolite groups.

THE MAGMA PLUMBING SYSTEM

When the petrogenetic scheme outlined above is related to the surface and near-surface distribution of the various magma types (Figure 101) and the timing of their emplacement, some conclusions can be reached concerning the evolution of the magma plumbing system beneath the Snowdon Centre.

The mantle-derived parental basaltic magmas were ponded, presumably as extensive sill-like reservoirs, at or below the base of the crust. To produce the various acidic fractions emplaced at the Snowdon Centre by 80–90 per cent fractional crystallisation of a basaltic parent, a minimum volume of 500 km³ basaltic magma would be necessary. For the 2nd Eruptive Cycle as a whole, the volume would be much greater and it is possible that the entire Snowdon graben was underlain, at subcrustal levels, by sill-like bodies of basaltic magma with an aggregate thickness of one to two kilometres.

The sub-LRTF basalt (island-arc tholeiite related) was the first magma to be emplaced at the Snowdon Centre, as sills about the developing Beddgelert pericline and as flows, with related hyaloclastite, to the west and north of the pericline. The A1 rhyolites and tuffs (Yr Arddu Tuffs), derived from a sub-LRTF basalt-type parent by fractional crystallisation and limited crustal assimilation, also developed during the pre-caldera phase (Figure 101). Probably no more than 2 km³ of A1 rhyolitic magma were erupted and the small, high-level magma chambers tapped were situated along the Yr Arddu Fracture.

The eruption of a minimum of 60 km³ of the A2 sub-alkaline rhyolitic magma, the most voluminous of the rhyolitic magmas, caused the collapse of the main Lower Rhyolitic Tuff Formation caldera (Figure 101). It was derived by fractional crystallisation of a basaltic parent, possibly formed by mixing of sub-LRTF type and Bedded Pyroclastic Formation-type basaltic magmas, with relatively minor crustal assimilation. Initially, the movement and accumulation of the A2 rhyolitic magma was controlled by the Beddgelert Fault

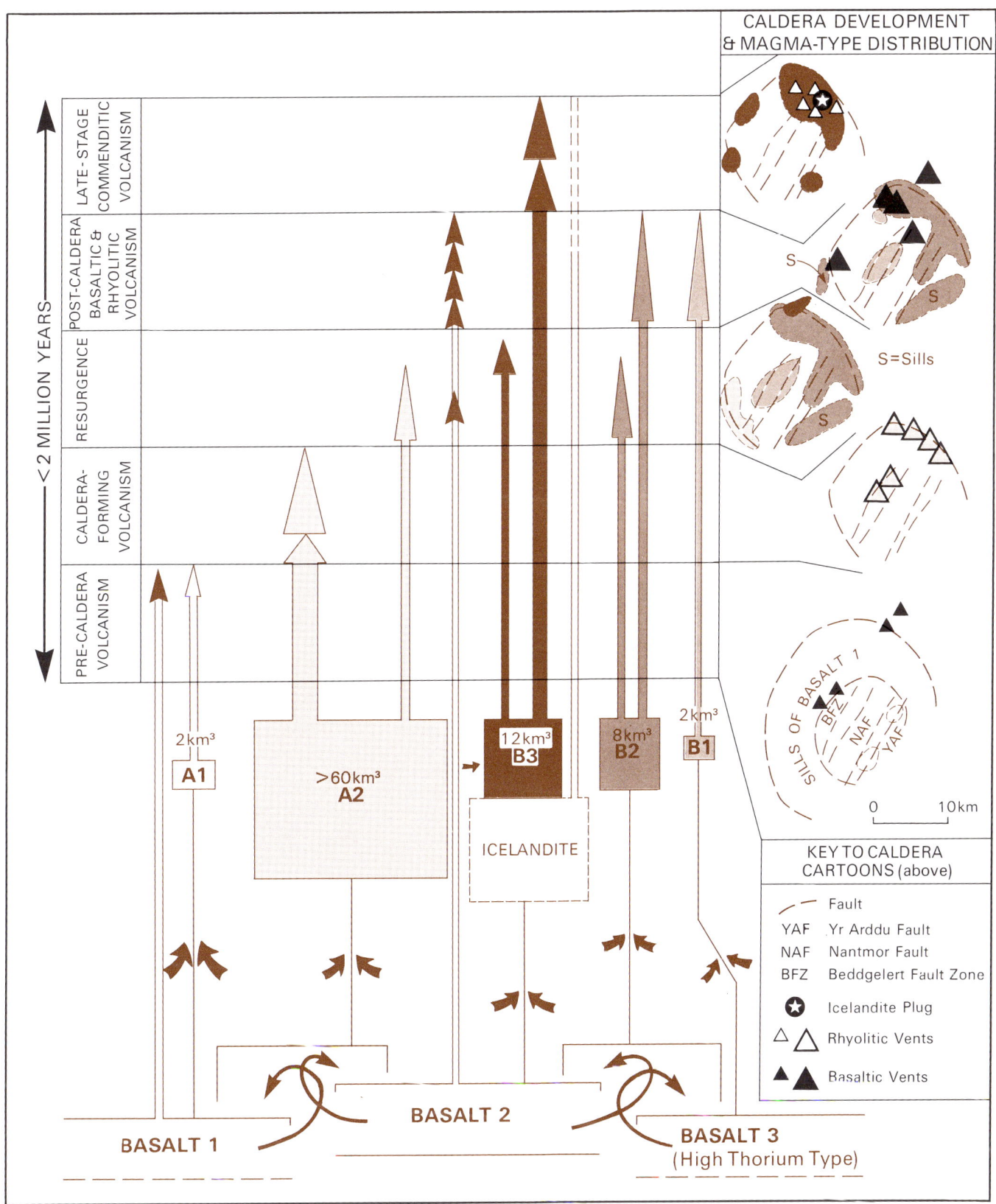

Figure 101 Snowdon Centre rhyolites, a model for their derivation from basaltic parents, their stratigraphical and areal distribution and the structural controls of emplacement within the Lower Rhyolitic Tuff Formation caldera.

Arrowheads in the upper part of the diagram represent the approximate time of emplacement of exogenous rhyolite domes, ash-flow tuffs (broad conduits from the A_2 and B_3 rhyolite reservoirs) and basalts. Arrows around the rhyolite reservoirs represent the relative degree of crustal contamination. Estimates of the volumes of the various rhyolite magma types erupted are given in the magma reservoirs.

Zone. However, by the time of its eruption, the magma chamber must have extended laterally to underlie the entire area of the subsequent Lower Rhyolitic Tuff Formation caldera. Furthermore, it must have been almost entirely evacuated of magma during these eruptions because subsequent rhyolitic magmas display a distinctly different basaltic parentage and both they, and the parent basalt, were emplaced across extensive areas of the caldera which was previously underlain by the A2 magma chamber. This chamber is envisaged as being disc-like (cf. Gudmundsson, 1988), about 600 m thick in the north and thinning to the south. The locus of eruption (Figure 101) and the site of maximum caldera subsidence was above the thickest part of the magma chamber, in the central and northern sectors of the caldera, and resulted in the asymmetry of the caldera as expressed in the thickness variations of the intracaldera tuffs (Figure 51). The eruptive pulses distinguished in the physical and chemical stratigraphy of the intracaldera tuffs indicate that some zoning of the A2 magma chamber, probably by element diffusion, was established prior to, and during pauses in its evacuation. A2 rhyolite intrusions are few and are limited to the south-east margin of the caldera. They were emplaced soon after the cessation of the caldera-forming eruptions and their small volume and restricted distribution supports the almost complete evacuation of the A2 rhyolitic magma chamber during caldera formation. The A2 rhyolites are the last manifestation of magma whose parentage can be traced, at least in part, to the sub-LRTF basalt type.

The emplacement of B1, B2 and B3 rhyolitic magmas was partly coeval (Figure 68), though the main eruptive phase of the B3 comenditic rhyolite was the final magmatic event of the Snowdon Centre. The marginal dolerite facies of the multiple intrusions in the south-east sector of the caldera indicates that some Bedded Pyroclastic Formation basaltic magma was emplaced prior to their B2a (transitional between B1 and B2) rhyolitic cores. B1 and B2 rhyolites occur in relatively small-scale intrusions and extrusive domes. B1 rhyolites are distributed along the Beddgelert Fault Zone, whereas B2 rhyolites occur at the north and north-east margin of the caldera and in the vicinity of the Nantmor Fault and Yr Arddu Fracture (Figure 101). Their partial contemporaneity but separate distribution, controlled by caldera-bounding and pre-existing fractures, suggests that they were derived

from relatively small, high-level magma chambers beneath different sectors of the caldera. Further evidence for the limited spatial extent of these rhyolitic magmas beneath the caldera is seen in the widespread coeval emplacement of Bedded Pyroclastic Formation basaltic magma (Figure 101) lacking any indication of significant mixing with the rhyolite magmas except, possibly, with B1 rhyolite.

The B3 rhyolitic magma was derived from the Bedded Pyroclastic Formation-type basalt via icelandite (Figure 101). A single plug of the latter intrudes the northern sector of the caldera but constraints on the timing of its emplacement are poor, though it certainly postdated the early Bedded Pyroclastic Formation volcanism. Analyses of the icelandite intrusion indicate a higher degree of crustal contamination than the B3 rhyolites, and suggest that it represents a relatively late intrusion that had a longer residence time in the crust. B3 rhyolites were largely emplaced along two arcuate fractures in the northern sector of the caldera (Figures 68 and 101). A relatively large volume ($> 10\,km^3$) of magma was erupted, partly as ash-flow tuffs, although restriction of outcrop as a result of recent erosion renders volume estimates imprecise. The arcuate fractures suggest localised nested-caldera formation above the main B3 magma chamber (Campbell et al., 1987).

In summary, the pattern of magma evolution at the Snowdon Centre was one of crystal fractionation from mixed basaltic parentage. The parental basaltic magmas resided at subcrustal levels and periodically ascended, together with their rhyolitic derivatives, to upper crustal levels through a system of deep crustal fractures. The denser intermediate derivatives, with the exception of a single intrusion of icelandite, did not reach upper crustal levels. The rhyolitic magma batches, derived from basaltic magams with relatively minor amounts of crustal contamination, occupied small and ephemeral high-level magma chambers beneath the Snowdon Centre. The largest rhyolite magma chamber (A2) was almost totally evacuated during caldera collapse. The model does not envisage either the existence of a single large basic to acidic zoned magma chamber beneath the Snowdon Centre or that wholesale crustal melting was the dominant factor in the production of the rhyolite magmas (cf. Kokelaar et al., 1984).

CHAPTER 5

Sedimentation and volcanism post-2nd Eruptive Cycle (post-Snowdon Volcanic Group)

The strata overlying the deposits of the 2nd Eruptive Cycle in central and south-western Snowdonia have been removed by erosion. However, a remnant of the strata is preserved overlying deposits of the Snowdon Centre in the core of the Dolwyddelan syncline in east Snowdonia. The lower part of the sequence is entirely of black graptolitic mudstone and in north-eastern Snowdonia there is a restricted development of sulphide deposition at the base.

Late-stage, exhalative ore deposition related to the Crafnant Centre

The Cae Coch massive pyrite deposit (Ball and Bland, 1985) locally overlies basaltic volcanic deposits (Dolgarrog Volcanic Formation) in north-eastern Snowdonia (Figure 102). Both dolerite and rhyolite intrusions occur in the vicinity. The deposit, up to 2 m thick, is stratabound and stratified (Figure 103). At its base, fragmental aggregates of finely crystallised pyrite, co-precipitated with silica, are common and have been interpreted as representing debris flows, remobilised from previously deposited ore and transported into shallow sea-floor depressions (Ball and Bland, 1985). The laminated, pyrite-quartz ore, forming the bulk of the deposit, suggests that the debris flows settled in a brine pool (Lydon, 1984). Intercalated mudstone bands have loaded bases and the subjacent banded pyrite ore shows evidence of soft-sediment deformation. The low energy, euxinic environment allowed the ore deposit to remain chemically and physically stable and to be slowly buried by pyrite-rich mud.

The main ore body thins to the north and south. At outcrop it is thin (<15 cm) and deeply weathered to an oxidic clay. The footwall basic rocks contain veins and patches of pyrite and pyrrhotite. In contrast to many stratiform massive sulphide ore deposits, there is no evidence of baryte or gypsum concentrations in the footwall although thin lenses of calcite concentrations do occur.

Tubes of pyrite, 5–10 mm in diameter, internal diameter 3.5 mm, infilled with radiating quartz, occur about 0.5 m below the top of the ore bed. Similar structures have been interpreted (Larter et al., 1981; Russell et al., 1981) as fossilised 'black smokers' and elsewhere it has been proposed (Boyce et al., 1983; Russell et al., 1984; Banks, 1985) that such features represent the 'chemical garden' growth of gel tubes and stalactites at the interface between two concentrated solutions of different compositions.

The upper contact of the ore body is locally irregular and draped by laminated mudstone. Pyrite tubes, 1–2 cm in diameter, which project into the overlying mudstone, were referred to as 'zebra ore' by the miners and were considered by Sherlock (1919) to be replaced worm casts.

The chemistry of the deposits reflects the simple mineralogy of pyrite and quartz with some carbonate. In the weath-

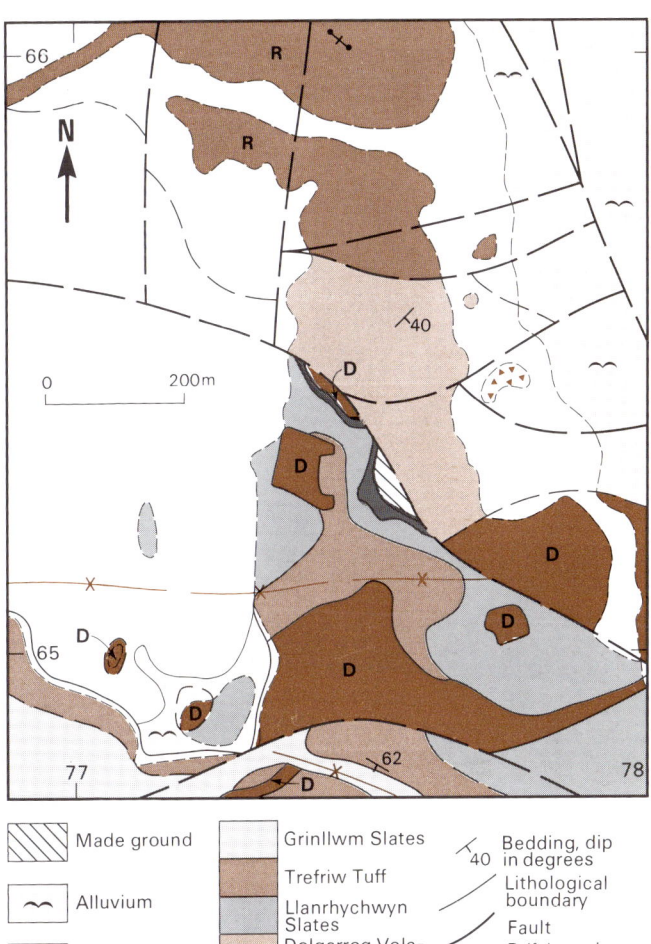

Figure 102 Map showing the distribution and context of the Cae Coch massive sulphide deposit in NE Snowdonia.

ered zone there is also an abundance of gypsum. The deposit shows zonal variation in traces of Ba, Cu, Zn, Ni and Co and contains detectable concentrations of Sn, Tl, Hg and Ag. On the basis of the ratio of Co/Ni, pyrite samples lie within the sedimentary field as defined by Bralia et al. (1979).

The footwall rocks have been highly altered. The basic rocks show relative increases in K, Ba, S and carbonate, and losses of Ca, V and Co, and probably LREE. A progressive gain in K, Ba and Pb and a depletion of Ga, Sr, Sn, Zn and REE are observed with increasing alteration of the rhyolite (see also Chapter 6).

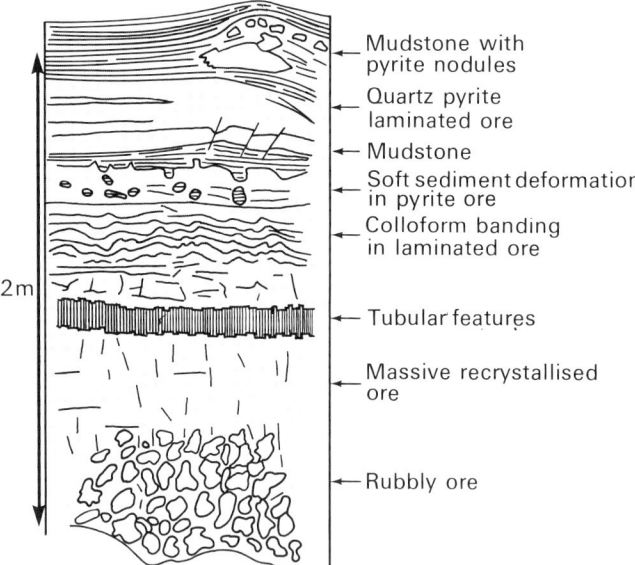

Figure 103 Section [SH 7752 6531] of the Cae Coch massive sulphide deposit (after Ball and Bland, 1985).

The Cae Coch ore body shows many features which are typical of the Kuroko type of deposit (Matsukuma and Horikoshi, 1970; Franklin et al., 1981; Ohmoto and Skinner, 1983; Urabe et al., 1983). Similar massive sulphide deposits, associated with bimodal, basalt/rhyolite volcanism, occur elsewhere in the Ordovician of the British Isles, e.g. Parys Mountain, Anglesey, and Avoca, SE Ireland (Leat et al., 1986; Jones, 1983). Here they are also interpreted to have been formed by hydrothermal exhalation on to the sea floor (Platt, 1977; Badham, 1978). The ore fluids were probably K- and carbonate-rich brines and it is possible that the source of the K lay in the extensive acidic ash-flow tuffs (Crafnant Volcanic Group) which underlie the deposit.

Black mud sedimentation

The euxinic environment, which facilitated local sulphide precipitation, is reflected in the widespread accumulation of black mudstone above the volcanic deposits of the 2nd Eruptive Cycle in eastern and north-eastern Snowdonia. In eastern Snowdonia, black 'sooty' graptolitic mudstones (Black Slates of Dolwyddelan) (Williams and Bulman, 1931; Howells et al., 1978) overlie the volcanic deposits of the Snowdon Centre in the core of the Dolwyddelan syncline. Pyrite in the mudstones is developed principally along bedding and cleavage planes. The phyllosilicate mineralogy of these mudstones is unusual in containing no chlorite (Merriman and Roberts, 1985). A *D. clingani* zone age is indicated by the contained graptolites.

In north and east Snowdonia, similar black graptolitic mudstones (the Llanrhychwyn Slates, Cadnant Shales; Howells et al., 1978, 1981) overlie the deposits of the Crafnant Centre. Black mudstone deposition at this time was widespread throughout North Wales (Cave, 1965), and in north-eastern Snowdonia the mudstones form the thickest known sequence, up to 450 m, in spite of the close association

with the two main centres of the 2nd Eruptive Cycle, the volcanic sequences show no indication of shallow-marine reworking. This evidence indicates extensive postvolcanic subsidence of the north-east–south-west-trending Snowdon graben which contained the Caradoc volcanic centres.

Local reactivation or uplift of a basaltic centre

The accumulation of black mudstone in the Snowdon graben was interrupted, in the vicinity of Trefriw, north-eastern Snowdonia, by the emplacement of basaltic tuff (Trefriw Tuff), up to 70 m thick, (Howells et al., 1981, 1985b). The distribution of the tuff was restricted and to the south it passes laterally into a thin sequence, about 15 m, of dark grey mudstone with impersistent sandstone intercalations.

The basaltic tuffs occur in massive and cross-laminated flaggy beds and locally, with increasing epiclastic debris, grade into tuffites with siltstone intercalations. Some beds are crowded with rounded clasts of altered vesicular basalt in a matrix of comminuted chloritised basaltic glass and albitised feldspar. The lithologies and bedforms of the tuffs indicate reworking and remobilisation of basic volcanics. The restriction of the tuffs to the vicinity of the earlier Dolgarrog Volcanic Formation centre suggests that this was the source, either from a brief recurrence of volcanic activity or from local uplift and erosion of the earlier deposits. However, the activity signalled a marked sedimentological change throughout the district.

Post-volcanic sedimentation

Subsequent to the emplacement of the basaltic tuffs, grey silty mudstones with irregular thin flaggy beds of rippled, cross-laminated sandstones accumulated and comprise the Conwy Mudstones (British Geological Survey, 1985a; Howells et al., 1985b). The Caradoc–Ashgill boundary has been determined close to the base of this sequence. In places, thicker beds of coarse, pebbly turbiditic sandstones occur and bioturbation is common. The general pattern of sedimentation and the lithologies bear a closer resemblance to the overlying Silurian than to the underlying Ordovician.

South-east of Betws y Coed there is a pronounced, though not markedly angular unconformity separating the Caradoc and Ashgill. The macrofaunas indicate the absence of late Caradoc (Marshbrookian–Onnian, Figure 9) and early Ashgill (Pusgillian–Cautleyan) strata over most of the area on the north-west side of the Bala fault. In addition, Longvillian strata are rarely represented in this area and locally, the uppermost Soudleyan is also absent. A more complete Caradoc and Ashgill sequence is again present south and south-east of the Harlech Dome; here the late Caradoc strata comprise black, graptolitic mudstones, similar to, but markedly thinner than those of north-eastern Snowdonia.

The nature and extent of the Caradoc–Ashgill unconformity is considered to reflect intra-Ashgill fault-block reactivation (Bassett et al., 1966; Campbell, 1983, 1984; Fitches and Campbell, 1987) and is a manifestation of an event of regional significance.

The Ordovician–Silurian boundary has been distinguished to the east of Snowdonia (Warren et al., 1985) within a conformable silty mudstone sequence.

CHAPTER 6

Post-emplacement modification of the volcanic sequence

The interpretation of ancient volcanic sequences is constrained by post-emplacement modification of the primary characters of the rocks. Their original composition may be altered by devitrification of primary glass, hydrothermal alteration and regional metamorphism. These processes also modify, or completely obliterate original textures. In addition, folding and associated cleavage formation may produce new textures, modify depositional thickness variations and affect the pattern of outcrop distribution. Here, the significance of these processes in the Caradoc sequence of North Wales is briefly assessed. Epigenetic, volcanogenic mineralisation is also discussed because it is intimately associated with hydrothermal alteration.

The alteration processes outlined above can be divided into two categories:

1 Volcanogenic alteration, including devitrification and the various episodes of hydrothermal alteration which were broadly contemporaneous with magmatism.

2 Regional metamorphism and deformation, which, in North Wales, extended to early Devonian times, some fifty million years after the cessation of volcanism.

Volcanogenic alteration

DEVITRIFICATION

Devitrification textures are clearly preserved in many of the silicic volcanic rocks of Snowdonia in spite of subsequent low-grade regional metamorphism. The original glassy character of most rhyolites of both major episodes of volcanism is indicated by the common preservation of perlitic fabrics (Plate 26B) though all are now devitrified. Spherulitic and snowflake devitrification fabrics (Plate 32A and D) are common and spherulites may vary in size and abundance in adjacent flow-bands (Plate 32C), most likely reflecting the variable water content of the original glass.

Devitrification of the welded tuffs was locally preceded or accompanied by vapour phase crystallisation, which either accentuated or, locally, obscured the original welding fabric. For example, the welding foliation in both the lower and upper flow units of the Pitts Head Tuff Formation is accentuated, on a mesoscopic scale, by the redistribution and crystallisation of silica into quartz-rich lenticles, parallel to the foliation (Plate 24A). In places, however, this crystallisation completely obliterates the original texture, which is preserved in less-altered parts of the flow. A more widespread destruction of original pyroclastic textures is seen in the locally thick, irregular lenses of silicified welded tuff near the base of the intracaldera facies of the Lower Rhyolitic Tuff Formation in the vicinity of the Beddgelert Fault Zone. In these rocks, thin (0.5–2.0 mm) and even quartz segregations, locally flow-folded, are normally the only visible texture.

These features, together with the apparent discordant relationship of the lenses with the enveloping cleaved tuffs, led to them being interpreted as rhyolite intrusions. Very rarely, however, thin sections reveal silicification fronts where the banding passes abruptly, but concordantly, into a eutaxitic welding fabric with clearly discernible flattened shards (Plate 17E). Here, devitrification and redistribution of silica was facilitated by vapour streaming at the base of the thick sequence of intracaldera tuffs in the vicinity of a postulated fissure vent.

Spherulitic and subspherulitic devitrification textures are also common in many of the welded tuffs (Plate 17F). Some of the large siliceous nodules in the subaerial facies of, for example the lower flow unit of the Pitts Head Tuff, may also have originated as spherulites. Perlitic fractures are preserved in some of the devitrified welded tuffs and, where devitrification and subsequent metamorphic recrystallisation have obliterated the original fabrics, they confirm the original welded character of the tuffs.

Studies of more recent volcanic rocks indicate that low temperature hydration and devitrification of rhyolite glass may result in changes in bulk rock chemistry, with significant variations in SiO_2, H_2O, the Fe_2O_3/FeO ratio, Na_2O and K_2O (Lipman, 1965; Noble, 1967; Lofgren, 1970). Enrichment in K_2O and depletion in Na_2O are commonly recorded. The chemical effects of vapour phase crystallisation and devitrification in the Caradoc volcanic rocks of Snowdonia cannot be distinguished from the effects of hydrothermal alteration and low-grade regional metamorphism (see below). However, the significance of the sum of these processes is illustrated by plots of $Na_2O + K_2O$ against $K_2O \times 100/Na_2O + K_2O$ (Figure 104) for rhyolites and welded ash-flow tuffs from both the Llewelyn and Snowdon Volcanic groups. These indicate that the majority of the samples analysed lie outside the normal igneous spectrum (Hughes, 1972) and are enriched in K_2O.

HYDROTHERMAL ALTERATION AND RELATED METALLIC MINERALISATION

Hydrothermal alteration includes all mineralogical, chemical and textural changes in rocks resulting from interaction with hot water of varying chemistry. So defined, the process may grade into regional, low-grade metamorphism. Here, however, we refer specifically to either the local effects of sea-water reacting with hot magma or the results of local geothermal activity related to specific volcanic centres.

It is reasonable to assume that magma–sea-water interaction, affecting lava and hot pyroclastic flows emplaced sub-aqueously and magma intruded into incompletely dewatered sediments, was common within the marine-dominated environments of Snowdonia. However, rigorous investigation is necessary to distinguish its effects from those of subsequent

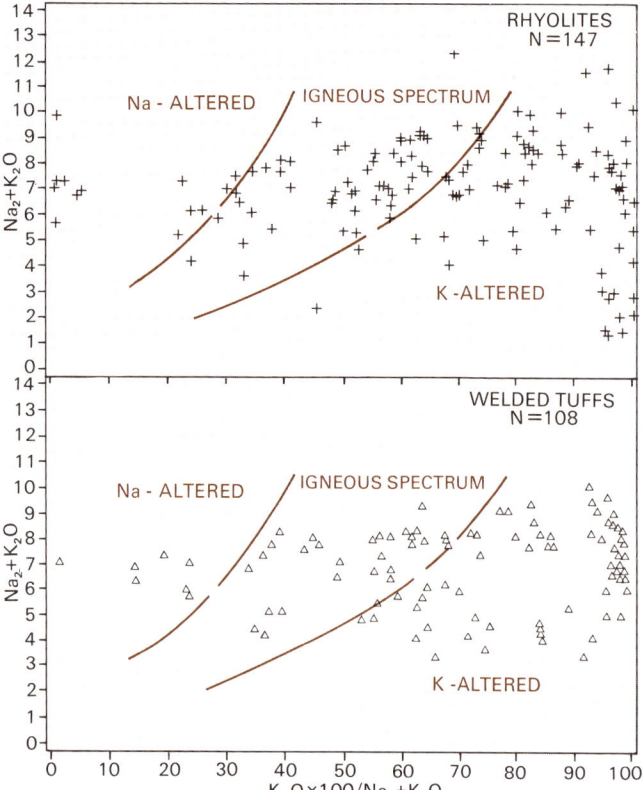

Figure 104   Rhyolites and welded ash-flow tuffs, Llewelyn and Snowdon Volcanic groups, $Na_2O + K_2O$ vs. $K_2O \times 100 / Na_2O + K_2O$ diagram.

The field of igneous spectrum (defined by Hughes, 1972) reflects the range of compositions attributable to primary magmatic variation. The areas outwith the field reflect compositions attributable to alteration processes.

events. In the Tal y Fan dolerite sill, for example, Bevins and Merriman (1988) considered the development of K-feldspar in the chilled contact and marginal zones to result from early alteration, K-feldspar having been reported from basic rocks altered by low-temperature reaction with sea-water in various ocean basins (Alt and Honnorez, 1984; Böhlke et al., 1984). Bevins and Merriman (1988) have also suggested that olivine was largely converted to saponite by hydration during this early hydrothermal phase and, with the highest fH_2O confined to the outer 2–3 m of the intrusion, hydrothermal alteration of other minerals would be expected but cannot be distinguished with certainty from later metamorphic assemblages.

Hydrothermal activity was probably commonplace in the vicinity of the major volcanic centres in Snowdonia. Alteration of the bulk chemistry of the volcanic rocks due to such activity has been investigated at the northernmost Conwy Rhyolite Formation centre, the Crafnant Centre and at the Lower Rhyolitic Tuff Formation caldera, where hydrothermal Cu/Pb/Zn mineralisation was also developed.

Conwy Rhyolite Centre

Widespread hydrothermal alteration of the rhyolites around Conwy Mountain, in the vicinity of the Conwy Rhyolite

centre (Figure 11), involved locally intense sericitisation with white mica completely replacing both K- and Na-feldspars. The presence of tridymite paramorphs and equant or doubly terminated β-quartz paramorphs, after α-quartz, confined within thin veins, and patches of white mica, supports a hydrothermal origin for the alteration (Ball and Merriman, 1989). These authors concluded that the hydrothermal process resulted in the loss of substantial quantities of Na, Ca and Fe, along with minor amounts of associated trace elements such as Sr and Co, and a slight, but significant increase in the LREE (Appendix 1, Table 3). Concentrations of Zr, Nb and the HREE were unaffected. In spite of this alteration, however, most of the samples analysed fall within the Igneous Spectrum (Hughes, 1972) on a plot of $Na_2O + K_2O$ vs. $K_2O \times 100 / Na_2O + K_2O$, in contrast to, for example, the low-Zr (<700 ppm Zr) rhyolites associated with the Lower Rhyolitic Tuff Formation caldera, the majority of which fall outside that range (Figure 105).

Crafnant Centre

In the vicinity of the Crafnant Centre, hydrothermal alteration has been studied in the basaltic tuffs below the Cae Coch massive sulphide deposit (Ball and Bland, 1985), and in a nearby rhyolite intrusion. The altered basaltic tuffs show significant enrichment in K_2O, Na_2O, Ba, Sr, and Rb, and depletion in MnO, CaO, Ce, Co, Ni and Pb. The Na enrichment contrasts with the depletion recorded in the basic rocks in the Lower Rhyolitic Tuff Formation caldera (see below and Appendix 1, Table 7). Light or middle REE are depleted only in rocks immediately underlying the ore deposit.

In the outer part of the rhyolite intrusion, microcline concentration indicates a high temperature hydrothermal addi-

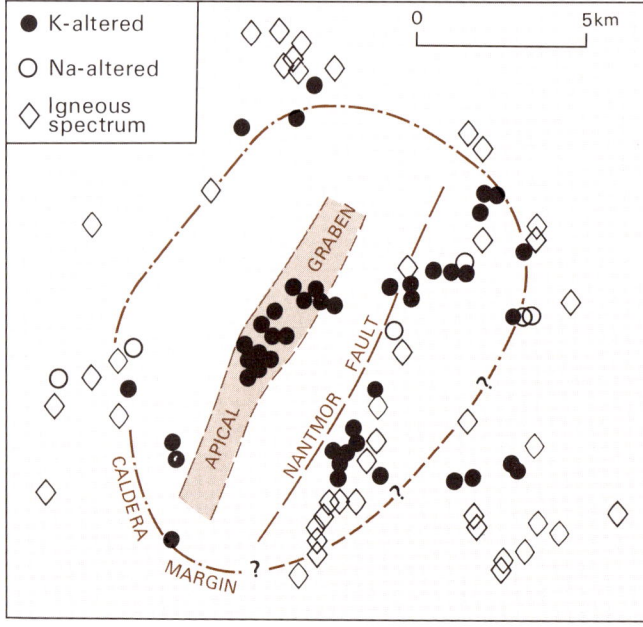

Figure 105   Sketch map showing the relationship of K-altered, Na-altered and unaltered rhyolites to the Lower Rhyolitic Tuff Formation caldera. Rhyolite types defined from the $Na_2O + K_2O$ vs. $K_2O \times 100 / Na_2 + K_2O$ plot of Figure 104.

tion of K_2O, accompanied by depletion in Na_2O, CaO, Fe_2O_3, and MgO, as compared to samples from the interior (Ball and Bland, 1985). Potassium-rich samples from the outer part of the intrusion contain much lower values of total REE.

Snowdon Centre: Lower Rhyolitic Tuff Formation caldera

Extensive hydrothermal alteration of the volcanic rocks within the Lower Rhyolitic Tuff Formation caldera is recognised from a comparison of their geochemistry with that of the same strata outside the caldera. For example, most low-Zr rhyolites (< 700 ppm Zr), which are defined as K-altered by reference to the Igneous Spectrum diagram of Hughes (1972), plot within or close to the margin of the Lower Rhyolitic Tuff Formation caldera (Figure 105), whilst most of those within the normal Igneous Spectrum lie outside the caldera. Typically, enrichment in Ba, K_2O and Rb and depletion in CaO, Na_2O and sometimes Sr has occurred within the caldera in most of the volcanic lithologies. In addition, the intracaldera tuffs and the low-Zr rhyolites, the latter related to resurgence, are relatively enriched in SiO_2. Slight depletion in TiO_2, Nb and Zr in the intracaldera tuffs of the Lower Rhyolitic Tuff Formation, not recorded in other rocks of similar composition within the caldera, may represent depositional processes rather than post-emplacement alteration.

Cu, Pb and Zn sulphides occur in thin veins, generally greater than 20 cm in width, and disseminations intimately associated with hydrothermal alteration within the caldera (Reedman et al., 1985). These deposits were mined on a small scale during the last century (Bick, 1982). In only a few cases, as at Brittania Mine (Colman and Laffoley, 1986), have individual veins been worked for more than 200 m along strike.

The mineralisation is mainly distributed along the apical graben and at the northern and southern margins of the caldera (Figure 106). Within the graben the mineralised veins, trending N40°E to N60°E, are normally very impersistent and are concentrated in groups. More persistent veins, up to several hundred metres long, with trends varying from N40°W to N80°W, occur at Brittania and Lliwedd mines in the north-west and at Moel Hebog and Cwm Bychan in the south (Figure 106). Most are fracture-fill veins displaying a variety of ore textures (Plate 47) and many contain clasts of wallrock. The fracture systems probably developed by hydraulic fracturing in areas of intense hydrothermal activity.

Five characteristic ore-mineral assemblages have been recognised (Reedman et al., 1985).

1 Quartz + pyrite + chalcopyrite + sphalerite
2 Quartz + pyrite ± pyrrhotite + chalcopyrite + sphalerite ± galena
3 Sphalerite + galena ± chalcopyrite ± pyrite
4 Calcite + marcasite ± haematite ± chalcopyrite ± sphalerite ± galena
5 Quartz + magnetite + haematite + pyrite

The mineral paragenesis is indicated in Figure 107. Assemblage 2, containing pyrrhotite, represents the highest temperature sulphide assemblage and is confined to the apical graben where the most extensive hydrothermal alteration also occurs. Assemblage 5 is not found in association with the other assemblages; it is restricted to a fault zone and

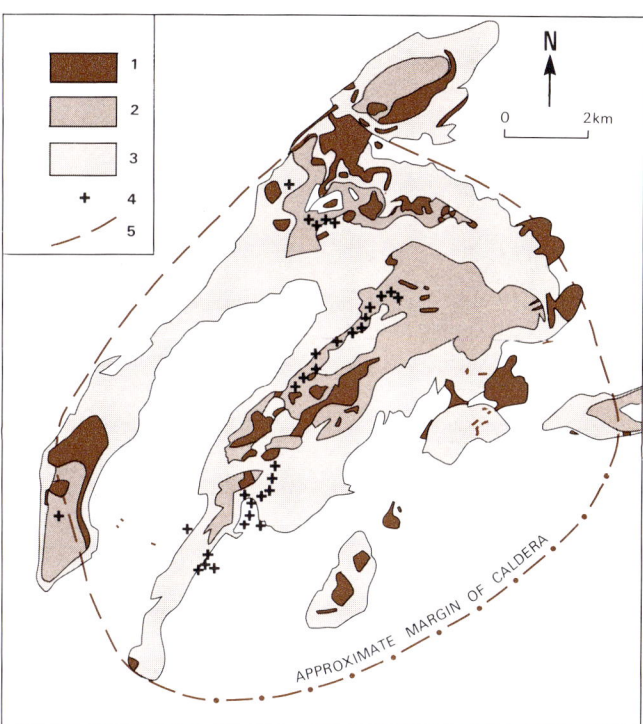

Figure 106 Sketch map showing relationship of mineralisation to the Lower Rhyolitic Tuff Formation caldera (modified after Reedman et al., 1985).

1: rhyolite; 2: Bedded Pyroclastic Formation; 3: Lower Rhyolitic Tuff Formation; 4: disused mineral working; 5: caldera margin.

breccia dyke cutting the base of the Lower Rhyolitic Tuff Formation in Cwm Tregalan.

The mineralised veins occur predominantly in the basic rocks of the Bedded Pyroclastic Formation and lie close to the Lower Rhyolitic Tuff Formation/Bedded Pyroclastic Formation boundary, estimated to have been at a depth of 1–2 km at the time of mineralisation (Reedman et al., 1985). This distribution reflects the influence of wallrock chemistry on the deposition of sulphides from the mineralising fluids and, possibly, the depth at which optimum P/T conditions for ore deposition existed. The relationship between the distribution of ore-minerals and wallrock composition is most clearly demonstrated by galena, which occurs in quantity only in assemblages 2 and 3, and exclusively in veins cutting basic rocks of the Bedded Pyroclastic Formation. Numerous barren quartz veins also occur within both the Lower Rhyolitic Tuff Formation and the Bedded Pyroclastic Formation in the caldera and are considered to have formed by hydraulic fracturing approximately contemporaneous with the mineralised veins (Fitches, 1987).

Wallrock chemistry has been investigated for samples taken at various distances from mineralised veins and compared with that of similar lithologies from nonmineralised localities (Colman, in press). Basic tuffs and basalts of the Bedded Pyroclastic Formation display a general increase in SiO_2, Fe_2O_3 and MnO content and a decrease in Al_2O_3, CaO and Na_2O content in the vicinity of mineralisation. K_2O content

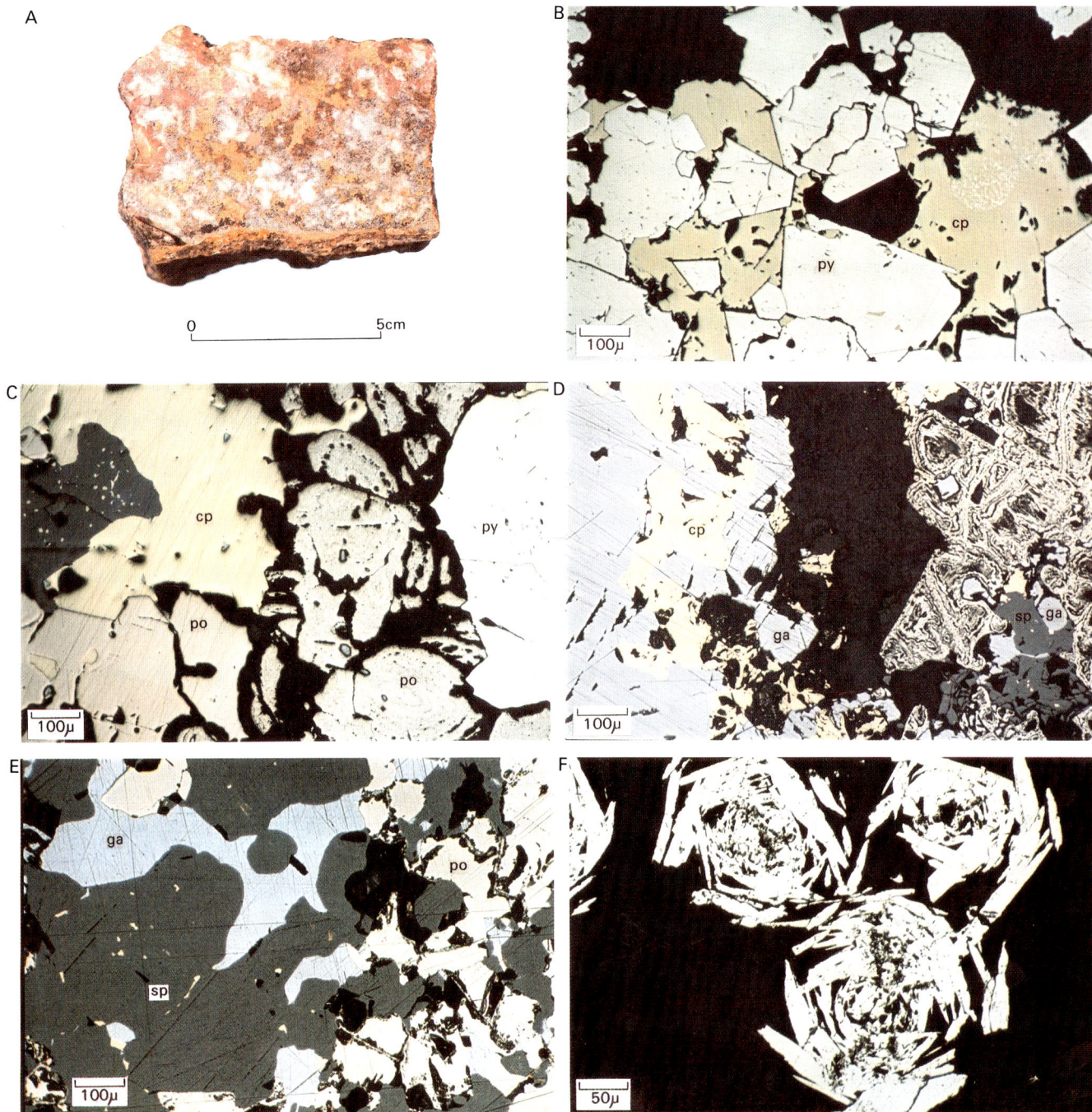

Plate 47 Sulphide mineralisation associated with hydrothermal alteration at the Snowdon Centre

A Sample of quartz-pyrite-chalcopyrite-sphalerite vein Cwm Llan [SH 622 520].
Photomicrographs:
B Assemblage 1; quartz + pyrite (py) + chalcopyrite (cp) + sphalerite from Cwm Bychan [SH 603 475].
C Assemblage 2; quartz + pyrite (py) ± pyrrhotite (po) + chalcopyrite (cp) + sphalerite ± galena from Cwm Llan [SH 622 520].
D Assemblage 2; colloform pyrite (py) in contact with wallrock fragments. Galena (ga), chalcopyrite (cp) and sphalerite (sp) also present. Cwm Llan [SH 622 520].
E Assemblage 2; quartz + pyrite ± pyrrhotite (po) + chalcopyrite + sphalerite (sp) ± galena (ga). Sphalerite contains chalcopyrite inclusions. Lliwedd [SH 633 533].
F Assemblage 4; spheroidal marcasite from the calcite + marcasite ± hematite ± chalcopyrite ± sphalerite ± galena assemblage, Britannia Mine [SH 613 548].

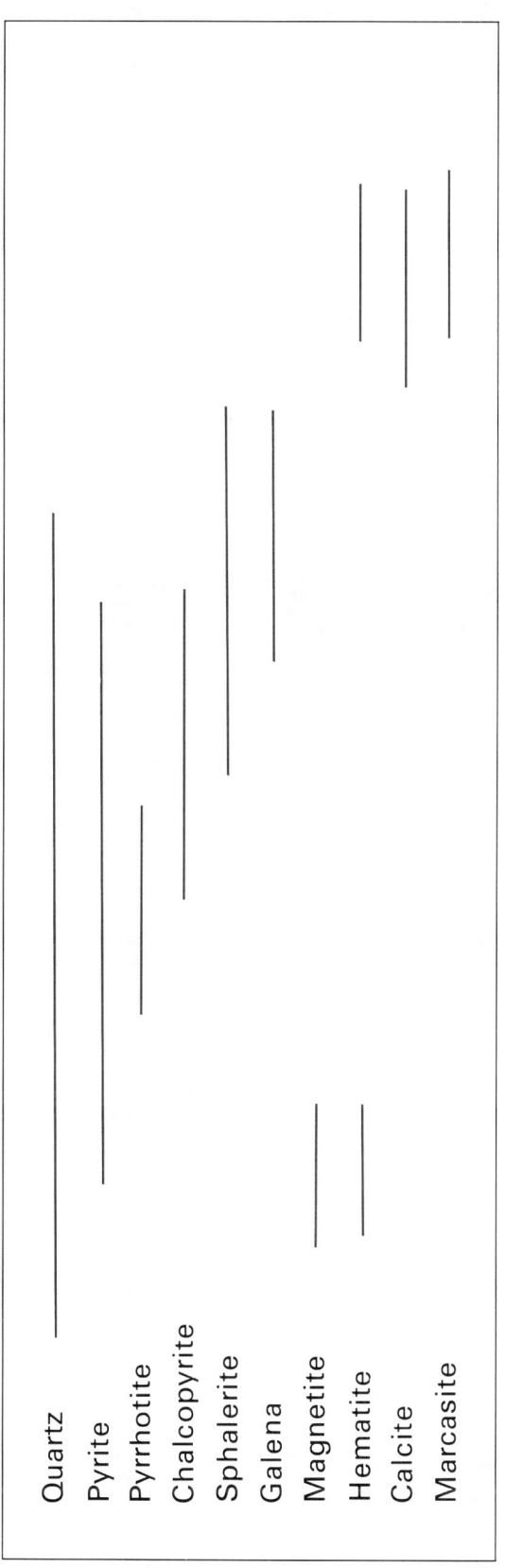

Figure 107   Mineral parageneses of vein assemblages within the Lower Rhyolitic Tuff caldera (after Reedman et al., 1985).

is erratic with a tendency to general enrichment. Similar trends are apparent in the acid tuff wallrocks of the Lower Rhyolitic Tuff Formation, though iron enrichment is much less marked. These major element variations broadly reflect those found around other volcanogenic vein-sulphide deposits in a variety of palaeogeothermal fields (e.g. Peterson and Lambert, 1979; Thurlow et al., 1975; Urabe et al., 1983). Amongst the trace elements, As, Cu, Pb, Rb, S, W and Zn are enriched in the basic (Bedded Pyroclastic Formation) wallrocks while Sr is strongly depleted. Similar trends are displayed in the acid (Lower Rhyolitic Tuff Formation) wallrocks except that Cu, Pb and Zn are depleted. Wallrocks from the Cwm Llan deposit demonstrate (Figure 108) some of the trends noted above, most notably the strong depletion in $Na_2O + CaO$. Enhanced levels of fluorine are also characteristic of many of the wallrocks, though fluorite is absent from the vein assemblages because of the generally low levels of calcium. The immobile trace elements such as Ti, Nb, Zr and the REE have not been affected by wallrock alteration. Mineralogically, the most conspicuous effect of alteration has been the development of chlorite at the expense of feldspars, leading to the increase in Fe, Mg, and K and the decrease in Ca and Na.

Employing the Alteration Index (AI = (MgO + K_2O)/($Na_2O + K_2O + CaO + MgO$) of Ishikara et al. (1976), values of 45 per cent for the basic rocks of the Bedded Pyroclastic Formation in the nonmineralised areas rise to values of greater than 90 per cent in the mineralised areas. Similarly, in acid rocks (Lower Rhyolitic Tuff Formation), background values of about 50 per cent rise to 80–90 per cent near the vein deposits.

In the absence of fluid inclusion and stable isotope studies, the likely source, chemistry and temperature of the mineralising fluids remains conjectural. However, as a result of the caldera setting, it is reasonable to infer that most of the metal content was derived by the leaching of volcanic rocks at high levels in the crust by a convecting hydrothermal cell that incorporated meteoric (sea) water and was driven by a residual subcaldera magmatic heat source (cf. Reedman et al., 1985). A magmatic input of fluids and metals is also possible, as indicated by the presence of F anomalies in wallrocks. Depletion of Cu, Pb and Zn in the acid tuffs of the Lower Rhyolitic Tuff Formation within the caldera suggests leaching of the metals from the tuffs and subsequent deposition in veins within the superjacent basic rocks.

The precise age of the mineralisation has not been proved, though there is evidence to indicate that it postdates the emplacement of the peralkaline rhyolites of the Upper Rhyolitic Tuff Formation and certainly predates cleavage formation. The distribution of the deposits is closely linked to the caldera structure, but Lipman et al. (1976) have demonstrated that, for example, although many of the mid-Tertiary base- and precious-metal deposits of the San Juan Mountains, Colorado, are associated with structures of the Silverton Caldera, they were emplaced 5–15 million years after the caldera formed. In the vicinity of the Lower Rhyolitic Tuff Formation caldera, magmatic activity may have continued for a similar period of time after caldera collapse. The nearby Mynydd Mawr microgranite intrusion yields a Rb/Sr age of 438 ± 4 Ma (Evans, 1989a,b) as compared to the Longvillian age (about 450 Ma) of the Lower Rhyolitic Tuff

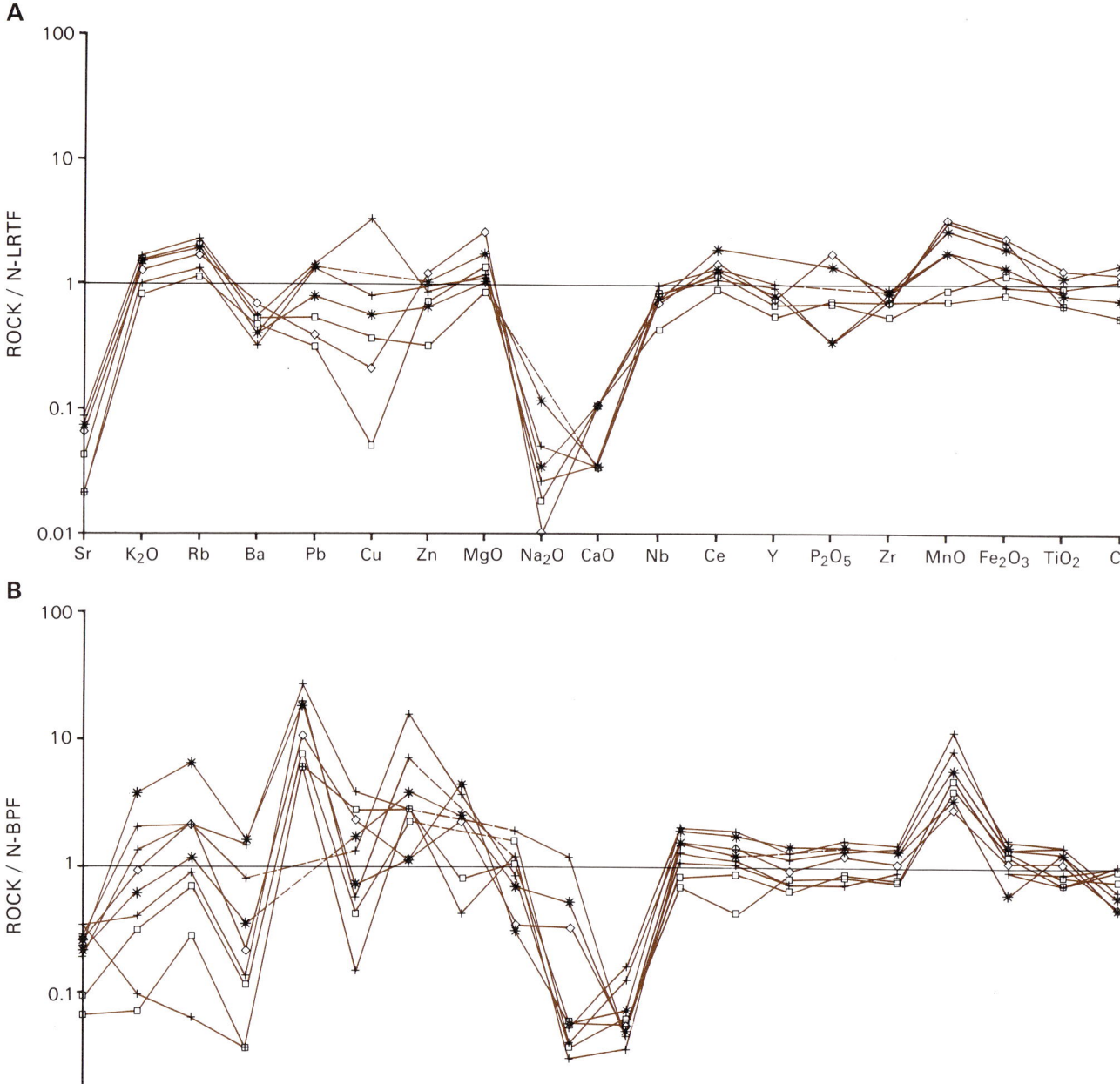

Figure 108   Wallrock alteration close to mineralised veins at Cwm Llan Mine [SH 6236 5226].

A. Element variation in wallrock samples of acid ash-flow tuffs of the Lower Rhyolitic Tuff Formation (LRTF)
normalised against average analyses of a suite of LRTF samples distant from mineralisation.
B. Element variation in wallrock samples of basalts and basaltic sandstones of the Bedded Pyroclastic
Formation (BPF) normalised against average analyses of a suite of BPF samples distant from mineralisation.

Formation caldera. Furthermore, small mineralised veins containing pyrite, chalcopyrite and fluorite in a gangue of quartz and calcite occur at the margins of the Mynydd Mawr intrusion.

Regional deformation and metamorphism

Rocks of the Lower Palaeozoic Welsh Basin have traditionally been assigned to the 'non-metamorphic' or 'paratectonic'

Caledonides (Dewey, 1969; Phillips et al., 1976). Nevertheless, they were subjected to extensive folding and low-grade regional metamorphism during the Acadian orogeny in early to mid-Devonian times. It follows, therefore, that the effects of thermal dewatering, deformation and metamorphic recrystallisation on element distributions, isotope systems and the geometry of individual units must be taken into account when interpreting the primary character of Caradoc volcanism.

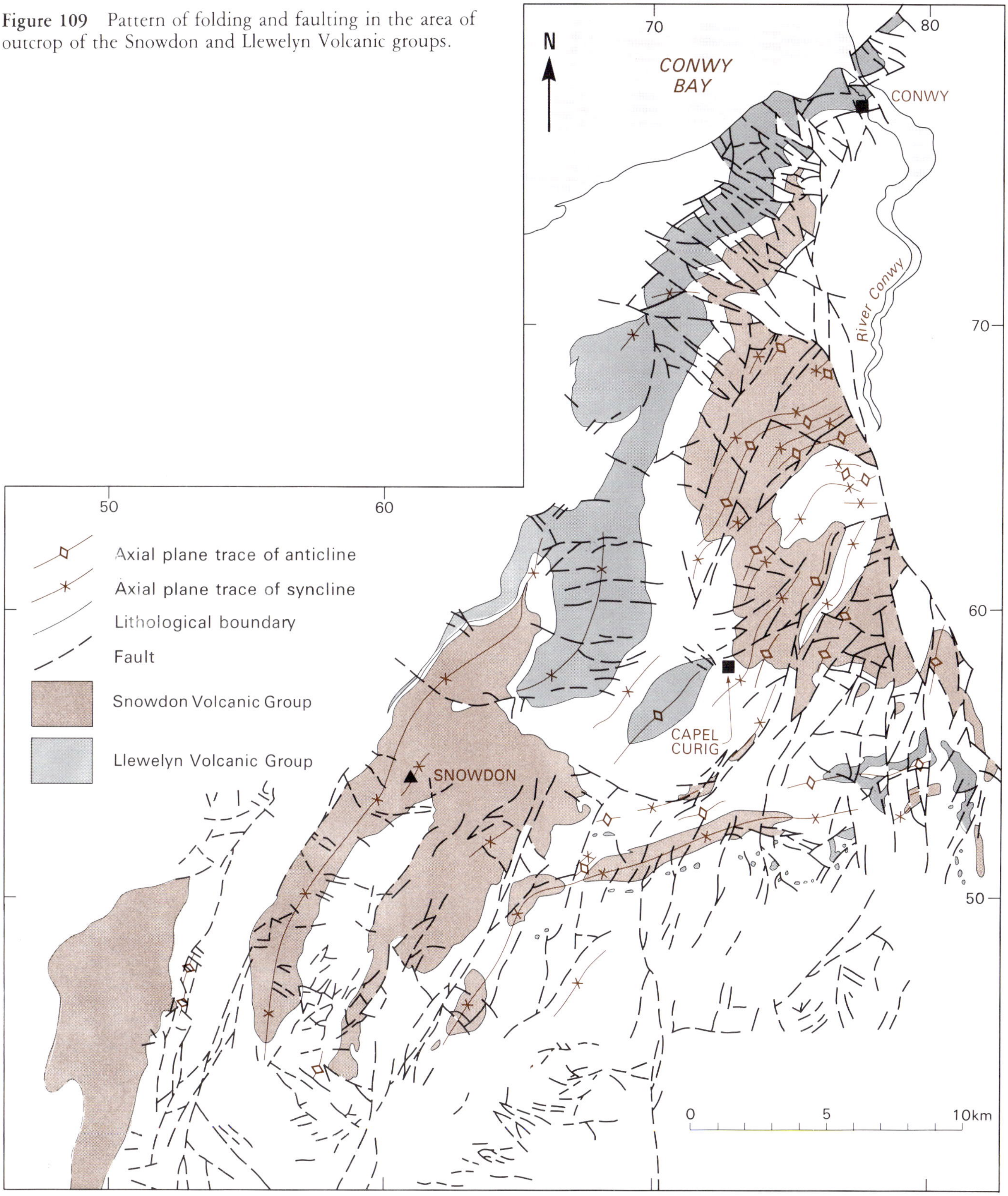

Figure 109   Pattern of folding and faulting in the area of outcrop of the Snowdon and Llewelyn Volcanic groups.

Deformation

A single period of regional deformation, of early to mid-Devonian (Acadian) age, resulted in the formation of major, south-east-verging periclinal folds and a predominantly axial-planar cleavage in North Wales (Coward and Siddans, 1979;

Campbell et al., 1985). An arcuate pattern of fold-axial traces and cleavage is developed across the region (Figure 109).

Medium-sized, upright to steeply inclined, periclinal folds with wavelengths of 2–3 km cause numerous repetitions of the upper volcanic horizons across the area, but the absence of such folds in the north-east has limited the outcrop of the

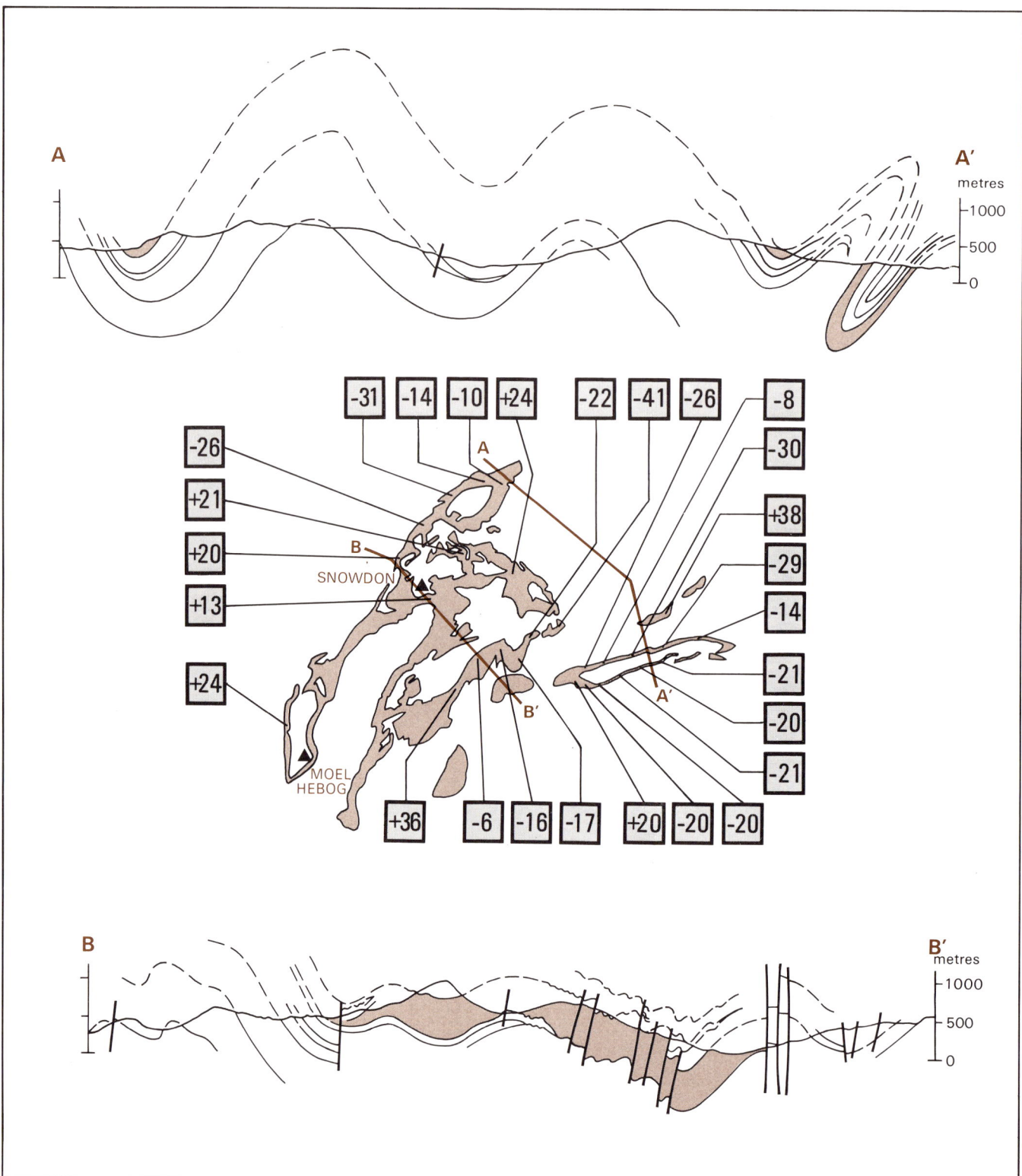

Figure 110 The effects of regional strain on original thicknesses of the Lower Rhyolitic Tuff Formation (brown) (after Wilkinson, 1988 and later modified, by him, with later unpublished data).

Sections (A-A[1], B-B[1]) show styles of folding; on the outcrop map, 27 localities are indicated where thickening (+) and thinning (−), expressed as a percentage of original thicknesses, have been determined.

lower volcanic horizons to a single strike section. Interlimb angles in the folds vary from 140° to 10°, tightness ranges from flattened concentric, with more or less constant limb thickness, to tight folds in which the hinges have been thickened and the limbs attenuated (Wilkinson, 1988; Figure 110). The axes of the major folds often coincide with zones of marked lateral thickness or facies changes within the folded sequences.

A penetrative cleavage is ubiquitous in all but the most competent units. Most commonly it is axial planar to the major folds and generally fans across them (Wilkinson, 1988). However, in the Moel Hebog–Cwm Idwal syncline the cleavage transects the fold axial trace (Wilkinson, 1988). Cleavage dips are normally 65–75° to the north-west, but locally may be as low as 30°. The dominant process during cleavage formation was pressure solution with some accompanying grain rotation and change of grain shape. In some of the nonwelded ash-flow tuffs, the effect of grain rotation and flattening imparts a 'pseudo-welding' fabric, easily mistaken for true welding in unoriented thin sections.

Faults are common throughout the area and trend predominantly north, north-east and west-north-west. Many that were active either during or subsequent to regional folding represent the reactivation of pre-existing fracture zones that had previously influenced basin development, sedimentation and volcanism (Rast, 1961; Webb, 1983; Campbell et al., 1987, 1988; Wilkinson and Smith, 1988;

Kokelaar, 1988). There is some evidence of lateral displacement on some of the faults, including the lateral offset of geophysical anomalies and local disruption of the regional cleavage arc. Both dextral and sinistral movements are postulated, but amounts are small (Figure 111), rarely more than 1–2 km, and interpretation of volcanic facies changes, based on their current relative positions, is not seriously affected. Perhaps most significant is the dextral movement (2–4 km) postulated along the Cwm Pennant fault (Figure 111; Smith, 1988), restoration of which would place the Pitts Head Tuff caldera (Llwyd Mawr) in closer alignment with the other Snowdon Volcanic Group calderas.

Wilkinson (1987, 1988) analysed strain markers at 250 sites in central Snowdonia to evaluate the regional strain variations and the effects of tectonic thickening or thinning on the original thickness variations. From these data it can be shown (Figure 110), for example, that the thickness variations in the Lower Rhyolitic Tuff Formation ash-flow tuffs used in reconstruction of the Lower Rhyolitic Tuff Formation caldera are predominantly depositional, and only locally is tectonic deformation a significant factor.

The recent studies of Wilkinson (1988) and Smith (1988) have refined the model of regional deformation in North Wales. They propose that the dominant process was 'resultant' transpression generated by the superimposition of regional simple shear on a basement/cover sequence in which older fractures divided the basement into a mosaic of relatively rigid blocks. Strain heterogeneity is reflected in the contrast between deformation of the cover over stable blocks and that above the intervening fractures. Deformation was focussed into zones of high shortening above the fractures, in which folding is associated with faulting, separated by broad zones of low shortening and undulose folding above the rigid blocks. Arcuate cleavage and fold axial-trace patterns are largely a result of deformation around rigid basement indentors, such as the raised block of the proto-Harlech Dome and, on a smaller scale, the Caradoc subvolcanic intrusions (Campbell et al., 1985).

Metamorphism

Studies of regional variations in the metamorphic grade of the Lower Palaeozoic rocks of North Wales (Figure 112) have been based on mineral alterations in the metabasites (Roberts, 1981) and on the white mica crystallinity of the less than 2 μm fraction of metapelites (Roberts and Merriman, 1985). The latter method has the advantage, because of the wide distribution of metapelites, of allowing a relatively uniform distribution of sampling points over the region. Three stages of metapelite recrystallisation, spanning high diagenetic zone through anchizone to epizone conditions, have been distinguished. These correspond approximately to the zeolite, prehnite-pumpellyite or prehnite-actinolite and low greenschist facies respectively in metabasites. The highest temperatures attained during metamorphism probably lay in the range 330°–380°C (Roberts and Merriman, 1985). The metamorphic grade of the pelites is closely related to their degree of deformation (Merriman and Roberts, 1985); the lowest grade, Stage 1, metapelites either lack or have only a very weak cleavage while the highest grade, Stage 3, metapelites are strongly cleaved (Roberts and Merriman, 1985). This conclusion accords with an earlier study based on the

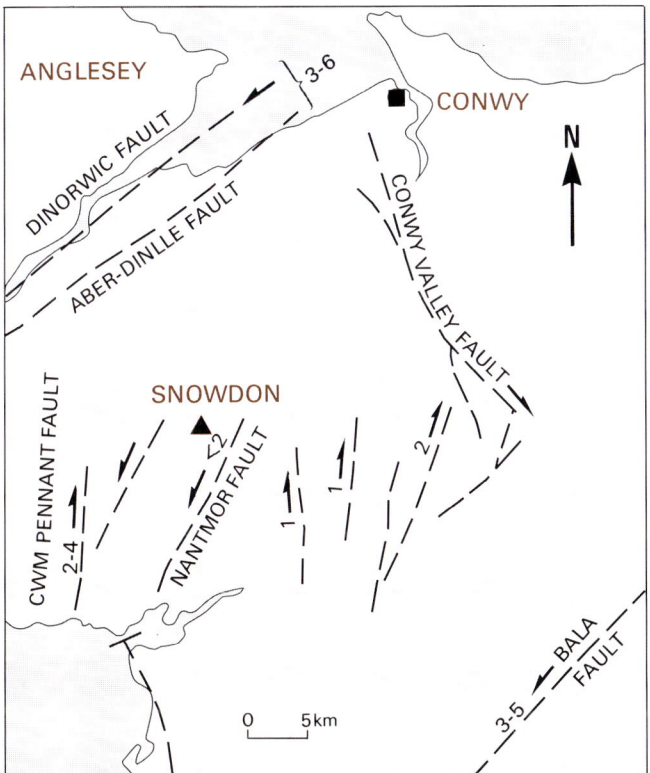

Figure 111 Estimates, in kilometres, of post-Caradoc cumulative strike-slip displacements on the faults within the area of the Snowdon and Llewelyn Volcanic groups outcrop and on the Dinorwic, Aber-Dinlle and Bala faults. Sea grey.

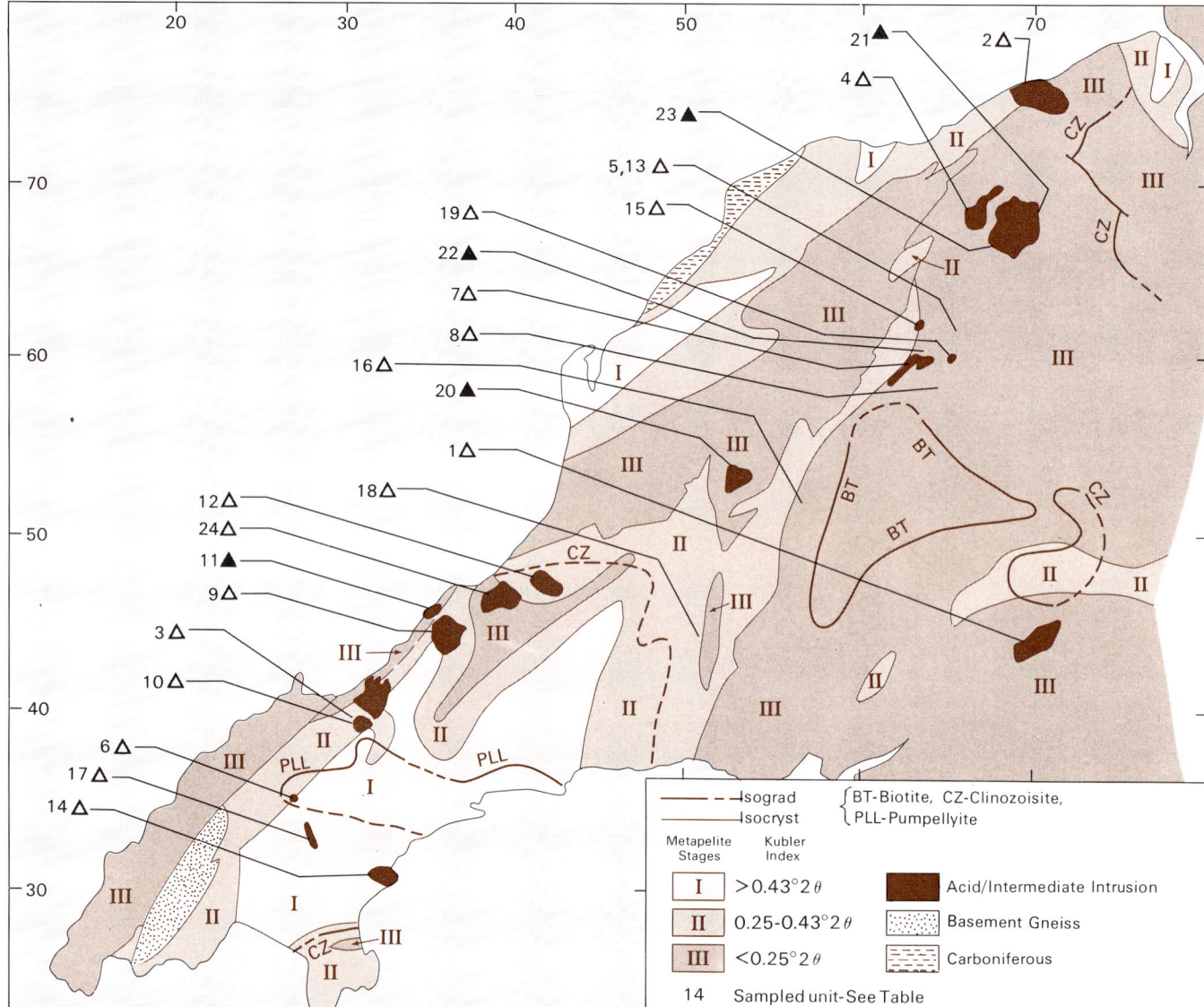

Figure 112 Map showing regional variations in metamorphic grade in Snowdonia and Lleyn (after Roberts and Merriman, 1985) and Ordovician intrusions dated by Rb/Sr whole-rock isochrons (after Evans, 1989a).

metabasites (Roberts, 1981) which indicated that the metamorphic recrystallisation was largely syn- and immediately post-tectonic. There is no consistent relationship between grade of metamorphism and stratigraphic level or, by inference, depth of burial, in either Ordovician or Silurian strata. Roberts and Merriman (1985) proposed, however, that Cambrian strata were subjected to weak burial metamorphism prior to Tremadoc uplift and the subsequent Arenig transgression.

Reviewing metamorphism across the whole of the Welsh Basin, Bevins and Rowbotham (1983) suggested that grades were generally higher in the centre of the basin and lower towards its margins. They concluded that this indicated that regional metamorphic grade was dominantly controlled by depth of burial. Robinson and Bevins (1986) suggested that

the depth-related regional metamorphism of the Welsh Basin was indicative of a low-pressure facies series. Subsequently Robinson (1987) and Bevins and Robinson (1988) argued that the low-pressure facies series metamorphism predated cleavage formation, occurring during a period of enhanced thermal flux accompanying early basin extension and magmatism, for which the term 'diastathermal metamorphism' was coined. Recent estimates of metamorphic field gradients across the Welsh Basin range from about $40°C/km$ (Robinson and Bevins, 1986), to about $50°C/km$ (Bevins and Merriman, 1988) and are consistent with a low-pressure facies series. However, in North Wales, areas of highest grade generally coincide with high strain zones (Roberts and Merriman, 1985; Wilkinson and Smith, 1988; Roberts et al., 1989) suggesting that final equilibration was reached during deformation.

Thus the regional metamorphism was syntectonic and immediately post-tectonic. It approximately coincided with widespread resetting of the Rb/Sr isotope systems of the Caradoc sedimentary rocks, extrusive volcanic rocks and subvolcanic intrusions in early Devonian times (Evans, 1990a,b; see also below).

Post-emplacement element mobility: the Tal y Fan dolerite sill

Element mobility, during post-emplacement cooling and/or low grade metamorphism, in one of the large basic intrusions of Snowdonia, was evaluated during a detailed study of the 110 m thick subconcordant Tal y Fan dolerite sill (Merriman et al., 1986).

The mode of emplacement and the primary textural, mineralogical and geochemical variations through the sill

have already been described (Chapter 5) and have also been discussed in detail by Merriman et al. (1986).

Two phases of post-emplacement mineral development can be distinguished; an early hydrothermal phase when K-feldspar replaced plagioclase microphenocrysts in the marginal and contact zones, and saponite replaced olivine in the central zone, and a later regional metamorphic phase which resulted in the development of prehnite-pumpellyite-epidote and prehnite-actinolite-epidote assemblages (Merriman et al., 1986; Bevins and Merriman, 1988). The regional metamorphism is estimated from the P-T-X grid of Liou et al. (1985) to have occurred at a temperature of 310°C at 1.85 kbar. This temperature estimate is comparable with that calculated (R Dearnley, personal communication, 1986) from a study of annealed plagioclase grain size profiles across the margins of the intrusion.

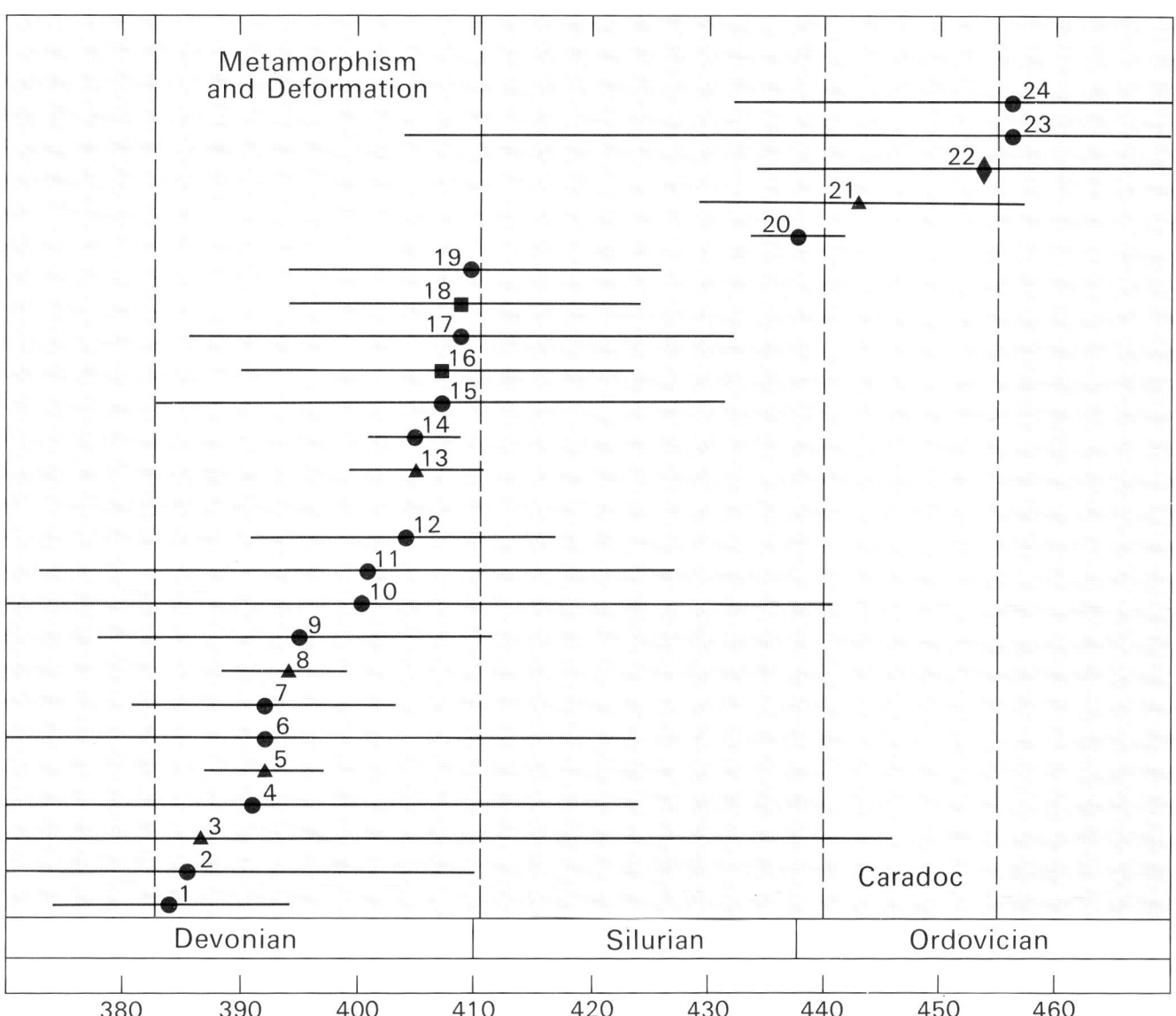

Figure 113 Range of Rb/Sr isotope ages (million years) in Snowdonia and Lleyn (after Evans, in press).
Circles—intrusions, triangles—extrusive volcanic rocks; squares—cleaved mudstones; diamond—hornfels. Time scale after Snelling (1987). Numbers 1 to 24 refer to Table 2.

Representative samples of the eight textural zones were analysed for a wide range of major and trace elements (see Merriman et al., 1986, for details). Ratios of major and trace elements to Zr have been plotted to distinguish the effects of secondary element mobility from primary geochemical variation (Figure 91). A high degree of mobility is recognised in the LIL elements Rb, Sr, K and Ba, particularly in the marginal and chilled contact zones. There is evidence for limited mobility of CaO, Al_2O_3 and Na_2O in the outermost 4–5 m of the intrusion while Mn, and probably Cr, display only local mobility limited to the scale of a hand specimen. In the central and inner marginal zones, MgO, Ni, Co and SiO_2 abundances can be related to the distribution of olivine phenocrysts which suggests that the mobility of these elements was insignificant in the central and inner marginal zones. However, more erratic behaviour in the outer marginal zones and contact zones suggests that, here, there has been limited mobility. The incompatible elements Ti, P, Zr, Y, Nb, Ta, Hf and the REE display strong linear correlations and appear to have been largely unaffected by alteration, even in the outer marginal and chilled contact zones. Amongst the REE, only Eu displays any evidence of mobility, with minor depletion at both margins related to redistribution of Ca. The study therefore indicates that, with the exception of the LIL

elements Rb, Sr, K and Ba, post-emplacement mobility was very limited, and samples from the central zone of the intrusion preserve primary compositions.

It has not yet been finally determined whether the chemical alteration recorded took place mainly during the cooling history of the sill or during the much later low-grade regional metamorphic event. However, the studies confirm that use of the 'immobile trace elements' in petrogenetic modelling is justified.

Metamorphic resetting of Rb/Sr isotope systems

An extensive study of the Rb/Sr geochronology of the extrusive acid volcanic rocks, acid to intermediate subvolcanic intrusions and cleaved metapelites (Figure 113) of northern Snowdonia has been made by Evans (1990a,b). The results clearly indicate that there was resetting of the Rb/Sr whole-rock systems during early Devonian metamorphism, at about 400 Ma (Roberts et al., 1989).

Regression results for strongly cleaved metapelites (Nantmor Group) of Ordovician age from Rhyd Ddu (Caradoc) and from Hendre Ddu (? Llanvirn), give ages of 407 ± 17 and 409 ± 15 Ma respectively (Evans, 1990a,b). Combined, these indicate an age for the regional metamorphism and cleavage formation of 408 ± 10 Ma, postdating the age of deposition

Table 2 Rb/Sr regression results for intrusive and extrusive volcanic rocks and cleaved mudstones, Snowdonia and Lleyn Peninsula

	Rock suite	Age	Initial ratio	MSWD	No. of samples
1	Tan y Grisiau granite	384 ± 10	0.7135 ± 0.0009	9.0	18
2	Penmaenmawr microdiorite	386 ± 24	0.7100 ± 0.0007	7.8	11
3	Boduan andesite	387 ± 59	0.7066 ± 0.0009	15.2	5
4	Aber-Drosgl microtonalite	391 ± 33	0.7080 ± 0.0008	11.0	10
5	Welded tuff, Braich tu du Volcanic Formation	392 ± 5	0.7093 ± 0.0003	1.7	6
6	Carn Fadryn microdiorite	392 ± 32	0.7065 ± 0.0004	4.8	11
7	Bwlch y Cwyion microgranite	392 ± 11	0.7112 ± 0.0013	7.6	10
8	Rhyolite, Lower Rhyolite Tuff Formation	394 ± 5	0.7068 ± 0.0004	2.2	8
9	Caergribin granophyre	395 ± 17	0.7137 ± 0.0040	1.6	10
10	Garn Boduan microgranodiorite	400 ± 41	0.7050 ± 0.0005	0.5	6
11	Outer Garnfor microgranodiorite	400 ± 27	0.7060 ± 0.0005	10.6	11
12	Bwlch Mawr microgranodiorite	404 ± 13	0.7081 ± 0.0003	0.8	3
13	Rhyolite, Braich tu du Volcanic Formation	405 ± 6	0.7089 ± 0.0003	1.9	8
14	Llanbedrog granophyre	405 ± 4	0.7125 ± 0.0008	2.4	14
15	Mynydd Perfedd microgranodiorite	407 ± 24	0.7075 ± 0.0004	2.6	5
16	Cleaved mudstone (Nantmor Group, Caradoc)	407 ± 17	0.7149 ± 0.0011	2.0	11
17	Nanhoron Quarry granophyre	409 ± 23	0.7246 ± 0.0146	7.6	11
18	Cleaved mudstone (Nantmor Group, ?Llanvirn)	409 ± 15	0.7161 ± 0.0011	6.3	9
19	Ogwen microgranite	410 ± 16	0.7100 ± 0.0028	3.2	10
20	Mynydd Mawr microgranite	438 ± 4	0.7040 ± 0.0021	2.1	5
21	Trachyandesite, Foel Fras Volcanic Complex	443 ± 14	0.7070 ± 0.0007	2.6	5
22	Hornfels of the Bwlch-y-Cwyion microgranite	454 ± 30	0.7090 ± 0.0003	1.5	5
23	Microdiorite, Foel Fras Volcanic Complex	456 ± 52	0.7052 ± 0.0009	1.6	9
24	Gurn Ddu microgranodiorite	456 ± 24	0.7045 ± 0.0004	1.8	8

Numbers in left hand column are those referred to in Figures 134 and 135.

by at least 50 Ma. This accords well with a K/Ar age of 399 ± 10 Ma obtained from metamorphic chlorite separated from strongly cleaved sandstone of Caradoc (Longvillian) age (recalculated from Fitch et al., 1969).

Rb/Sr whole rock isochrons obtained from extrusive rhyolites of the Braich tu du Volcanic Formation (Soudleyan) and the Lower Rhyolitic Tuff Formation (Longvillian), and from a welded tuff of the Braich tu Du Volcanic Formation (Evans, 1990a,b) give ages of 405 ± 6 Ma, 394 ± 5 Ma and 392 ± 5 Ma respectively (Table 2). The apparent ages of these Caradoc extrusive volcanic rocks correspond to those of metamorphism and slaty cleavage formation in the enclosing sediments and indicate that the Rb/Sr isotopic systems have been reset.

Similar late Silurian and Devonian ages, ranging from 384 ± 10 Ma to 410 ± 16 Ma, have been obtained from many of the subvolcanic intrusions (Table 2), indicating resetting of their Rb/Sr systems several tens of millions of years after intrusion. However, two of the intrusions, a microdiorite of the Foel Fras Volcanic Complex and the Mynydd Mawr microgranite, and a possibly extrusive andesite of the Foel Fras Volcanic Complex, yield ages of 456 ± 52 Ma, 438 ± 4 Ma and 443 ± 13 Ma respectively; these appear not to have been reset during the late Caledonian metamorphic event. Similarly, contact metamorphosed mudstones from the metamorphic aureole of the Bwlch y Cywion microgranite yield an isochron age of 454 ± 20 Ma (Evans, 1990a,b), indicating the age of emplacement of the intrusion, whereas the microgranite itself gives a reset age of 392 ± 11 Ma.

The Rb/Sr isotope results from the Caradocian extrusive volcanic rocks and associated subvolcanic intrusions in northern Snowdonia and the adjacent Lleyn peninsula (Figure 112) indicate that there is no correlation between those having reset ages and their grade of metamorphism. The temperature reached at any particular locality during regional metamorphism does not appear to be a major factor in rehomogenisation of the isotopic systems. However, Evans (1990a,b) has shown that all the rocks with 'reset' Devonian ages display extensive replacement by secondary minerals resulting from hydration reactions during low-grade regional metamorphism. This suggests that it is the extent to which the rocks have reacted with water that effects the Rb/Sr isotopic systems. The rocks that retain their Ordovician isotopic ages show only a limited development of secondary hydrous mineralogy. However, in the case of the Bwlch y Cwyion microgranite hornfels zone, hydrated silicates, particularly mica and chlorite, had developed during contact metamorphism, prior to the regional metamorphism, and remained metastable during the latter event.

The broad range of reset ages obtained (412 Ma to 384 Ma), with the igneous rocks yielding ages the same or younger than those of the cleaved metapelites, is shown in Figure 113. This supports the contention that the source of water for the hydration reactions in the extrusive volcanic rocks and the intrusions was metamorphic water released from the surrounding sediments as the latter were deformed and partially dehydrated (Evans, 1990a,b). The influx of water into the igneous rocks and the closed system behaviour of their isotopes during metamorphism (Evans, 1990a,b) suggests that they acted as sumps for the water lost from the adjacent sediments.

CHAPTER 7

Conclusions

In the introduction to this book reference was made to Harker (1889), who recognised in the Ordovician volcanic rocks of Snowdonia the potential for understanding the processes of volcanic activity. In central and northern Snowdonia, this volcanism was contained entirely within two stages of the Caradoc Series (about 2–3 Ma) and in the uplifted and deeply dissected sequence it is possible to consider its activity from beginning to end. Such an opportunity is normally denied to those who study recent volcanic activity and relatively young deposits.

For the most part, the volcanic sequence is associated with marine sediments. It was this association that led Sedgwick (1843) to conclude that the extrusive volcanic rocks were emplaced in a submarine environment. In recent years, research in modern marine environments (e.g. the Deep Sea Drilling Project) has made major advances. However, because of the constraints on direct observation in such research, it can be argued that consideration of uplifted, ancient marine strata may be as productive.

Clearly, there are disadvantages in considering ancient sequences for detailed interpretation. For them to have been uplifted would necessitate their involvement in an orogeny and, as a result, folding, faulting and metamorphism. However, in north-west Wales the folding is relatively simple and the metamorphism is low grade. As a result, original surfaces can be reconstructed without too many reservations and the original mineralogies and fabrics of the rocks can generally be distinguished or inferred with confidence.

The restriction of outcrop caused by the deep dissection of a folded sequence during the erosion of the landscape to its present level could result in distribution patterns of elements of the geology which do not accurately reflect their original disposition. However, in Snowdonia such difficulties have been mitigated by detailed mapping, and the resultant lithostratigraphical and biostratigraphical framework enables an accurate, three-dimensional, regional picture to be drawn.

During the Caradoc (Soudleyan and Longvillian stages) the area of central and northern Snowdonia was the locus of bimodal (basalt-rhyolite) volcanism. The associated sediments were dominantly marine (inner- to outer-shelf) with only occasional expressions of subaerial and shoreline depositional environments. The siting of the volcanic activity was not fortuitous; it was controlled by the regional tectonic framework established during the late Precambrian and dominated by fractures with a broadly north-east–south-west trend, such as the Menai Straits Fault System and associated faults to the south-east, which affected sedimentation and volcanism in early Cambrian times (British Geological Survey, 1988). Reactivation of these deep-seated faults during Caradoc times formed a north-east–south-west-trending Snowdon graben which controlled magma movement at depth and restricted the distribution of the intrusive and most of the extrusive volcanic rocks. However, the surface manifestation of these faults is not always well defined and

this suggests that their upward propagation to shallow crustal levels was mainly dissipated in unconsolidated cover. Even so, contemporaneous faults are recognised within the graben; for example, during the 1st Eruptive Cycle (Llewelyn Volcanic Group), small-scale faults developed block and basin topography which affected distribution of the extrusive basalts.

The bounding fracture zones of the Snowdon graben have been interpreted (Kokelaar, 1988) to represent an extension and deflection of the north–south-trending Barmouth Fracture Zone and Rhobell Fracture Zone to the south which had been controlling influences on the earlier, Tremadoc–Llanvirn volcanism in southern Snowdonia. However, in the Lleyn peninsula, the broadly north-east–south-west alignment of Ordovician intrusions and the occurrence of a Caradoc eruptive centre (Fitch, 1967), together with the system of faults, suggest that the Snowdon graben extends to the south-west and crosscuts the earlier north–south structures, reflecting a change from the Tremadoc–Llanvirn regional stress patterns (Campbell et al., 1988).

Similarly orientated structures, which affected both sedimentation and volcanism, are a prominent feature of the Welsh Lower Palaeozoic Basin, for example the Bala Fault profoundly influenced volcanism and sedimentation in Arenig–Llanvirn times (Dunkley, 1979). The Llanvirn to Caradoc volcanism of the Builth and Shelve inliers may also have been constrained in a narrow graben (cf. Kokelaar, 1988) and the Caradoc volcanism of the Berwyn Hills (Brenchley, 1978) is sited close to the extension of this structure. Further reactivation of these structures occurred during later phases of crustal extension (e.g. Carboniferous, Permo-Triassic) and compression (e.g. late Silurian–early Devonian). For example, the reactivation of the extension of the Builth–Shelve graben is graphically displayed by gravity data, which reflects the Permo-Triassic sedimentary infill of the Cheshire Basin.

Within the Snowdon graben, the nature of the eruptive centres, and the form and size of associated volcanotectonic structures, was to a large extent controlled by magma composition and the character of the substrate. The most voluminous eruptions, 20 to 60 km³, were acidic and produced substantial and widespread pyroclastic flows. The basic eruptions, mainly of basalt, hyaloclastite and basic tuff, were probably an order of magnitude smaller in volume. The eruptive centres have been determined from the distribution and variations in thickness and facies of individual extrusive volcanic units and associated near-surface intrusions. The complex Caradoc sequence within the graben indicates that it was the site of intense volcanism, during which the eruptive centres shifted temporally and spatially, and the deposits of successive eruptions interdigitated within a sequence of predominantly marine sediments.

Two eruptive cycles have been defined and these mainly comprise basalts, basic tuffs, rhyolites and acidic ash-flow tuffs, with a few intermediate, trachyandesite lavas and tuffs.

The two cycles differ fundamentally in their relative proportions of extrusive basic, intermediate and acidic magma. The 1st Eruptive Cycle contains only small volumes of basic magma, but significant intermediate lavas, whereas the 2nd Eruptive Cycle is essentially bimodal (basic and acidic) with virtually no intermediate magma. The absence of significant volumes of intermediate magma, together with the lack of geophysical evidence for the existence of large bodies of major mafic residua, have led previous authors (Allen, 1982; Kokelaar et al., 1984) to favour partial melting of the crust to generate the large volumes of acidic magma observed. However, Croudace (1982) interpreted the granitoid intrusions of Caradoc age on the Lleyn peninsula as having been derived by crystal fractionation from andesitic magmas, on the basis of trace element and REE geochemistry. Similarly, it is concluded here that the magmatic evolution of both eruptive cycles was dominated by crystal fractionation consistent with the closed-system Rayleigh Fractionation Law, and that crustal assimilation, though recognisable, was a subordinate process.

In the 1st Eruptive Cycle (Llewelyn Volcanic Group), two distinct fractionation suites are represented, although the parental basaltic compositions for only one, the Foel Grach Basalt Formation of calcalkaline, volcanic-arc affinity, has been identified in outcrop. The genesis of residual rhyolite compositions (e.g. Conwy Rhyolite Formation) can be achieved by a two-stage crystal fractionation model involving initial removal of plagioclase, olivine and ilmenite from a basaltic parent (cf. Foel Grach Basalt Formation) to produce 'trachyandesite' and then the removal of plagioclase, augite, ilmenite and apatite to produce rhyolite. Petrochemical and experimental data (Ball and Merriman, 1989) suggest that the rhyolites evolved at relatively shallow crustal levels (P_{H_2O} < 500 bar).

In the 2nd Eruptive Cycle (Snowdon Volcanic Group), two distinct basaltic parental compositions have been identified and a third is proposed. The sub-LRTF basalts represent one end-member, of island-arc tholeiite affinity, and the Bedded Pyroclastic Formation basalts, representing transitional volcanic-arc to ocean-island basalts, another. A third basaltic parent, similar to the sub-LRTF basalts but enriched in Th, is proposed. Dolerites of both the Snowdon and Crafnant centres are predominantly of Bedded Pyroclastic Formation affinity but some are transitional between the Bedded Pyroclastic Formation and sub-LRTF basalt compositions. Such variations of basaltic composition, tholeiitic, within-plate and volcanic-arc like, are typical of back-arc marginal basin environments (Saunders and Tarney, 1984). Recently, Kokelaar et al. (in press) have attributed such variability, in a study of Tremadoc to Llandeilo basaltic volcanism in the Welsh Basin, to mantle heterogeneity unrelated to subduction and to variable partial melting of the mantle source. Leat and Thorpe (1989) explain the transitional character of the Bedded Pyroclastic Formation basalts, in particular, in terms of mixing of two discrete end-member partial melts derived from two distinct mantle sources. In addition, they argue that since ocean-island basalt is not thought to erupt above well-established subduction zones (Fitton and Dunlop, 1985), then the eruption of ocean-island basalts during Bedded Pyroclastic Formation times establishes a minimum age for the end of subduction of oceanic lithosphere beneath

Snowdonia. The relationships are similar to those distinguished by Ormerod et al. (1988), Fitton et al. (1988) and Leat et al. (1988) in the Basin and Range province in western USA where subduction is considered to have ceased about 2 ma before the eruption of ocean-island basalt. If this is also the case in Snowdonia, the subduction here probably ceased during or before the 1st Eruptive Cycle. The parental basaltic magmas are assumed to have been derived from a garnet-free, spinel- (or plagioclase-) lherzolite mantle source, with up to 60 per cent fractional crystallisation of plagioclase, olivine, pyroxene and Fe-Ti-oxide, leading to intermediate compositions.

Crystal fractionation is again the dominant process in the genesis of rhyolites associated with the 2nd Eruptive Cycle, but it also involves limited crustal assimilation. The rhyolites are interpreted as having been derived, via a suite of intermediate magmas, from basaltic parents of varying composition, reflecting mixed basalt types, namely sub-LRTF, Bedded Pyroclastic Formation and high-Th. The mixing of these parental basalts is presumed to have taken place at deep crustal or subcrustal levels. The derivation of the B3 rhyolite (high Zr) from icelandite did not involve significant zircon fractionation and in this respect differs from that of the other rhyolites.

In general terms, there is no consensus regarding the petrogenesis of bimodal volcanic suites (see review in Doe et al., 1982). Partial melting of heterogeneous upper mantle is commonly invoked, as here, to explain temporal and spatial variations in basalt chemistry (e.g. Thorpe et al., 1984b). However, the genesis of the silicic magmas is more contentious, with strong bodies of opinion favouring either partial melting of the upper crust or crystal fractionation of a basaltic parent, or, as here, a combination of both processes. Some experimental data have been cited to support rhyolite derivation by partial melting of available crust, for example by Conrad et al. (1988) for the voluminous rhyolitic magmas of the Taupo Volcanic Zone. However, much of the most convincing evidence in support of a partial crustal melt origin is based on isotopic studies. In view of the paucity of reliable Sr initial ratios and Pb isotopic data for the Caradoc volcanic rocks in North Wales, it is necessary to consider the evidence from analogous sequences elsewhere. For example, the late Cenozoic bimodal volcanism of the western USA is characterised by basalt and rhyolite with significantly different Pb and Sr isotopic compositions (Lipman et al., 1978; Doe et al., 1982; Bacon et al., 1984). Similarly, Davies and Macdonald (1987) present Pb isotope data for the bimodal basalt-comendite Naivasha Complex (Kenya) which suggests that the basalts and comendites are not cogenetic. They argue, therefore, for comendite derivation by crustal melting of a heterogeneous source region, despite the fact that trace-element data might be interpreted as indicating their derivation from basalt by fractional crystallisation, dominated by feldspar, with some crustal assimilation (mass assimilation/mass crystallisation (Ma/Mc) − about 0–0.2).

In contrast to the continental-rift settings of the western USA and Kenya, however, and in a tectonic regime probably more akin to that occurring during Caradoc times in northwest Wales, Sr isotopic data from the bimodal Kikai volcano, in the Ryukyu island-arc setting of southern Japan, and the Usu volcano in north-east Japan, yield a narrow scatter of

[87]Sr/[86]Sr ratios (Notsu et al., 1987). These data imply a common source for both the basalts and the rhyolites. Their favoured interpretation is for derivation of the rhyolites by fractional crystallisation of the local basaltic magma, and this is supported by the major- and trace-element data (Ujike et al., 1986). Of the Sr initial ratio data for acid magmas in the Caradoc of North Wales, that for the Mynydd Mawr microgranite, whose Rb-Sr system was not reset during Caledonian metamorphism, yields a low initial ratio of 0.7040. This is comparable to values for Kikai and Usu volcanoes (respectively 0.705 and 0.704; Notsu et al., 1987) and is consistent with a mantle origin rather than derivation by crustal fusion.

Grunder and Mahood (1988) modelled the genesis of large volumes of silicic magma by fractional crystallisation of a basaltic parent. They consider that genesis by fractional crystallisation of the three voluminous (>1000 km), zoned ash-flow tuffs of the Pleistocene Calabozos complex in central Chile explains major element variations in phenocryst and whole-rock analyses. However, as with the rhyolites of the Snowdon Centre, enrichment of certain trace elements requires a degree of crustal assimilation (Ma/Mc = about 0.1–0.3). Similarly, Thorpe et al. (1984b), in their model for the Cerro Galan volcanic centre (Argentina), invoke derivation of the large-volume ignimbrites from complex assimilation-fractional crystallisation and magma mixing processes within the continental crust.

Inevitably, the petrogenesis of the two Caradocian eruptive cycles in general, and the Snowdon rhyolites in particular, was considerably more complex than the model presented here. Several authors stress the open-system nature of many high-level silicic magma systems (e.g. Macdonald et al., 1987; Lipman, 1988). Indeed, all of the processes invoked by the latter author for the open-system evolution of the mid-Tertiary Questa caldera and cogenetic volcanic and plutonic rocks were likely to have operated to some extent at least during the evolution of the Snowdon and other centres. These include protracted crystal fractionation, replenishment by mantle and lower crustal melts of varying chemical and isotopic character, mixing of evolved with more primitive magmas, upper crustal assimilation and possibly volatile transfer processes.

The volcanic sequences dominated by acidic ash-flow tuffs and volumes of the major eruptions are as follows: Members 1 and 2 of the climactic phase of the 1st Eruptive Cycle (Capel Curig Volcanic Formation), each greater than 20 km[3], the initial phase of the 2nd Eruptive Cycle (Pitts Head Tuff Formation) >25 km[3], the main phase at the Snowdon Centre (Lower Rhyolitic Tuff Formation) about 60 km[3], and the Crafnant Centre (Crafnant Volcanic Group) about 50 km[3]. Such volumes would alone suggest that they were caldera-forming events. On a plot of caldera diameter vs. erupted volume (cf. Cas and Wright, 1987, p.233, fig. 8.13) these volumes crudely relate to caldera diameters of at least 10–20 km. As has recently been observed (Lipman, 1976, 1984; Walker, 1984), not all calderas conform to the cylindrical block subsidence model (Clough et al., 1909; Williams, 1941; Smith and Bailey, 1968). Within the Snowdon graben three different caldera types have been distinguished, which can be related to three different settings; subaerial, transitional shoreline and submarine.

Subaerial calderas are inferred for the climactic phase of the 1st Eruptive Cycle (1st and 2nd members of the Capel Curig Volcanic Formation) and recognised for the initial phase of the 2nd Eruptive Cycle (Pitts Head Tuff Formation). In the former, the caldera is postulated to lie beyond the current coastline. Its subaerial setting is inferred from the subaerial, alluvial fan-type sediments underlying the most proximal facies of the two outflows. In the latter, the eruptive centre was located in a narrow rift (possibly offset by later faulting from the median rift of the Snowdon graben) which includes a volume of about 34 km[3] of primary ash-flow tuffs intruded by late-stage rhyolite domes. The caldera form was probably similar to a small graben (cf. Walker, 1984, fig. 1b). As in the former case, the proximal outflow tuffs from this centre are underlain by subaerial, alluvial fan conglomerates and sandstones. In both cases, the outflows passed distally into marine environments; in the former to the south, and in the latter to the north (see below).

A distinctive feature of both these examples of subaerial ash-flow tuff volcanism in the Caradoc sequence is the absence of any significant Plinian fall-out deposits. This is in marked contrast with the numerous examples of recent subaerial ash-flow tuffs which are preceded by a Plinian fall-out phase and followed by a fall-out from the attendant cloud. It suggests that only very low eruptive columns were developed and that the eruptions were characterised by a 'boiling-over' style as described at Mount Lamington, Papua, New Guinea (Taylor, 1958).

The transitional, shoreline, caldera type is most clearly represented by the Lower Rhyolitic Tuff Formation caldera which developed sequentially above the Beddgelert Fault. In the early stages the eruptions were fissure-controlled, subaerial and sited near the centre of the subsequent caldera. They occurred close to the shoreline, which was established during the initial phase of the 2nd Eruptive Cycle activity and modified by updoming caused by upward magma migration. During the main phase of caldera subsidence, the locus of eruption shifted northwards to the vicinity of the northern margin of the caldera. This spatial shift contributed greatly to the eventual asymmetric downsag character of the caldera. Its form can be compared to either a normal-faulted graben or a downsagged graben, or a combination of both (Walker, 1984, fig. 1). The occurrence of fault-controlled caldera development within a larger graben is similar to those in the Taupo Volcanic Zone (Wilson et al., 1984) and about the Vulsini volcano, Italy (Varekamp, 1980).

The intracaldera facies of the Lower Rhyolitic Tuff Formation caldera overlies shallow-marine sandstones and siltstones and the outflow facies extends distally into an outer-shelf setting. The top of the intracaldera primary tuffs was reworked in a shallow-marine environment. The absence of any profound change in the sedimentary setting during caldera development indicates that the caldera collapsed incrementally and subsidence broadly kept pace with tuff accumulation in the submarine environment. Any subaerial edifice was limited in extent.

During the main eruptive phase, the caldera was drowned by the sea but, particularly in the south, not to any great depth. Drowned Quaternary calderas are common and particularly distinctive are those which occur in the rifted volcanic arcs in Southern Japan such as Kikai (Notsu et al.,

1987) and Aira (Aramaki, 1984) in the rifted Ryuku arc. Similarly, Rabaul, New Britain (Walker et al., 1981), Santorini (Druitt, 1985) and Kos and Yali (Stadlbauer et al., 1986) in the Aegean Sea are islands which are fragments of the walls of drowned, collapsed calderas, although in these instances the nature of the intracaldera facies is not known.

The submarine caldera is typified by the Crafnant Centre in north-eastern Snowdonia. Here, the volcanic cycle was dominated by explosive acidic eruptions in an outer-shelf environment of silt and mud deposition. The most clearly determined pyroclastic facies is that of the outflows. The caldera is marked by proximal thickening of the volcanic sequence and corresponding disturbance of both tuffs and the intercalated sediments, caused by repeated secondary sloughing into and away from the vent area. The margins of the caldera are not clearly defined and the effects of any of the basement faults, which probably controlled the siting of the caldera, were dissipated in the unconsolidated cover. The volcanic sequence shows no evidence of shallow-water reworking, no significant edifice was constructed above wave-base, and the sequence was overlain by black graptolitic muds.

Details of modern examples of entirely submarine calderas are extremely rare in the literature which is largely due to the practical difficulties of their recognition. A substantial caldera, in water depth of greater than 1 km has been recognised (written communication, Dr R S Fiske, 1989) in the Okinawa back-arc trough. Also, the northerly extension of the Taupo Volcanic Zone in the Bay of Plenty contains a magnetic anomaly to the west of White Island (Cole, 1984; Lewis and Pantin, 1984) which possibly reflects a caldera similar to those within the zone on North Island, New Zealand. The most intensely studied examples of submarine calderas are those in the Tertiary of the Hukuroko Basin, Japan (Ohmoto and Skinner, 1983), although the stages of their physical evolution, unlike the associated massive sulphide deposits, were not clearly defined.

The downsag character and lack of a significant edifice are common to both the transitional shoreline and submarine calderas. The development of such features would be facilitated both by the extensional setting and the unlithified character of the substrate. As in the subaerial eruptions of the sequence, there is little evidence of Plinian fall-out ash; however, in the deeper marine setting, the suppression of the eruption column would be further enhanced by the hydrostatic pressure.

The recognition of the calderas in the Snowdon graben is not based solely on consideration of the pyroclastic rocks. Evidence for the shoreline caldera (Lower Rhyolitic Tuff Formation, Snowdon Centre) is corroborated by the distribution of the intimately associated rhyolite intrusions and extrusions and the spatially associated sulphide mineralisation. In the deeper marine setting of the Crafnant Centre, bedded sulphide ores accumulated in brine pools above the caldera in a setting similar to that recently described in the central part of the Okinawa trough (Halbach et al., 1989) and in the Hukuroko Basin, Japan (Ohmoto and Takahashi, 1983).

From a consideration of the temporal and spatial relationships of the intrusive and extrusive rhyolites and basalts at the Snowdon Centre, it is possible to draw some conclusions regarding the magma plumbing system and its relationship to caldera development. Most conventional models of large

volume silicic ash-flow eruptions and attendant caldera collapse (e.g. Smith and Bailey, 1968) envisage a single, large, high-level magma chamber, possibly zoned (Hildreth, 1979, 1981), which is partially evacuated (<10 per cent, Smith, 1979) to cause caldera collapse in one or more stages (e.g. Druitt and Sparks, 1984). However, the relationships at the Snowdon Centre indicate that the rhyolitic magmas were contained in small and ephemeral chambers located beneath different parts of the caldera. The A2 rhyolite magma, from which the main volume of silicic ash-flows was erupted, with consequent caldera collapse, briefly underlaid the whole of the caldera. However, it must either have been largely evacuated during caldera collapse or cooled sufficiently quickly to crystallise completely soon after, as basaltic magmas were erupted 'through' the magma chamber without any evidence of significant mixing, in the immediate post-caldera phase of volcanism. During the main-phase caldera-forming eruptions therefore, a relatively thin, disc-like magma chamber is envisaged, as proposed for icelandic calderas by Gudmundsson (1988). During the pre-caldera phase, the post-caldera resurgence and the late-stage comenditic ash-flow eruptions, small batches of rhyolitic magma of contrasting composition and parentage occupied small chambers largely controlled either by pre-caldera faults or caldera-bounding fractures. B3 rhyolitic magma may have underlain the whole caldera during the final phase of volcanism.

The basic eruptions which occurred intermittently during both the eruptive cycles were generally small and short-lived. In central Snowdonia, basalt effusion late in the interval between the two eruptive cycles and early in the second cycle (sub-LRTF basalt, see above) probably developed from a number of centres. Later, in the same area, following the main caldera-forming phase, the most extensive expression of basaltic volcanism occurred (Bedded Pyroclastic Formation). The sequence, dominated by basaltic sediments, represents shallow-marine reworking of small basaltic island volcanoes in a tectonically active environment, which was further complicated by resurgent activity within the caldera. Locally, pillowed-basalt flows, pillow-breccias and hyaloclastites are preserved and, close to the centres, a network of shallow-level, basaltic sills. In places, cross-cutting vents infilled with agglomerates have been distinguished.

In contrast, the basic volcanism associated with the Crafnant Centre in north-eastern Snowdonia is mainly expressed in local thick accumulations, up to 500 m. These show little evidence of reworking and corroborate the deep, outer-shelf environment established for the acidic eruptions (see above). Their relationships with the outflow tuffs from the acidic centre indicate that they accumulated in small basins, the development of which was facilitated by loading of the relatively dense basic lavas and tuffs on unlithified sediments in an extensional environment. Though not strictly analogous to the calderas, these small subsiding basins had the same general effect of entrapping and preserving thick primary volcanic sequences within the stratigraphic column.

The formation of subaqueous volcanic centres by explosive eruptions has been the subject of much controversy in recent years (see Cas and Wright, 1987). From work on the Miocene Green Tuff, Japan, Fiske and Matsuda (1964) proposed a model for submarine eruption involving vesiculation and fragmentation of magma and the submarine emplacement of related pyroclastic flows. However, McBirney (1963, 1971) argued that extensive vesiculation of acidic magmas could only occur in water depths of tens of metres or less.

More recently Burnham (1983), in a controversial paper, has argued that in the Hukuroko district, Japan, the vesicularity of juvenile acid pyroclasts is consistent with their eruption on the sea floor at water depths of 3.5 ± 0.5 km from a magma chamber 1.1 ± 0.3 km deeper. These depths are closely comparable with depths, estimated independently from a consideration of the influence of volatile contents of associated basalts, of up to 4000 m (Dudas, 1983), depth assemblages of foraminifera in associated mudstones, >3000 m (Guber and Merrill, 1983), and studies of fluid inclusions in ores, greater than 2000 m (Pisutha-Arnoud and Ohmoto, 1983). Similarly, Lonsdale and Hawkins (1985) have described in-situ rhyodacitic pumice erupted at depths of 2600–3600 m within the Mariana trough. Thus the problem of vesiculation at great water depths can no longer be considered an insurmountable obstacle in proposing subaqueous eruptions.

The order of depths to be considered in interpreting the subaqueous explosive eruption of magma within the Snowdon graben is considerably less than those considered in the Hukuroko district. The stratigraphical relationships of the Snowdon and Crafnant centres is such that the interpretation of them as being largely contained within a submarine environment is far easier to sustain than them being interpreted as the eroded remnants of subaerial centres. The shoreline development of the Snowdon (Lower Rhyolitic Tuff Formation) caldera is similar to that invoked by Busby-Spera (1984, 1986) in her description of the Triassic–early Jurassic sequence in the roof pendant at Mineral King, in the southern Sierra Nevada, which Cas and Wright (1987) considered a 'feasible' explanation.

With regard to the Crafnant Centre, the disturbed and disrupted pyroclastic and epiclastic (mud) elements indicate water depths sufficient to contain the eruptive column but not to suppress explosivity. The process envisaged is similar to that proposed by Kokelaar et al. (1985) for the eruption and emplacement of the Cader Rhwydog Tuff (Llanvirn) in southwest Wales.

The acidic ash-flow tuffs of the Caradoc sequence are characteristically fine-grained and, in spite of the post-emplacement devitrification, recrystallisation and low-grade metamorphism, their primary fabrics are remarkably well preserved. The consistently fine grade of the ash-flow tuffs reflects the intensity of the eruptive explosivity. This was probably enhanced by magma interaction with water in the vent region, perhaps derived in part from the unconsolidated cover. The dominantly comagmatic character of the components of the ash-flow tuffs, combined with the homogeneity and fineness of grade, has resulted in them being unexpectedly receptive to detailed geochemical correlation. This is a field in which future work can profitably develop, although with due cognisance taken of heterogeneity related to sorting within the flows and to incorporation at their bases, and more generally, distally, of sediment.

A major problem in the ash-flow tuff interpretations is that of welding in a subaqueous environment. Implicit in the interpretations in this book is the conclusion that an acidic ash-flow tuff can be emplaced in a submarine environment and

yet retain sufficient heat to weld. Also, the evidence of the sedimentary context of some tuffs, the lowest members of the Capel Curig Volcanic Formation and the Pitts Head Tuff Formation, indicates that the ash-flows were erupted subaerially. They then transgressed a shoreline and the tuffs were eventually welded in both subaerial and submarine environments. Characteristic fabrics within the tuffs in each environment do occur (Figure 114). Although none is solely diagnostic of the environment of deposition, considered together, as a facies, they can be regarded as such.

The irregular bases displayed by the ash-flow tuffs in the shoreface and subaqueous environments, with downward protrusion of tuff and intruded flames of sediment, clearly indicate emplacement on unlithified substrate. The scale of these features suggests that they were not just the product of emplacement on an irregular surface. Following the original discussion (e.g. Osmaston in Francis and Howells, 1973), Kokelaar (1982) has shown that such a process was facilitated by fluidisation of the sediment at the hot tuff–sediment interface. The locally developed micaceous replacement of glass shards at basal contacts may also reflect reaction with steam and resultant silica migration may have enhanced the development of siliceous nodules. Vapour streaming from the basal contact could also rheomorphose the welding fabric. Such possibilities were theoretically considered by Sparks et al. (1980) who concluded that welding might take place very readily in a subaqueous environment.

The relationships of the ash-flow tuffs in the Caradoc sequence are difficult, if not impossible to interpret without accepting that the ash-flows transgressed a shoreline. Additionally, they retained their integrity, displaced the shoreline and continued to flow further under water and, in some instances, retained sufficient heat to weld on emplacement. These proposals, outlined by Cas and Wright (1987), have caused much debate in recent years. After considering the various reported instances of subaqueous pyroclastic flows, they present (Cas and Wright, 1987, fig. 9.13) a schematic model for the passage of pyroclastic flows into water which clearly reflects their antipathy to the proposals above. They accept that a large hot pumiceous pyroclastic flow can be deposited in shallow water but argue that on further encroachment into the subaqueous environment it chills, disintegrates and continues as a water-saturated debris flow, which may in turn transform into an ash-turbidity flow. They state that 'the simplest, and undoubtedly the most important, way in which volcanic material is transported into ocean basin as flows is by epiclastic processes and slumping'. In the instances quoted, marginal to extensive, subaerial, island volcanoes in the Lesser Antilles, there can be little doubt that such processes are dominant. However, where the volcanism becomes progressively more contained within the subaqueous environment these processes become simultaneously less important.

In the Caradoc sequence, deposits derived by reworking of pyroclastic debris, slumping and remobilisation into debris flows and tuff-turbidites are common. They are characterised by their bed forms and their variable contamination with unconsolidated epiclastic debris. They can be distinguished from the repeated intercalations of background epiclastic sediments and the completely preserved non-welded and welded primary ash-flow tuffs. It is probably because of this

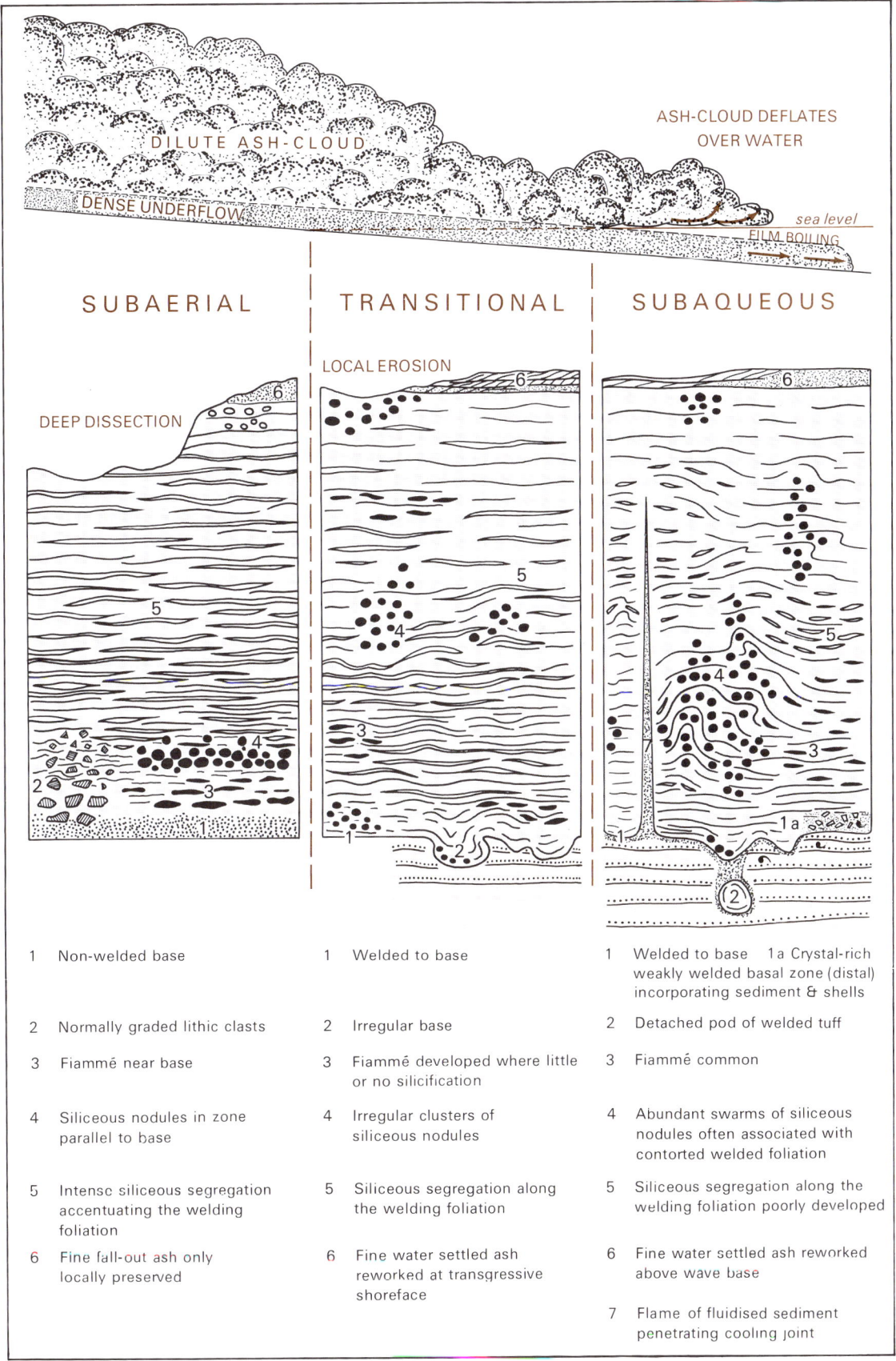

DILUTE ASH-CLOUD

ASH-CLOUD DEFLATES OVER WATER

DENSE UNDERFLOW

sea level

FILM BOILING

SUBAERIAL | **TRANSITIONAL** | **SUBAQUEOUS**

LOCAL EROSION

DEEP DISSECTION

1 Non-welded base	1 Welded to base	1 Welded to base 1a Crystal-rich weakly welded basal zone (distal) incorporating sediment & shells
2 Normally graded lithic clasts	2 Irregular base	2 Detached pod of welded tuff
3 Fiammé near base	3 Fiammé developed where little or no silicification	3 Fiammé common
4 Siliceous nodules in zone parallel to base	4 Irregular clusters of siliceous nodules	4 Abundant swarms of siliceous nodules often associated with contorted welded foliation
5 Intense siliceous segregation accentuating the welding foliation	5 Siliceous segregation along the welding foliation	5 Siliceous segregation along the welding foliation poorly developed
6 Fine fall-out ash only locally preserved	6 Fine water settled ash reworked at transgressive shoreface	6 Fine water settled ash reworked above wave base
		7 Flame of fluidised sediment penetrating cooling joint

Figure 114 Model of ash-flow emplacement, and of contact and internal facies relations of welded ash-flow tuffs with respect to environment of deposition.

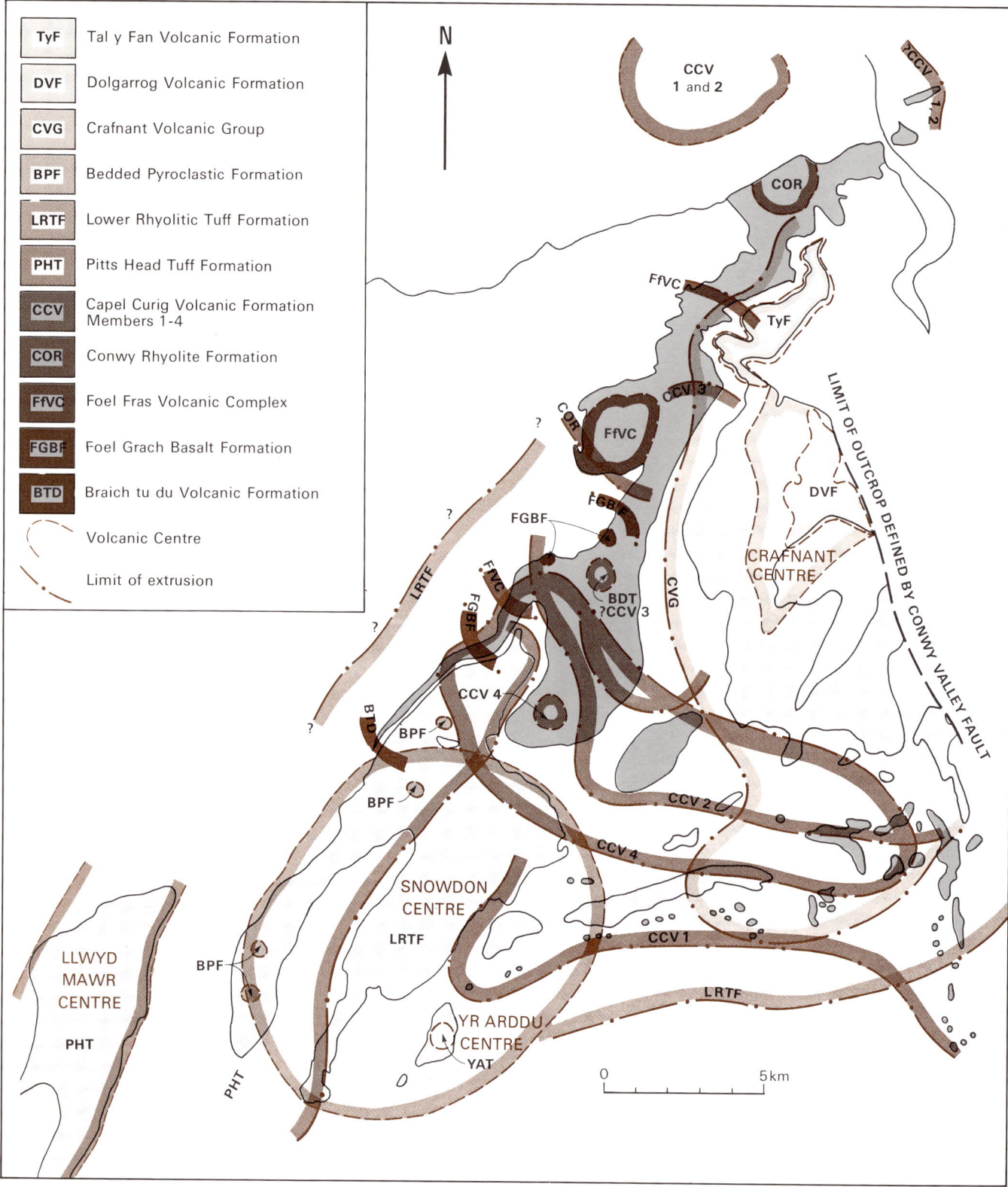

Figure 115 Sketch map showing distribution of Caradoc eruptive centres, related intrusions and limits of associated extrusions.

that the primary origin of the latter is so difficult to dispute.

We hope that in this book we have demonstrated, at least to some extent, the perspicacity of Harker's final statement (1889). A common factor to both Harker's work and ours is that the interpretations are based on the geological maps available. Harker (1889, p.2) acknowledged 'the admirable maps of the Geological Survey, supplemented by Sir Andrew Ramsay's Memoir'. These maps were produced on the scale of 1 inch to 1 mile. We have been involved in the primary 1:10 000 sheet survey of the Bangor (106) and Snowdon (119) 1:50 000 geological sheets and acknowledge the contribution of our colleagues in this work. As a result of this common approach to our interpretation it would be remiss of us if we did not attempt to emulate Harker's final sketch map (Figure 2) showing the distribution of intrusions and 'limits of the groups of lavas' in north-west Wales. In doing so (Figure 115) we were made all too well aware of how indebted subsequent generations of geologists should be to the extraordinary pioneering work of Sedgwick, Murchison, Ramsay and Harker in this classic area of Lower Palaeozoic geology.

REFERENCES

AIGNER, T. 1985. Storm depositional systems—dynamic stratigraphy in modern and ancient shallow-marine sequences. *Lecture notes in Earth Sciences*, Vol. 3. (Berlin: Springer-Verlag.)

ALLEGRE, C J, TREVOR, M, MINSTER, J F, MINSTER, B, and ALBAREDE, F. 1977. Systematic use of trace elements in igneous processes. *Contributions to Mineralogy and Petrology*, Vol. 60, 55–75.

ALLEN, P M. 1982. Lower Palaeozoic volcanism in Wales, the Welsh Borderland, Avon and Somerset. 65–91 in *Igneous rocks of the British Isles*. SUTHERLAND, D S (editor). (London: Wiley.)

— and JACKSON, A A. 1985. Geology of the country around Harlech. *Memoir of the British Geological Survey*, Sheet 135 with part of 149. (England and Wales.)

ALT, J C, and HONNOREZ, J. 1984. Alteration of the upper oceanic crust, DSDP site 417: mineralogy and chemistry. *Contributions to Mineralogy and Petrology*, Vol. 87, 149–169.

ANDERSON, T, and FLETT, J S. 1903. Report on the eruptions of the Soufrière in St. Vincent and on a visit to Montagne Pelé in Martinique. Pt. I. *Royal Society of London Philosophical Transactions*, Series A, Vol. 200, 353–553.

ARAMAKI, S. 1984. Formation of the Aira Caldera, Southern Kyushu, ~22,000 years ago. *Journal of Geophysical Research*, Vol. 89, B10, 8485–8501.

ARTHUR, M J. 1982. Investigations of geophysical anomalies in the Hereford area of the Welsh Borderland. *Publication of the Applied Geophysical Research, Institute of Geological Sciences*, No. 122.

BACON, C R, KURASAWA, H, DELEVAUX, M H, KISTLER, R W, and DOE, B R. 1984. Lead and strontium isotopic evidence for crustal interaction and compositional zonation in the source regions of Pleistocene basaltic and rhyolitic magmas of the Coso volcanic field, California. *Contributions to Mineralogy and Petrology*, Vol. 85, 366–375.

BADHAM, J P N. 1978. Slumped sulphide deposits at Avoca, Ireland and their significance. *Transactions of the Institution of Mining and Metallurgy*, Vol. B87, 21–26.

BAILEY, J C. 1981. Geochemical criteria for a refined tectonic discrimination of orogenic andesites. *Chemical Geology*, Vol. 32, 139–154.

BALL, T K, and BLAND, D J. 1985. The Cae Coch volcanogenic massive sulphide deposit, Trefriw, North Wales. *Journal of the Geological Society of London*, Vol. 142, 889–898.

— and MERRIMAN, R J. 1989. The petrology and geochemistry of the Ordovician Llewelyn Volcanic Group, Snowdonia, North Wales. *British Geological Survey Research Report*, No. SG/89/1.

BALLY, A W. 1983. A picture and work atlas, seismic expression of structural styles. *Publication of the American Association of Petroleum Geologists*.

BANCROFT, B B. 1928. On the unconformity at the base of the Ashgillian in the Bala District. *Geological Magazine*, Vol. 65, 484–493.

— 1933. Correlation tables of the stages Costonian–Onnian in England and Wales. Privately printed and published by the author, Blakeney, Gloucestershire, 1–4.

— 1945. The brachiopod zonal indices of the stages Costonian to Onnian in Britain. *Journal of Palaeontology*, Vol. 19, 181–252.

BANKS, D A. 1985. A fossil hydrothermal worm assemblage from the Tynagh lead-zinc deposit, Ireland. *Nature, London*, Vol. 313, 128–131.

BASSETT, D A. 1969. Some of the major structures of early Palaeozoic age in Wales and the Welsh Borderland: An historical essay. 67–116 in *The Precambrian and Lower Palaeozoic rocks of Wales*. WOOD, A (editor). (Cardiff: University of Wales Press.)

— WHITTINGTON, H B, and WILLIAMS, A. 1966. The stratigraphy of the Bala District, Merionethshire. *Quarterly Journal of the Geological Society of London*, Vol. 122, 219–271.

BEAVON, R V. 1963. The succession and structure east of the Glaslyn river, North Wales. *Quarterly Journal of the Geological Society of London*, Vol. 119, 479–512.

— 1980. A resurgent cauldron in the early Palaeozoic of Wales, U.K. *Journal of Volcanology and Geothermal Research*, Vol. 7, 157–174.

BECKLEY, A J. 1987. Basin development in North Wales during the Arenig. *Geological Journal*, Vol. 22, 19–30.

BEVINS, R E. 1979. The geology of the Strumble Head—Fishguard region, Dyfed, Wales. Unpublished PhD thesis, University of Keele.

— 1982. Petrology and geochemistry of the Fishguard Volcanic Complex, Wales. *Geological Journal*, Vol. 17, 1–21.

— and MERRIMAN, R J. 1988. Compositional controls on coexisting prehnite-actinolite and prehnite-pumpellyite facies assemblages in the Tal y Fan metabasite intrusion, N. Wales: implications for Caledonian metamorphic field gradients. *Journal of Metamorphic Geology*, Vol. 6, 17–39.

— and ROACH, R A. 1979a. Pillow lava and isolated pillow breccia of rhyodacitic composition from the Fishguard Volcanic Group, Lower Ordovician S.W. Wales, United Kingdom. *Journal of Geology*, Vol. 87, 193–201.

— — 1979b. Early Ordovician volcanism in Dyfed, S.W. Wales. 603–609 in *The Caledonides of the British Isles—reviewed*. HARRIS, A L, HOLLAND, C H, and LEAKE, B E (editors). *Special Publication of the Geological Society of London*, Vol. 8, 603–609.

— and ROBINSON, D. 1988. Low grade metamorphism of the Welsh Basin Lower Palaeozoic succession: an example of diastathermal metamorphism. *Journal of the Geological Society of London*, Vol. 145, 363–366.

— and ROWBOTHAM, G. 1983. Low grade metamorphism within the Welsh sector of the paratectonic Caledonides. *Geological Journal*, Vol. 18, 141–167.

— KOKELAAR, B P, and DUNKLEY, P N. 1984. Petrology and geochemistry of lower to middle Ordovician igneous rocks in Wales: a volcanic arc to marginal basin transition. *Proceedings of the Geologists' Association*, Vol. 95, 337–347.

BICK, D. 1982. The old copper mines of Snowdonia. Publishers. The Pound House.

BÖHLKE, J K, ALT, J C, and MUEHLENBACHS, K. 1984. Oxygen isotope water relations in altered deep sea basalts: low temperature mineralogical control. *Canadian Journal of Earth Sciences*, Vol. 21, 66–77.

BOYCE, A J, COLEMAN, M L, and RUSSELL, M J. 1983. Formation of fossil hydrothermal chimneys and mounds from Silvermines, Ireland. *Nature, London*, Vol. 306, 545–550.

BRALIA, A, SABATINI, G, and TROJA, F. 1979. A revaluation of the Co/Ni ratio in pyrite as a geochemical tool in ore genesis problems. *Mineralium Deposita*, Vol. 14, 353–374.

BRANNEY, M J. 1986. Discussion on Isolated pods of subaqueous welded ash-flows tuff: a distal facies of the Capel Curig Volcanic Formation (Ordovician) N. Wales. *Geological Magazine*, Vol. 123, 589–590.

BRENCHLEY, P J. 1964. Ordovician ignimbrites in the Berwyn Hills, North Wales. *Geological Journal*, Vol. 4, 43–54.

— 1972. The Cwm Clwyd Tuff, North Wales: a palaeogeographical interpretation of some Ordovician ash-shower deposits. *Proceedings of the Yorkshire Geological Society*, Vol. 39, 199–224.

— 1978. The Caradocian rocks of the north and west Berwyn Hills, North Wales. *Geological Journal*, Vol. 13, 137–164.

BRITISH GEOLOGICAL SURVEY. 1985a. Bangor. England and Wales Sheet 106. Solid. 1:50 000. (Southampton: Ordnance Survey for British Geological Survey.)

— 1985b. The Passes of Nant Ffrancon and Llanberis. Solid and Drift. 1:25 000, SH65/66. Classical areas of British geology. (Southampton: Ordnance Survey for British Geological Survey.)

— 1986. Bethesda and Foel Fras. Solid and Drift. 1:25 000. SH66/67. Classical areas of British geology. (Southampton: Ordnance Survey for British Geology Survey.)

— 1988. Llyn Padarn. Solid and Drift. 1:25 000. SH55/56. Classical areas of British geology. (Southampton: Ordnance Survey for British Geological Survey.)

— 1989. Snowdon Solid and Drift. 1:25 000, SH64/65. Classical areas of British geology (Southampton: Ordnance Survey for British Geological Survey.)

BROMLEY, A V. 1965. Intrusive quartz latites in the Blaenau Ffestiniog area, Merioneth. *Geological Journal*, Vol. 4, 247–256.

— 1969. Acid plutonic igneous activity in the Ordovician of North Wales. 387–408 in *The Precambrian and Lower Palaeozoic rocks of Wales*. WOOD, A (editor). (Cardiff: University of Wales Press.)

BURNHAM, C W. 1983. Deep submarine pyroclastic eruptions. 142–148 in The Kuroko and related volcanogenic massive sulphide deposits. OHMOTO, H, and SKINNER, B J (editors). *Economic Geology Monograph*, No. 5.

BUSBY-SPERA, C J. 1984. Large-volume rhyolite ash-flow eruptions and submarine caldera collapse in the Lower Mesozoic Sierra Nevada, California. *Journal of Geophysical Research*, Vol. 89, 8417–8428.

— 1986. Depositional features of rhyolitic and andesitic volcanic rocks of the Mineral King Submarine Caldera Complex, Sierra Nevada, California. *Journal of Volcanology and Geothermal Research*, Vol. 27, 43–76.

CAMPBELL, S D G. 1983. The geology of an area between Bala and Betws y coed, North Wales. Unpublished PhD thesis, University of Cambridge.

— 1984. Aspects of dynamic stratigraphy (Caradoc–Ashgill) in the northern part of the Welsh marginal basin. *Proceedings of the Geologists' Association*, Vol. 95, 390–391.

— REEDMAN, A J, and HOWELLS, M F. 1985. Regional variations in cleavage and fold development in North Wales. *Geological Journal*, Vol. 20, 43–52.

— — — and MANN, A C. 1987. The emplacement of geochemically distinct groups of rhyolites during the evolution of the Lower Rhyolitic Tuff Formation caldera (Ordovician), N. Wales, U.K. *Geological Magazine*, Vol. 124, 501–511.

— HOWELLS, M F, SMITH, M, and REEDMAN, A J. 1988. A Caradoc failed-rift within the Ordovician Marginal Basin of Wales. *Geological Magazine*, Vol. 125, 257–266.

CAREY, S, and SIGURDSSON, H. 1984. A model of volcanogenic sedimentation in marginal basins. 37–58 in Marginal basin geology: volcanic and associated sedimentology and tectonic processes in modern and ancient marginal basins. KOKELAAR, B P, and HOWELLS, M F (editors). *Special Publication of the Geological Society of London*, No. 16.

CAS, R A F, and WRIGHT, J V. 1987. *Volcanic successions: modern and ancient: a geological approach to processes, products and successions*. (London: Allen and Unwin (Publishers) Ltd.)

CATTERMOLE, P, and JONES, A. 1970. The geology of the area around Mynydd Mawr, Nantlle, Caernarvonshire. *Geological Journal*, Vol. 7, 111–128.

CAVE, R. 1965. The Nod Glas sediments of Caradoc age in North Wales. *Geological Journal*, Vol. 4, 279–298.

CHAPIN, C E, and ELSTON, W E. 1979. Ash-flow tuffs. *Special Paper of the Geological Society of America*, No. 180.

— and LOWELL, G R. 1979. Primary and secondary flow structures in ash-flow tuffs of the Gribbles Run palaeovalley, central Colorado. 137–154 in Ash-flow tuffs. CHAPIN, C E, and ELSTON, W E (editors). *Special Paper of the Geological Society of America*, No. 180, 137–154.

CLOUGH, C T, MAUFE, H B, and BAILEY, E B. 1909. The cauldron subsidence of Glen Coe and the associated igneous phenomena. *Quarterly Journal of the Geological Society of London*, Vol. 65, 611–676.

COCKS, L R M, and FORTEY, R A. 1982. Faunal evidence for oceanic separations in the Palaeozoic of Britain. *Journal of the Geological Society of London*, Vol. 139, 467–480.

COLE, J W. 1984. Taupo-Rotorua Depression: an ensialic marginal basin of North Island, New Zealand. 109–120 in Marginal basin geology: Volcanic and associated sedimentary and tectonic processes in modern and ancient marginal basins. KOKELAAR, B P, and HOWELLS, M F (editors). *Special Publication of the Geological Society of London*, No. 16. (Oxford: Blackwell.)

COLMAN, T B, and LAFFOLEY, N,D'A. 1986. Brittania or Snowdon Mine. *Bulletin of the Peak District Mines Historical Society*, Vol. 9, 313–331.

COLMAN, T B. In press. The geochemistry of wall-rock alteration associated with volcanogenic sulphide vein mineralisation Snowdonia, North Wales.

CONRAD, W K, NICHOLLS, I A, and WALL, V J. 1988. Water-saturated and -undersaturated melting of metaluminous and peraluminous crustal compositions at 10kb: evidence for the origin of silicic magmas in the Taupo Volcanic Zone, New Zealand, and other occurrences. *Journal of Petrology*, Vol. 29, 765–803.

CORNWELL, J D, PATRICK, D J, and TAPPIN, R J. 1980. Geophysical evidence for a concealed extension of the Tanygrisiau microgranite and its possible relation to mineralisation. *Mineral Reconnaissance Programme Report, Institute of Geological Sciences*, No. 38.

COWARD, M P, and SIDDANS, A W B. 1979. The tectonic evolution of the Welsh Caledonides. 187–198 in The

Caledonides of the British Isles—reviewed. HARRIS, A L, HOLLAND, C H, and LEAKE, B E (editors). *Special Publication of the Geological Society of London*, No. 8.

CROUDACE, I W. 1982. The geochemistry and petrogenesis of the Lower Palaeozoic granitoids of the Lleyn Peninsula, North Wales. *Geochimica et Cosmochimica Acta*, Vol. 46, 609–622.

DAKYNS, J R, and GREENLY, E. 1905. On the probable Peléan origin of the felsitic slates of Snowdon and their metamorphism. *Geological Magazine*, Vol. 42, 541–549.

DAVIES, D A B. 1936. Ordovician rocks of the Trefriw district (North Wales). *Quarterly Journal of the Geological Society of London*, Vol. 92, 62–90.

DAVIES, G R, and MACDONALD, R. 1987. Crustal influences in the petrogenesis of the Naivasha basalt-comendite complex: combined trace element and Sr-Nd-Pb isotope constraints. *Journal of Petrology*, Vol. 28, 1009–1031.

DAVIES, R G. 1959. The Cader Idris Granophyre and its associated rocks. *Quarterly Journal of the Geological Society of London*, Vol. 115, 189–216.

DAVIES, W, and CAVE, R. 1976. Folding and cleavage determined during sedimentation. *Sedimentary Geology*, Vol. 15, 89–133.

DEAN, W T. 1958. The faunal succession in the Caradoc Series of South Shropshire. *Bulletin of the British Museum of Natural History (Geology)*, Vol. 3, No. 6, 191–231.

— 1965. A shelly fauna from the Snowdon Volcanic Series at Twll Du, Caernarvonshire. *Geological Journal*, Vol. 4, 301–314.

DELANEY, P T, and POLLARD, D D. 1982. Solidification of basaltic magma during flow in a dike. *American Journal of Science*, Vol. 282, 856–885.

DEWEY, J F. 1969. Evolution of the Appalachian/Caledonian Orogen. *Nature, London*, Vol 222, 124–129.

— 1982. Plate tectonics and the evolution of the British Isles. *Journal of the Geological Society of London*, Vol. 139, 371–412.

DIGGENS, J N, and ROMANO, M. 1968. The Caradoc rocks around Llyn Cowlyd, North Wales. *Geological Journal*, Vol. 6, 31–48.

DOE, B R, LEEMAN, W P, CHRISTIANSEN, R L, and HEDGE, C E. 1982. Lead and strontium isotopes and related trace elements as genetic tracers in the upper Cenozoic rhyolite-basalt association of the Yellowstone plateau volcanic field. *Journal of Geophysical Research*, Vol. 87, 4785–4806.

DRUITT, T H. 1985. Vent evolution and lag breccia formation during the Cape Riva eruption of Santorini, Greece. *Journal of Geology*, Vol. 93, 439–454.

— and SPARKS, R S J. 1982. A proximal ignimbrite facies on Santorini, Greece. *Journal of Volcanology and Geothermal Research*, Vol. 13, 137–172.

— — 1984. On the formation of calderas during ignimbrite eruptions. *Nature, London*, Vol. 310, 679.

DUDAS, F Ö. 1983. The effect of volatile content on the vesiculation of submarine basalts. 134–141 *in* The Kuroko and related volcanogenic massive sulphide deposits. OHMOTO, H, and SKINNER, B J (editors). *Economic Geology Monograph*, No. 5.

DUNKLEY, P N. 1979. Ordovician volcanicity of the SE Harlech Dome. 597–601 *in* The Caledonides of the British Isles — reviewed. HARRIS, A L, HOLLAND, C H, and LEAKE, B E (editors). *Special Publication of the Geological Society of London*, No. 8.

EVANS, J A. 1990a. Resetting of the Rb-Sr whole-rock system of an Ordovician microgranite during low-grade metamorphism. *Geological Magazine*, Vol. 126, 675–679.

— 1990b. Resetting of Rb/Sr whole-rock isotope systems during low-grade metamorphism, North Wales. Unpublished PhD thesis, University of London.

EVANS, R B, and GREENWOOD, P G. 1988. Outcrop magnetic susceptibility measurements as a means of differentiating rock types and their mineralization. *Asian Mining '88*. Papers presented at the Asian Mining '88 Conference. (London: The Institution of Mining and Metallurgy.).

FENNER, C N. 1923. The origin and mode of emplacement of the great tuff deposit in the Valley of Ten Thousand Smokes. *National Geographic Society, Contributed Technical Papers, Katmai Series*, No. 1.

FISHER, R V. 1983. Flow transformations in sediment gravity flows. *Geology*, Vol. 11, 273–274.

— 1984. Submarine pyroclastic rocks. 5–28 *in* Marginal basin geology: volcanic and associated sedimentary and tectonic processes in modern and ancient marginal basins. KOKELAAR, B P, and HOWELLS, M F (editors). *Special Publication of the Geological Society of London*, No. 16.

— and SCHMINCKE, H-U. 1984. *Pyroclastic rocks*. (Berlin: Springer-Verlag.)

FISKE, R S, and MATSUDA, T. 1964. Submarine equivalent of ash-flows in the Tokiwa Formation, Japan. *American Journal of Science*, Vol. 262, 76–106.

FITCH, F J. 1967. Ignimbrite volcanism in North Wales. *Bulletin Volcanologique*, Vol. 30, 199–219.

— MILLER, J A, EVANS, A L, GRASTY, R L, and MENEISY, M Y. 1969. Isotopic age determinations on rocks from Wales and the Welsh Borders. 23–45 *in* The Precambrian and Lower Palaeozoic rocks of Wales. WOOD, A (editor). (Cardiff: University of Wales Press.)

FITCHES, W R. 1987. Aspects of veining in the Welsh Lower Palaeozoic Basin. 325–342 *in* Deformation of sediments and sedimentary rocks. JONES, M E, and PRESTON, R M F (editors). *Special Publication of the Geological Society of London*, No. 29.

— and CAMPBELL, S D G. 1987. Tectonic evolution of the Bala Lineament in the Welsh Basin. *Geological Journal*, Vol. 22, 131–153.

FITTON, J G, and DUNLOP, H M. 1985. The Cameroon Line, West Africa, and its bearing on the origin of oceanic and continental alkali basalt. *Earth and Planetary Science Letters*, Vol. 72, 23–38.

— and HUGHES, D J. 1970. Volcanism and plate tectonics in the British Ordovician. *Earth and Planetary Science Letters*, Vol. 8, 223–228.

— THIRLWALL, M F, and HUGHES, D J. 1982. Volcanism in the Caledonian orogenic belt in Britain. 611–636 *in Andesites*. THORPE, R S (editor). (London: John Wiley & Sons.)

— JAMES, D, KEMPTON, P D, ORMEROD, D S, and LEEMAN, P. 1988. The role of lithospheric mantle in the generation of late Cenozoic basic magmas in the western United States. 331–349 *in* Oceanic and continental lithosphere. MENZIES, M A, and COX, K (editors). *Journal of Petrology, Lithosphere Issue*.

FLORES, R M. 1975. Shortheaded stream deltas: Model for Pennsylvanian Haymond Formation. *Bulletin of the American Association of Petroleum Geologists*, Vol. 59, 2288–2301.

FRANCIS, E H, and HOWELLS, M F. 1973. Transgressive welded ash-flow tuffs among the Ordovician sediments of N.E. Snowdonia, N. Wales. *Journal of the Geological Society of London*, Vol. 129, 621–641.

FRANKLIN, J M, LYDON, J W, and SANGSTER, D F. 1981. Volcanic-associated massive sulfide deposits. 485–627 in *Economic Geology 75th Anniversary Volume*.

FRITZ, W J, and AXELROD, R B. 1987.   Dynamics of some large wave ripples, Ordovician, Capel Curig Volcanic Formation, N. Wales. Abstract, Society of Economic Paleontologists and Mineralogists, Annual Meeting, Denver.

— HOWELLS, M F, REEDMAN, A J, and CAMPBELL, S D G. 1990. Volcaniclastic sedimentation in an Ordovician subaqueous caldera, Lower Rhyolitic Tuff Formation, North Wales. *Bulletin of the Geological Society of America*, Vol. 96, 1246–1256.

GEOLOGICAL SURVEY OF GREAT BRITAIN. 1965.   Aeromagnetic map of Great Britain. Sheet 2. England and Wales (South of National Grid Line 500 km N). 1:625 000. (Chessington, Surrey: Ordnance Survey for Geological Survey of Great Britain.)

GIBBONS, W. 1983.   Stratigraphy, subduction and strike-slip faulting in the Mona Complex of North Wales—a review. *Proceedings of the Geologists' Association*, Vol. 94, 147–163.

— 1985.   Geology and paleobiology of islands in the Ordovician Iapetus Ocean: review and implications: discussion. *Bulletin of the Geological Society of America*, Vol. 96, 1225–1226.

— and GAYER, R A. 1985.   British Caledonian terranes.   3–16 *in* The tectonic evolution of the Caledonide-Appalachian Orogen. GAYER, R A (editor). *Earth Evolution Sciences, Monograph Series*, No. 1.

GRUNDER, and MAHOOD, G A. 1988.   Physical and chemical models of zoned silicic magmas: The Loma Seca Tuff and Calabozos caldera, Southwest Andes. *Journal of Petrology*, Vol. 29, 831–867.

GUBER, A L, and MERRILL, S. 1983.   Palaeobathymetric significance of foramanifera from the Hukoroku district.   55–70 *in* The Kuroko and related volcanogenic massive suldhide deposits. OHMOTO, H, and SKINNER, B J (editors). *Economic Geology Monograph*, No. 5.

GUDMUNDSSON, A. 1988.   Formation of collapse calderas. *Geology*, Vol. 16, 808–810.

HAHN, G A, ROSE, W J, and MEYERS, T. 1979.   Geochemical correlation of genetically related rhyolitic ash-flow and air-fall ashes, central and western Guatemala and equatorial Pacific. 101–112 *in* Ash-flow tuffs. CHAPIN, C E, and ELSTON, W E (editors).   *Special Paper of the Geological Society of America*, No. 180.

HALBACH, P, NAKAMURA KO-ICHI, WAHSNER, M, LANGE, J, SAKAI, H, KÄSELITZ, L, HANSEN, R-D, YAMANO, M, POST, J, PRAUSE, B, SEIFERT, R, MÄICHAELIS, W, TEICHMANN, F, KINOSHITA, M, MÄRTEN, A, ISHIBASHI, J, CZERWINSKI, S, and BLUM, N. 1989.   Probable modern analogue of Kuroko-type massive sulphide deposits in the Okinawa Trough back-arc basin. *Nature, London*, Vol. 36, 496–498.

HARKER, A. 1889.   *The Bala Volcanic Series of Caernarvonshire.* (Cambridge: University Press.)

HART, S R, and DAVIS, K C. 1978.   Nickel partitioning between olivine and silicate melt.   *Earth and Planetary Science Letters*, Vol. 40, 203–219.

HARVEY, P K, and ATKIN, B P. 1982.   Automated X-Ray Fluorescence Analysis.   1–10 *in* Sampling and analysis for the mineral industry. *Institute of Mining and Metallurgy*.

HILDRETH, W. 1979.   The Bishop Tuff: evidence for the origin of compositional zonation in silicic magma chambers.   43–75 *in* Ash-flow tuffs. CHAPIN, C E, and ELSTON, W E (editors). *Special Paper of the Geological Society of America*, No. 180.

— 1981.   Gradients in silicic magma chambers. implications for lithospheric magmatism.   *Journal of Geophysical Research*, Vol. 86, 10153–10192.

HOWELLS, M F, ALLEN, P M, ADDISON, R, WEBB, B C, LYNAS, B D T, and JACKSON, A. 1977.   Folding and cleavage determined during sedimentation—comments.   *Sedimentary Geology*, Vol. 17, 333–338.

— CAMPBELL, S D G, and REEDMAN, A J. 1985a.   Isolated pods of subaqueous welded ash-flow tuff: a distal facies of the Capel Curig Volcanic Formation (Ordovician), North Wales.   *Geological Magazine*, Vol. 122, 175–180.

— — — and TUNNICLIFF, S P. 1987.   An acidic fissure-controlled volcanic centre (Ordovician) at Yr Arddu, N. Wales. *Geological Journal*, Vol. 21, 133–149.

— FRANCIS, E H, LEVERIDGE, B E, and EVANS, C D R. 1978. *Classical area of British geology: Capel Curig and Betws y Coed: Description of 1:25 000 sheet SH75.* (London: HMSO for Institute of Geological Sciences.)

— and LEVERIDGE, B E. 1980.   The Capel Curig Volcanic Formation.   *Report of the Institute of Geological Sciences*, No. 80/6.

— — ADDISON, R, EVANS, C D R, and NUTT, M J C. 1979. The Capel Curig Volcanic Formation, Snowdonia, North Wales; variations in ash-flow tuffs related to emplacement environment. 611–618 *in* The Caledonides of the British Isles — reviewed. HARRIS, A, HOLLAND, C H, and LEAKE, B E (editors). *Special Publication of the Geological Society of London*, No. 8.

— — — and REEDMAN, A J. 1983.   The lithostratigraphical subdivision of the Ordovician underlying the Snowdon and Crafnant volcanic groups, North Wales.   *Report of the Institute of Geological Sciences*, No. 83/1.

— — and EVANS, C D R. 1973.   Ordovician ash-flow tuffs in eastern Snowdonia.   *Report of the Institute of Geological Sciences*, No. 73/3.

— — — NUTT, M J C. 1981a.   *Classical areas of British geology: Dolgarrog: Description of 1:25 000 Geological Sheet SH76.* (London: HMSO for Institute of Geological Sciences.)

— —and REEDMAN, A J. 1981b.   *Snowdonia—rocks and fossils.* (London: Unwin Paperbacks.)

— REEDMAN, A J, and CAMPBELL, S D G. 1986.   The submarine eruption and emplacement of the Lower Rhyolitic Tuff Formation (Ordovician), N. Wales.   *Journal of the Geological Society of London*, Vol. 143, 411–424.

— — and LEVERIDGE, B E. 1985b.   Geology of the country around Bangor. Explanation for 1:50 000 Sheet 106 (England and Wales). (London: HMSO for British Geological Survey.)

HUGHES, C J. 1972.   Spilites, keratophyres and the igneous spectrum.   *Geological Magazine*, Vol. 109, 513–527.

— 1982.   *Igneous petrology. Developments in petrology 7.* (Amsterdam: Elsevier.)

HUIJSMANS, J P P. 1985.   Calc-alkaline lavas from the volcanic complex of Santorini, Aegean Sea, Greece.   *Geologica Ultraiectina*, Vol. 41.

HURST, J M. 1979.   The stratigraphy and brachiopods of the upper part of the type Caradoc of South Salop.   *Bulletin of the British Museum of Natural History (Geology)*, Vol. 32, 183–304.

HUTTON, D H W. 1987.   Strike-slip terranes and a model for the evolution of the British and Irish Caledonides.   *Geological Magazine*, Vol. 124, 405–425.

INSTITUTE OF GEOLOGICAL SCIENCES. 1976.   Capel Curig and Betws y Coed. Solid and Drift. 1:25 000. SH75. Classical areas of British geology. (Southampton: Ordnance Survey for Institute of Geological Sciences.)

— 1978a. Liverpool Bay. Sheet 53°N–04°W. Solid geology. 1:250 000 (Southampton: Ordnance Survey for Institute of Geological Sciences.)

— 1978b. Liverpool Bay. Sheet 53°N–04°W. Bouguer gravity anomaly map (Provisional edition). 1:250 000 (Southampton: Ordnance Survey for Institute of Geological Sciences.)

— 1981a. Anglesey. Sheet 53°N–06°W. Solid geology. 1:250 000. (Southampton: Ordnance Survey for Institute of Geological Sciences.)

— 1981b. Dolgarrog. Solid and Drift. 1:25 000. SH76. Classical areas of British geology. (Southampton: Ordnance Survey for Insitute of Geological Sciences.)

— 1982a. Harlech. England and Wales Sheet 135 and part of 149. Solid and Drift. 1:50 000. (Southampton: Ordnance Survey for Institute of Geological Sciences.)

— 1982b. Cardigan Bay. Sheet 52°N–06°W. Solid geology. 1:250 000 (Southampton: Ordnance Survey for Institute of Geological Sciences.)

ISHIKARA, Y, SAWAGUCHI, T, IWAYA, S, and HORIUCHI, M. 1976. Delineation of prospecting targets for Kuroko deposits based on modes of volcanism of underlying dacite and alteration haloes. *Mining Geology*, Vol. 26, 105–117 [in Japanese].

JACKSON, M P A, and TALBOT, C J. 1989. Anatomy of mushroom-shaped diapirs. *Journal of Structural Geology*, Vol. 11, 211–230.

JONES, O T. 1938. On the evolution of a geosyncline. *Quarterly Journal of the Geological Society of London*, Vol. 94, lx–cx.

— and PUGH, W J. 1949. An early Ordovician shoreline in Radnorshire, near Builth Wells. *Quarterly Journal of the Geological Society of London*, Vol. 105, 65–99.

JONES, W B. 1983. The geological association of sulphide mineralisation at Avoca, County Wicklow—a new interpretation based on field evidence. *Journal of Earth Sciences of the Royal Dublin Society*, Vol. 5, 145–152.

KOKELAAR, B P. 1979. Tremadoc to Llanvirn volcanism on the southeast side of the Harlech Dome (Rhobell Fawr), N. Wales. 591–596 *in* The Caledonides of the British Isles—reviewed. HARRIS, A L, HOLLAND, C H, and LEAKE, B E (editors). *Special Publication of the Geological Society of London*, No. 8.

— 1982. Fluidization of wet sediments during the emplacement and cooling of various igneous bodies. *Journal of the Geological Society of London*, Vol. 139, 21–33.

— 1986. Petrology and Geochemistry of the Rhobell Volcanic Complex: amphibole dominated fractionation at an early Ordovician arc volcano in North Wales. *Journal of Petrology*, Vol. 27, 887–914.

— 1988. Tectonic controls of Ordovician arc and marginal basin volcanism in Wales. *Journal of the Geological Society of London*, Vol. 145, 759–775 .

— HOWELLS, M F, BEVINS, R E, ROACH, R A, and DUNKLEY, P N. 1984. The Ordovician marginal basin of Wales. 245–269 *in* Marginal basin geology, volcanic and associated sedimentary and tectonic processes in modern and ancient marginal basins. KOKELAAR, B P, and HOWELLS, M F (editors). *Special Publication of the Geological Society of London*, No. 16.

— BEVINS, R E, and ROACH, R A. 1985. Submarine silicic volcanism and associated sedimentary and tectonic processes, Ramsay Island, S.W. Wales. *Journal of the Geological Society of London*, Vol. 142, 591–614.

LAPWORTH, C. 1879. On the tripartite classification of the Lower Palaeozoic rocks. *Geological Magazine*, Vol. 16, 1–15.

LARTER, R C L, BOYCE. A J, and RUSSELL, M J. 1981. Hydrothermal pyrite chimneys from the Ballynoe Barite Deposit, Silvermines, County Tipperary, Ireland. *Mineralium Deposita*, Vol. 16, 309–318.

LEAT, P T, and THORPE, R S. 1986. Geochemistry of an Ordovician basalt-trachybasalt-subalkaline/peralkaline rhyolite association from the Lleyn Peninsula, North Wales, U.K. *Geological Journal*, Vol. 21, 29–43.

— —1989. Snowdon Volcanic Group basalts date end of south-directed Caledonian subduction by the Longvillian. *Journal of the Geological Society of London*, Vol. 146, 965–970.

— JACKSON, S E, THORPE, R S, and STILLMAN, C J. 1986. Geochemistry of bimodal basalt-subalkaline/peralkaline rhyolite provinces associated with volcanogenic mineralization within the Southern British Caledonides. *Journal of the Geological Society of London*, Vol. 143, 259–274.

— THOMPSON, R N, MORRISON, M A, HENDRY, G L, and DICKIN, A P. 1988. Compositionally diverse Miocene-Recent rift related magmatism in NW Colorado: partial melting, and mixing of mafic magmas from 3 different asthenospheric and lithospheric mantle sources. *Journal of Petrology*, special lithosphere issue, 351–377.

LE BAS, M J, LE MAITRE, R W, STRECKEISEN, A, and ZANETTIN, B. 1986. A chemical classification of volcanic rocks based on the Total Alkali-Silica diagram. *Journal of Petrology*, Vol. 27, 745–750.

LEGGETT, J K, McKEWAN, W S, and SOPER, N J. 1983. A model for the crustal evolution of southern Scotland. *Tectonics*, Vol. 2, 187–210.

LEWIS, K B, and PANTIN, H M. 1984. Intersection of a marginal basin with a continent: structure and sediments of the Bay of Plenty, New Zealand. 121–135 *in* Marginal basin geology: volcanic and associated sedimentary and tectonic processes in modern and ancient marginal basins. KOKELAAR, B P, and HOWELLS, M F (editors). *Special Publication of the Geological Society of London*, No. 16.

LIOU, J G, MARYUMA, S, and CHO, M. 1985. Phase equilibria and mineral paragenesis of metabasites in low grade metamorphism. *Mineralogical Magazine*, Vol. 49, 321–333.

LIPMAN, P W. 1965. Chemical comparison of glassy and crystalline volcanic rocks. *Bulletin of the United States Geological Survey*, No. 1201-D.

— 1976. Caldera-collapse breccias in the western San Juan Mountains, Colorado. *Bulletin of the Geological Society of America*, Vol. 87, 1397–1410.

— 1984. The roots of ash-flow calderas in western North America: windows into granitic batholiths. *Journal of Geophysical Research*, Vol. 89, 8801–8841.

— 1988. Evolution of silicic magma in the upper crust: the mid-Tertiary Latir volcanic field and its cogenetic granitic batholith, northern New Mexico, U.S.A. *Transactions of the Royal Society of Edinburgh: Earth Sciences*, Vol. 79, 265–288.

— FISHER, F S, MEHNERT, H H, NAESER, C W, LUEDKE, R G, and STEVEN, T A. 1976. Multiple ages of mid-Tertiary mineralization in the western San Juan Mountains, Colorado. *Economic Geology*, Vol. 71, 571–588.

— DOE, B R, HEDGE, C E, and STEVEN, T A. 1978. Petrologic evolution of the San Juan volcanic field, southwestern Colorado: Pb and Sr isotope evidence. *Bulletin of the Geological Society of America*, Vol. 89, 59–82.

LOCKLEY, M G. 1980. The Caradoc faunal associations of the area between Bala and Dinas Mawddwy, North Wales. *Bulletin* of the British Museum of Natural History (Geology), Vol. 33, 165–235.

LOFGREN, G. 1970. Experimental devitrification rates of rhyolitic glass. Bulletin of the Geological Society of America, Vol. 81, 553–560.

LONSDALE, P F, and HAWKINS, J W K. 1985. An off-axis geothermal field with silicic volcanism in the Mariana Trough back-arc basin. Bulletin of the Geological Society of America.

LOWMAN, R D W, and BLOXAM, T W. 1981. The petrology of the Lower Palaeozoic Fishguard Volcanic Group and associated rocks E of Fishguard, N Pembrokeshire (Dyfed), South Wales. Journal of the Geological Society of London, Vol. 138, 47–68.

LYDON, J W. 1984. Some observations on the mineralogical and chemical zonation pattern of volcanogenic sulphide deposits of Cyprus. Geological Survey of Canada Papers, No. 84-1A, 611–616.

LYNAS, B D T. 1983. Two new Ordovician volcanic centres in the Shelve inlier, Powys, Wales. Geological Magazine, Vol. 120, 535–542.

McBIRNEY, A R. 1963. Factors governing the nature of submarine volcanism. Bulletin Volcanologique, Vol. 26, 455–469.

— 1971. Oceanic volcanism: a review. Geophysical Space Physics Review, Vol. 9, 523–556.

MACDONALD, G A. 1972. Volcanoes. (Englewood Cliffs, New Jersey: Prentice-Hall, Inc.)

MACDONALD, R. 1974. Nomenclature and petrochemistry of the peralkaline oversaturated extrusive rocks. Bulletin Volcanologique, Vol. 38, 498–516.

— DAVIS, G R, BLISS, C M, LEAT, P T, BAILEY, D K, and SMITH, R L. 1987. Geochemistry of high-silica peralkaline rhyolites, Naivasha, Kenya Rift Valley. Journal of Petrology, Vol. 28, 979–1008.

McKERROW, W S, and COCKS, L R M. 1976. Progressive faunal migration across the Iapetus Ocean. Nature, London, Vol. 263, 304–306.

MAHOOD, G A. 1984. Pyroclastic rocks and calderas associated with strongly peralkaline magmatism. Journal of Geophysical Research, Vol. 89, 8540–8552.

MANN, A C. 1983. Trace element geochemistry of high alumina basalt–andesite–dacite–rhyodacite lavas of the Main Volcanic Series of Santorini Volcano, Greece. Contribution to Mineralogy and Petrology, Vol. 84, 43–57

MARSHALL, P. 1935. Acid rocks of the Taupo-Rotorua district. Transactions of the Royal Society of New Zealand, Vol. 64, 323–366.

MATSUKUMA, T, and HORIKOSHI, E. 1970. Kuroko deposits in Japan, a review. 153–179 in Volcanism and ore genesis. TATSUMI, T (editor). (Tokyo.)

MERRIMAN, R J, BEVINS, R E, and BALL, T K. 1986. Geochemical variations within the Tal y Fan intrusion: implications for element mobility during low-grade metamorphism. Journal of Petrology, Vol. 27, 1409–1436.

— and ROBERTS, B. 1985. A survey of white mica crystallinity and polytypes in pelitic rocks of Snowdonia and Lleyn, North Wales. Mineralogical Magazine, Vol. 49, 305–319.

MURCHISON, R I. 1835. On the Silurian System of rocks. London and Edinburgh Philosophical Magazine. Series 3.7, 46–52.

MURPHY, F C, and HUTTON, D H W. 1986. Is the Southern Uplands of Scotland really an accretionary prism? Geology, Vol 14, 354–357.

NEUMANN VAN PADANG, M. 1933. De uitbarsting van den Merapi (Midden Java) in de jaren 1930–1931: Ned. Indies. Dienst Mijnbouwk. Vulkan. Seism. Mededel, Vol. 12. [With English summary.]

NOBLE, D C. 1967. Sodium, potassium and ferrous iron contents of some secondarily hydrated natural silicic glasses. American Mineralogist, Vol. 52, 230–285.

— VOGEL, T A, PETERSON, P A, LANDIS, G P, GRANT, N K, JEZEK, P, and McKIE, E H. 1984. Rare element enriched S type ash-flow tuffs containing phenocrysts of muscovite, andalusite and sillimanite, S.E. Peru. Geology, Vol. 12, 35–39.

NOCKOLDS, S R. 1938. Acmite in the riebeckite-microgranite of Mynydd Mawr. Mineralogical Magazine, Vol. 25, 35.

NOTSU, K, ONO, K, and SOYA, T. 1987. Strontium isotopic relations of bimodal volcanic rocks at Kikai volcano in the Ryúkú arc, Japan. Geology, Vol. 15, 345–348.

NYSTROM, J O. 1984. Rare earth element mobility in vesicular lava during low-grade metamorphism. Contribution to Mineralogy and Petrology, Vol. 88, 328–331.

O'BRIEN, C, PLANT, J A, SIMPSON, P R, and TARNEY, J. 1985. The geochemistry, metasomatism and petrogenesis of the granites of the English Lake District. Journal of the Geological Society of London, Vol. 142, 1139–1158.

OHMOTO, H, and SKINNER, B J. 1983. The Kuroko and related volcanogenic massive sulfide deposits: introduction and summary of new findings. 1–8 in The Kuroko and related volcanogenic massive sulphide deposits. OHMOTO, H, and SKINNER, B J (editors). Economic Geology Monograph, No. 5.

— and TAKAHASHI, T. 1983. Geological setting of the Kuroko deposits, Japan. III Submarine calderas and Kuroko genesis. 39–54 in The Kuroko and related volcanogenic massive sulphide deposits. OHMOTO, H, and SKINNER, B J (editors). Economic Geology Monograph, No. 5.

OLIVER, R L. 1954. Welded tuffs in the Borrowdale Volcanic Series, English Lake District, with a note on similar rocks in Wales. Geological Magazine, Vol. 91, 473–483.

ORMEROD, D S, HAWKESWORTH, C J, ROGERS, N W, LEEMAN, W P, and MENZIES, M A. 1988. Tectonic and magmatic transitions in the Western Great Basin, USA. Nature, London, Vol. 333, 349–353.

ORTON, G. 1987. Discussion of Submarine eruption and emplacement of the Lower Rhyolitic Tuff Formation, Ordovician, N Wales. Journal of the Geological Society of London, Vol. 144, 523–525.

— 1988a. Ordovician volcaniclastic sedimentation, North Wales. Unpublished DPhil thesis, University of Oxford.

— 1988b. A spectrum of mid-Ordovician fan and braidplain deltaic sequences, North Wales: a consequence of varying fluvial input. 23–49 in Fan deltas: tectonic setting, sedimentology and recognition. NEMEX, W, and STEEL, R J (editors). (Glasgow: Blackie.)

PEARCE, J A. 1982a. Trace element characteristics of lavas from destructive plate boundaries. 525–548 in Andesites. THORPE, R S (editor). (New York: John Wiley.)

— 1982b. Role of the sub-continental lithosphere in magma genesis at active continental margins. 230–249 in Continental basalts and mantle xenoliths. HAWKESWORTH, C J, and NORRY, M J (editors). (Nantwich: Shiva.)

— and CANN, J R. 1973. Tectonic setting of basic volcanic rocks determined by using trace element analyses. Earth and Planetary Science Letters, Vol. 19, 290–300.

— HARRIS, N B W, and TINDLE, A G. 1984a. Trace element discrimination diagrams for the tectonic interpretation of granitic rocks. *Journal of Petrology*, Vol. 25, 956–983.

— LIPPARD, S J, and ROBERTS, S. 1984b. Characteristics and tectonic significance of supra-subduction zone ophiolites. 77–94 *in* Marginal basin geology: volcanic and associated sedimentary and tectonic processes in modern and ancient marginal basins. KOKELAAR, B P, and HOWELLS, M F (editors). *Special Publication of the Geological Society of London*. No. 16.

— and NORRY, M J. 1979. Petrogenetic implications of Ti, Zr, Y and Nb variations in volcanic rocks. *Contributions to Mineralogy and Petrology*, Vol. 69, 33–47.

PERRET, F A. 1935. The eruption of Mt. Pelée 1929–1932. *Carnegie Institute of Washington Publication*, No. 458.

PETERSON, M D, and LAMBERT, I B. 1979. Mineralogical and chemical zonation around the Woodlawn Cu-Pb-Zn ore deposit, south-eastern New South Wales. *Journal of the Geological Society of Australia*, Vol. 26, 169–186.

PHILLIPS, W E A, STILLMAN, C J, and MURPHY, T. 1976. A Caledonian plate tectonic model. *Journal of the Geological Society of London*, Vol. 132, 579–609.

PICKERILL, R K. 1977. Trace fossils from the Upper Ordovician (Caradoc) of the Berwyn Hills, Central Wales. *Geological Journal*, Vol. 12, 1–16.

— and BRENCHLEY, P J. 1979. Caradoc marine benthic communities of the south Berwyn Hills, North Wales. *Palaeontology*, Vol. 22, 229–264.

PICKERING, K T. 1987. Wet sediment deformation in the Upper Ordovician Point Leamington Formation: an active thrust–imbricate system during sedimentation, Notre Dame Bay, north-central Newfoundland. 213–219 *in* Deformation of sediments and sedimentary rocks. JONES, M E, AND PRESTON, R M F (editors). *Special Publication of the Geological Society of London*, No. 29.

PISUTHA-ARNOUD, V, and OHMOTO, J. 1983. Thermal history, and chemical and isotopic composition of the ore-forming fluids responsible for the Kuroko massive sulphide deposits in the Hukuroko district of Japan. 523–558 *in* The Kuroko and related volcanogenic massive sulphide deposits. OHMOTO, H, and SKINNER, B J (editors). *Economic Geology Monograph*, No. 5.

PLATT, J W. 1977. Volcanogenic mineralisation at Avoca, Co. Wicklow, Ireland and its regional implications. 163–170 *in* Volcanic processes in ore genesis. JONES, M (editor). *Institution of Mining and Metallurgy and Geological Society of London*.

RAMSAY, A C. 1866. The geology of North Wales. *Memoir of the Geological Survey of Great Britain*, 1st edition.

— 1881. The geology of North Wales. *Memoir of the Geological Survey of Great Britain*, No. 3, 2nd edition.

RAST, N. 1961. Mid-Ordovician structures in south-western Snowdonia. *Liverpool and Manchester Geological Journal*, Vol. 2, 645–651.

— 1969. The relationship between Ordovician structure and volcanicity in Wales. 303–335 in *The Pre-cambrian and Lower Palaeozoic rocks of Wales*. WOOD, A (editor). (Cardiff: University of Wales Press.)

— BEAVON, R V, and FITCH, F J. 1958. Sub-aerial volcanicity in Snowdonia. *Nature, London*, Vol. 181, 508.

REEDMAN, A J, WEBB, B C, ADDISON, R, LYNAS, B D T, LEVERIDGE, B E, and HOWELLS, M F. 1983. The Cambrian–Ordovician boundary between Aber and Betws Garmon, Gwynedd, North Wales. *Report of the Institute of Geological Sciences*, No. 83/1.

— LEVERIDGE, B E, and EVANS, R B. 1984. The Arvon Group ('Arvonian') of North Wales. *Proceedings of the Geologists' Association*, Vol. 95, 313–321.

— COLMAN, T B, CAMPBELL, S D G, and HOWELLS, M F. 1985. Volcanogenic mineralization related to the Snowdon Volcanic Group (Ordovician), North Wales. *Journal of the Geological Society of London*, Vol. 142, 875–888.

— HOWELLS, M F, ORTON, G, and CAMPBELL, S D G. 1987a. The Pitts Head Tuff Formation: a subaerial to submarine welded ash-flow tuff of Ordovician age, North Wales. *Geological Magazine*, Vol. 124, 427–439.

— PARK, K H, MERRIMAN, R J, and KIM, S E. 1987b. Welded tuff infilling a volcanic vent at Weolsong, Republic of Korea. *Bulletin of Volcanology*, Vol. 49, 541–546.

REID, F, and FROSTICK, L E. 1985. Beach orientation, bar morphology and the concentration of metalliferous placer deposits: a case study, Lake Turkana, N. Kenya. *Journal of the Geological Society of London*, Vol. 192, 837–848.

RIDGWAY, J. 1975. The stratigraphy of Ordovician volcanic rocks on the southern and eastern flanks of the Harlech Dome in Merionethshire. *Geological Journal*, Vol. 10, 87–106.

— 1976. Ordovician palaeogeography of the southern and eastern flanks of the Harlech Dome, Merionethshire, North Wales. *Geological Journal*, Vol. 11, 121–136.

ROBERTS, B. 1969. The Llwyd Mawr Ignimbrite and its associated volcanic rocks. 337–356 in *The Pre-Cambrian and Lower Palaeozoic rocks of Wales*. WOOD, A (editor). (Cardiff: University of Wales Press.)

— 1981. Low grade and very low grade regional metabasic Ordovician rocks of Llyn and Snowdonia, Gwynedd, North Wales. *Geological Magazine*, Vol. 118, 189–200.

— EVANS, J A, MERRIMAN, R J, and SMITH, M. 1989. Discussion on low grade metamorphism of the Welsh Basin Lower Palaeozoic succession: an example of diastathermal metamorphism? *Journal of the Geological Society of London*, Vol. 146, 885–890.

— and MERRIMAN, R J. 1985. The distinction between Caledonian burial and regional metamorphism in metapelites from North Wales: an analysis of isocryst patterns. *Journal of the Geological Society of London*, Vol. 142, 615–624.

— — 1990. Cambrian and Ordovician metabentonites and their relevance to the origins of associated mudrocks in the northern sector of the Lower Palaeozoic Welsh marginal basin. *Geological Magazine*, Vol. 126, 31–43.

— and SIDDANS, A W B. 1971. Fabric studies in the Llwyd Mawr Ignimbrite, Caernarvonshire, North Wales. *Tectonophysics*, Vol. 12, 283–306.

ROBINSON, D. 1987. The transition from diagenesis to metamorphism in extensional and collision settings. *Geology*, Vol. 15, 966–969.

— and BEVINS, R E. 1986. Incipient metamorphism in the Lower Palaeozoic marginal basin of Wales. *Journal of Metamorphic Geology*, Vol. 4, 101–113.

ROMANO, M, and DIGGENS, J N. 1969. Longvillian shelly faunas from the Dolwyddelan area, North Wales. *Geological Magazine*, Vol. 106, 603–606.

ROSS, G S, and SMITH, R L. 1961. Ash-flow tuffs, their origin, geological relation and identification. *United States Geological Survey Professional paper*, No. 366.

RUSSELL, M J, BOYCE, A J, LARTER, R C L, and SAMSON, I M. 1981. The significance of hydrothermal chimneys in the Silvermines deposit. In *Mineral exploration in Ireland, progress and developments 1971 – 1981.* Brown, A G (editor). (Irish Association for Economic Geologists.)

— Hall, A J, and Dullar, P. 1984. Implications of chemical garden growth to the understanding of certain ore morphologies in the Navan orebody. (Abstract). *Bulletin of the Mineralogical Society of London.*

Sato, H. 1977. Nickel content of basaltic magmas: identification of primary magmas and a measure of the degree of olivine fractionation. *Lithos*, Vol. 10, 113 – 120.

Saunders, A D, and Tarney, J. 1984. Geochemical characteristics of basaltic volcanism within back-arc basins. 59 – 76 *in* Marginal basin geology: volcanic and associated sedimentary and tectonic processes in modern and ancient marginal basins. Kokelaar, B P, and Howells, M F (editors). *Special Publication of the Geological Society of London*, No. 16.

Schiffman, P, and Lofgren, G E. 1982. Dynamic crystallization studies on the Grande Ronde pillow basalts, central Washington. *Journal of Geology*, Vol. 90, 49 – 78.

Sedgwick, A. 1843. Outline of the geological structure of North Wales. *Proceedings of the Geological Society of London*, Vol. 4, 212 – 224.

Shackleton, R M. 1954. The structural evolution of North Wales. *Liverpool and Manchester Geological Journal*, Vol. 1, 261 – 296.

— 1959. The stratigraphy of the Moel Hebog district between Snowdon and Tremadoc. *Liverpool and Manchester Geological Journal*, Vol. 2, 216 – 252.

Sheridan, M F. 1979. Emplacement of pyroclastic flows: A review. 125 – 136 *in* Ash-flow tuffs. Chapin, C E, and Elston, W E (editors). *Special Paper of the Geological Society of America*, No. 180.

Sherlock, R L. 1919. The geology and genesis of the Trefriw pyrites deposit. *Quarterly Journal of the Geological Society of London*, Vol. 74, 106 – 115.

Smith, M. 1987. The Tremadoc 'Thrust' Zone in southern central Snowdonia. *Geological Journal*, Vol. 22, 119 – 129.

— 1988. The tectonic evolution of the Cambro-Ordovician rocks of southern central Snowdonia. Unpublished PhD thesis, University of Wales, Aberystwyth.

Smith, R L. 1960a. Ash-flows. *Bulletin of the Geological Society of America*, Vol. 71, 795 – 842.

— 1960b. Zones and zonal variations in welded ash-flows. *Professional Paper of the United States Geological Survey*, No. 354-F, 149 – 159.

— 1979. Ash-flow magmatism. 5 – 27 *in* Ash-flow tuffs. Chapin, C E, and Elston, W E (editors). *Special Paper of the Geological Society of America*, No. 180.

— and Bailey, R A. 1968. Resurgent cauldrons. *Memoir of Geological Society of America*, No. 116, 613 – 662.

Snelling, N J. 1987. A geological time scale. *Modern Geology*, Vol. 11, 365 – 374.

Soper, N J, and Hutton, D H W. 1984. Late Caledonian sinistral displacements in Britain: implications for a three-plate collision model. *Tectonics*, Vol. 3, 781 – 794.

— Webb, B C, and Woodcock, N H. 1988. Late Caledonian (Acadian) transpression in north-west England: timing, geometry and geotectonic significance. *Proceedings of the Yorkshire Geological Society*, Vol. 46, 175 – 192.

Sparks, R S J, Self, S, and Walker, G P L. 1973. Products of ignimbrite eruptions. *Geology*, Vol. 1, 115 – 118.

— Sigurdsson, H, and Carey, J N. 1980. The entrance of hot pyroclastic flows into the sea, II. Theoretical considerations on subaqueous emplacement and welding. *Journal of Volcanology and Geothermal Research*, Vol. 7, 97 – 105.

Stadlbauer, E, Bohla, M, and Keller, J. 1986. The Kos Plateau-Tuff (Greece): a major ignimbrite eruption that crossed the open sea. Abstracts of International Volcanological Congress New Zealand, 1986, p.75.

Stillman, C J, and Francis, E H. 1979. Caledonide volcanism in Britain and Ireland. 557 – 588 *in* The Caledonides of the British Isles — reviewed. Harris, A L, Holland, C H, and Leake, B E (editors). *Special Publication of the Geological Society of London*, No. 8.

Taylor, G A. 1958. The 1951 eruption of Mont Lamington, Papua. *Geological and Geophysical Bulletin of the Australia Bureau of Mineral Resources*, Vol. 38, 1 – 117.

Taylor, S R, and McLennan, S M. 1985. *The continental crust: its composition and evolution.* (Oxford: Blackwell Scientific.)

Thirlwall, M F, and Jones, N W. 1983. Isotope geochemistry and contamination mechanics of Tertiary lavas from Skye, northwest Scotland. 186 – 208 *in Continental basalts and mantle xenoliths.* Hawkesworth, and Norry, M J (editors). (Nantwich: Shiva.)

Thomas, J E, Dodson, M H, Rex, D C, and Ferrara, G. 1966. Caledonian magmatism in North Wales. *Nature, London*, Vol. 209, 866 – 868.

Thompson, R N, Morrison, M A, Hendry, G L, and Parry, S J. 1984. An assessment of the relative roles of crust and mantle in magma genesis: an elemental approach. *Philosophical Transactions of the Royal Society of London*, No. A310, 549 – 590.

Thorpe, R S. 1979. Late Precambrian igneous activity in Southern Britain. 579 – 584 *in* The Caledonides of the British Isles — reviewed. Harris, A L, Holland, C H, and Leake, B E (editors). *Special Publication of the Geological Society of London*, No. 8.

— Beckinsale, R D, Patchett, P J, Piper, J D A, Davies, G R and Evans, J A. 1984a. Crustal growth and late Precambrian — early Palaeozoic plate tectonic evolution of England and Wales. *Journal of the Geological Society of London*, Vol. 141, 521 – 536.

— Francis, P W, and O'Callaghan, L. 1984b. Relative roles of source composition and fractional crystallization in the petrogenesis of Andean Volcanic rocks. *Philosophical Transactions of the Royal Society of London*, No. A310, 675 – 692.

— Leat P T, Bevins, R E, and Hughes, D J. 1989. Late-orogenic alkaline/subalkaline Silurian volcanism of the Skomer Volcanic Group in the Caledonides of south Wales. *Journal of the Geological Society of London*, Vol. 146, 125 – 132.

— — Mann, A, Campbell, S D G, Evans, J, Reedman, A J, and Howells, M F. In press. The petrogenesis of the Ordovician Snowdon Volcanic Centre, North Wales.

Thurlow, J G, Swanson, E A, and Strong, D F. 1975. Geology and lithogeochemistry of the Buchans polymetallic sulphide deposits, Newfoundland. *Economic Geology*, Vol. 70, 130 – 144.

Trythall, R J B, Eccles, C, Molyneux, S G, and Taylor, W E G. 1987. Age and controls of ironstone deposition (Ordovician) North Wales. *Geological Journal*, Vol. 22 (Thematic Issue), 31 – 43.

Ujike, O, Soya, T, and Ono, K. 1986. Major element, Rb, Sr, Y and Zr composition and origin of volcanic rocks from Kikai caldera, south of Kyushu. *Journal of the Japanese Association of*

Mineralogy, Petrography and Economic Geology, Vol. 81, 105–115.

URABE, T, SCOTT, S D, and HATTORI, K. 1983. A comparison of footwall-rock alteration and geothermal systems beneath some Japanese and Canadian volcanogenic massive sulfide deposits. 345–364 *in* The Kuroko and related volcanogenic massive sulphide deposits. OHMOTO, H, and SKINNER, B J (editors). *Economic Geology Monograph*, No. 5.

VAREKAMP, J C. 1980. The geology of the Vulsinian area, Lazio, Italy. *Bulletin Volcanologique*, Vol. 43, 487–503.

VIDAL, Ph, COCHERIE, A, and LE FORT, P. 1982. Geochemical investigation of the origin of the Manaslu leucogranite (Himalayas, Nepal). *Geochemica Cosmochemica Acta*, Vol. 46, 2279–2292.

WALKER, G P L. 1984. Downsag calderas, ring faults, caldera sizes and incremental caldera growth. *Journal of Geophysical Research*, Vol. 89, 8407–8416.

— HEMING, R F, SPROD, T J, and WALKER, H R. 1981. Latest major eruptions of Rabaul Volcano. 181–193 *in* Cooke-Davian Volume of volcanological papers. JOHNSON, K W (editor). *Memoir of the Geological Survey of Papua New Guinea*, No. 10.

WALSH, J N, BUCKLEY, F, and BARKER, J. 1981. The simultaneous determination of the rare earth elements in rocks using inductively coupled plasma source spectrometry. *Chemical Geology*, Vol. 33, 141–153.

WARREN, P T, PRICE, D, NUTT, M J C, and SMITH, E G. 1985. Geology of the country around Rhyl and Denbigh. *Memoir of the British Geological Survey*, sheets 95 and 107.

WEBB, B C. 1983. Caledonian structures in the Cambrian Slate Belt, Gwynedd, North Wales. *Report of the Institute of Geological Sciences*, No. 83/1.

WHITTINGTON, H B, and WILLIAMS, A. 1955. The fauna of the Derfel Limestone of the Arenig District. *Philosophical Transactions of the Royal Society*, No. B238, 397–427.

WILKINSON, I. 1987. A finite strain study of the Ordovician volcanic rocks of Snowdonia, North Wales, and its implications for a regional strain model. *Geological Journal*, Vol. 22, 95–105.

— 1988. The deformation of the Ordovician volcanic rocks of Snowdonia, North Wales. Unpublished PhD thesis, University of Wales, Aberystwyth.

— and SMITH, M. 1988. Basement fractures in North Wales: their recognition and control on Caledonian deformation. *Geological Magazine*, Vol. 125, 301–306.

WILLIAMS, A. 1963. The Caradocian brachiopod fauna of the Bala District. *Bulletin of the British Museum of Natural History (Geology)*, Vol. 8, 327–421.

— 1969. Ordovician faunal provinces with reference to brachiopod distribution. 117–154 in *The Precambrian and*

Lower Palaeozoic rocks of Wales. WOOD, A (editor). (Cardiff: University of Wales Press.)

— 1976. Plate tectonics and biofacies evolution as factors in Ordovician correlation. 29–66 in *The Ordovician System*. BASSETT, M G (editor). (Cardiff: University of Wales Press.)

WILLIAMS, D. 1930. The geology of the country between Nant Peris and Nant Ffrancon (Snowdonia). *Quarterly Journal of the Geological Society of London*, Vol. 96, 191–233.

WILLIAMS, H. 1927. The geology of Snowdon (North Wales). *Quarterly Journal of the Geological Society of London*, Vol. 87, 346–431.

— 1941. Calderas and their origin. *University of California Publications Bulletin of the Department of Geological Sciences*, No. 25, 239–346.

— and BULMAN, O M B. 1931. The geology of the Dolwyddelan Syncline (North Wales). *Quarterly Journal of the Geological Society of London*, Vol. 87, 425–458.

WILSON, C J N, ROGAN, A M, SMITH, I E, NORTHEY, D J, NAIRN, J A, and HOUGHTON, B F. 1984. Caldera volcanoes of the Taupo Volcanic Zone, New Zealand. *Journal of Geophysical Research*, Vol. 89, 8463–8488.

WINCHESTER, J A, and FLOYD, P A. 1977. Geochemical determination of different magma series and their differentiation products using immobile elements. *Chemical Geology*, Vol. 20, 325–343.

WOOD, D A. 1980. The application of a Th-Hf-Ta diagram to problems of tectonomagmatic classification and to establishing the nature of crustal contamination of basaltic lavas of the British Tertiary volcanic province. *Earth and Planetary Science Letters*, Vol. 50, 11–30.

WOODCOCK, N H. 1984a. Early Palaeozoic Sedimentation and Tectonics in Wales. *Proceedings of the Geologists' Association*, Vol. 95, 323–335.

— 1984b. The Pontesford lineament, Welsh Borderland. *Journal of the Geological Society of London*, Vol. 141, 1001–1014.

— and GIBBONS, W. 1988. Is the Welsh Borderland Fault System a terrane boundary? *Journal of the Geological Society of London*, Vol. 145, 915–924.

WRIGHT, D. 1979. Palaeoecology of the Caradocian succession of Snowdonia, North Wales. Unpublished PhD thesis, Kingston Polytechnic.

WRIGHT, J V, and COWARD, M P. 1977. Rootless vents in welded ash-flow tuffs from northern Snowdonia, N. Wales, indicating deposition environment of shallow water. *Geological Magazine*, Vol. 114, 133–140.

APPENDIX 1

Tables of representative analyses

All numbers without prefixes at the top of columns relate to KB prefixed numbers in the table explanations and the British Geological Survey database. X-ray fluorescence (XRF) analyses were provided by the Department of Geology, University of Nottingham (analysts, Drs Peter Harvey and Brian Atkins) except those which are indicated in the table explanations. Instrumental neutron activation analyses (INAA) were provided by Open University (analyst, Dr Richard Thorpe) and Goldsmith's College, University of London (analyst, Dr Allan Mann) and inductivity coupled plasma (ICP) analyses by King's College, University of London (analyst, Dr Nick Walsh) and University of Leeds/Royal Holloway and Bedford New College, University of London (analyst, Anwen Hughes).

Table 3 Early activity, 1st Eruptive Cycle (Llewelyn Volcanic Group)

	168A	196A	202	222B	012	154	059	083	072	TL072	TL074	087	174	178	180
SiO_2	70.13	69.89	70.96	71.36	65.73	63.98	60.38	62.91	59.05	57.93	54.02	50.32	72.97	71.58	73.21
Al_2O_3	14.69	13.78	14.34	12.85	14.56	14.16	14.50	14.55	14.89	15.93	15.77	16.05	13.35	13.37	13.26
TiO_2	0.29	0.37	0.38	0.75	1.21	1.24	1.47	1.33	1.54	2.49	2.33	2.05	0.32	0.39	0.29
FeO	2.54	2.21	2.41	4.84	5.57	5.97	7.47	6.56	7.81	7.57	8.62	12.72	1.74	3.09	1.98
MgO	0.42	0.44	0.39	1.76	1.54	1.44	2.07	2.14	2.87	1.82	4.51	5.87	0.73	1.73	0.47
CaO	1.67	2.29	0.61	1.18	1.11	2.41	3.76	2.49	5.11	2.36	2.93	3.41	1.47	0.91	1.02
Na_2O	3.56	2.61	5.70	4.21	4.44	4.39	3.24	5.41	3.92	6.25	4.71	4.94	3.42	3.75	2.92
K_2O	3.33	4.83	3.28	1.36	3.15	3.58	3.46	2.75	2.59	2.27	1.03	0.24	3.10	2.15	5.06
MnO	0.08	0.12	0.08	0.08	0.15	0.17	0.15	0.14	0.15	0.13	0.21	0.21	0.08	0.07	0.04
P_2O_5	0.07	0.10	0.08	0.15	0.30	0.35	0.40	0.33	0.39	0.64	0.63	0.37	0.08	0.08	0.04
LOI	2.54	3.72	1.26	1.94	1.95	2.62	2.36	1.35	1.69	2.42	4.69	3.53	2.23	2.12	1.44
Total	99.32	100.34	99.48	100.48	99.71	100.31	99.26	99.96	100.01	99.79	99.45	99.71	99.49	99.24	99.73
As	1	0	1	0	4	1	1	4	2	2	0	3	1	3	2
Ba	739	547	851	359	743	721	567	706	359	712	462	82	960	438	804
Co	3	0	6	17	18	14	14	16	19	22	0	36	5	13	0
Cr	13	0	3	50	14	7	16	11	49	12	7	94	6	24	17
Cu	7	4	2	5	21	10	14	12	26	4	3	23	4	2	6
Ga	17	16	15	15	16	20	20	18	19	18	21	25	16	15	16
Ni	3	28	0	78	12	8	12	10	23	2	6	38	7	6	3
Nb	18	17	15	11	12	22	18	19	20	24	20	14	17	15	22
Pb	20	18	21	12	11	16	17	17	15	8	16	8	17	9	24
Rb	139	191	105	76	71	61	112	70	91	51	203	4	102	85	34
Se	8		6		19	13	14	17	21	29	6	23	9	9	4
Sn	0	2	0	4	7	6	1	3	0	6	12		0	0	5
Sr	89	60	64	101	152	211	265	298	236	165	48	264	79	60	54
Th	13	9	19	13	9	12	9	9	11	6	27	2	16	14	27
U	3	4	3	0	2	2	1	1	3	3	3	2	2	2	4
V	8	14	19	70	96	99	115	109	149	220	15	227	14	37	13
W	3	3	3	3	4	2	1	2	2	7	4		6	3	3
Y	71	57	50	45	64	56	51	54	51	49	62	39	57	55	68
Zn	141	34	44	58	64	83	84	77	79	54	19	108	33	46	27
Zr	225	222	394	209	305	419	350	349	330	411	393	235	216	158	353
La	31.90*	27.20*	43	20	34.00*	41.80*	29	44.50*	27	38.20*	66	24.70*	25	21	37
Ce	66.40*	58.20*	75	29	70.70*	86.20*	81	90.50*	81	82.90*	106	56.20*	56	64	92
Pr*	7.60	6.80			8.40	9.90		10.60		10.10		7.20			
Nd*	34.20	31.10			39.50	44.70		47.80		47.60		34.40			
Sm*	8.60	7.80			9.10	9.60		10.10		10.80		8.10			
Eu*	1.54	1.64			1.99	2.35		2.56		2.73		2.37			
Gd*	9.20	8.10			9.60	9.30		9.80		10.60		8.50			
Dy*	9.94	8.63			10.05	8.64		8.76		9.13		7.93			
Ho*	2.05	1.78			2.05	1.75		1.75		1.82		1.59			
Er*	6.24	5.37			5.97	5.21		5.11		5.26		4.58			
Yb*	6.28	5.28			5.66	5.14		4.89		4.90		4.08			
Lu*	1.00	0.84			0.91	0.84		0.78		0.80		0.66			
S	62	0	0	0	44	45	59	127	63	11	0	0	249	228	112
Ag			0		1	1	1	0	0	0	0	0			
Cd			3		1	0	0	0	0	0	0	0	5		1
Cs			0		3	0	3	4	1	0		1	0		1
Sb			0		1	3	2	1	1	0		0			
Bi			0		1	2	0	0	2	0		0			
Mo			0		1	7	3	2	2	0		0			

Conwy Rhyolite Formation

KB 168A	Anafon valley	7010 6870
KB 196A	Conwy Mountain	7555 7792
KB 202	Penmaenbach	7500 7845
KB 222B	Foel Lwyd	7168 7274
KB 012	Bera Bach	6700 6792

Foel Fras Volcanic Complex

KB 154	Ysgolion Duon	6770 6360
KB 059	Llwytmor	6862 6943
KB 083	Braich tu du	6519 6250
KB 072	Bera Bach (intrusive)	6772 6836

Foel Grach Basalt Formation

TL 072	Carnedd Llewelyn	6867 6488
TL 074	Foel Grach	6897 6570
KB 087	Braich tu du	6517 6251

Braich tu du Volcanic Formation, type section 6490 6220

KB 174	No. 1 unit, rhyolite
KB 178	No. 2 unit, ash-flow tuff
KB 180	No. 3 unit, rhyolite

* ICP at King's College, University of London

Table 4

	TL069	1219	1204	TL084	112	TL009	569	118	TL012	121	128	TL094	TL097	TL090
SiO_2	73.74	75.24	76.80	75.89	80.76	76.41	79.06	74.76	74.98	12.79	74.38	75.47	75.09	73.76
Al_2O_3	13.28	12.75	10.83	12.72	10.62	11.80	11.49	12.85	11.96	13.61	12.72	11.30	11.28	13.28
TiO_2	0.19	0.20	0.20	0.18	0.16	0.20	0.20	0.26	0.18	0.28	0.27	0.28	0.29	0.26
FeO	2.97	2.50	2.70	1.87	2.04	2.12	2.33	2.68	2.17	2.63	2.41	3.20	3.50	2.72
MgO	0.42	0.33	0.57	0.20	0.75	0.14	0.45	1.47	0.12	2.16	2.21	0.56	0.75	0.42
CaO	0.76	1.63	1.63	1.19	0.19	0.43	0.14	0.56	0.71	0.13	0.34	1.37	1.11	0.61
Na_2O	4.37	3.09	2.17	5.91	0.64	1.31	2.70	1.35	1.11	2.20	0.59	4.59	4.41	4.32
K_2O	1.60	1.58	2.04	0.99	3.40	6.93	1.54	3.63	6.61	3.97	4.74	1.35	1.35	4.34
MnO	0.05	0.16	0.19	0.10	0.05	0.10	0.05	0.08	0.14	0.03	0.06	0.18	0.20	0.10
P_2O_5	0.26	0.27	0.07	0.09	0.12	0.10	0.08	0.07	0.12	0.09	0.09	0.03	0.05	0.07
LOI	1.81	2.45	2.48	1.10	1.59	0.71	1.38	2.30	1.29	1.93	2.19	1.89	1.64	0.56
Total	99.35	100.19	99.46	100.23	100.32	100.25	99.41	100.01	99.38	99.82	99.99	100.21	99.68	100.44
As	3	0	0	3	0	2	0	1	2	1	1	2	4	0
Ba	560	386	279	254	801	1123	284	604	2747	889	763	389	399	646
Co	10	6	0	0	6	0	0	16	0	2	3	0	4	5
Cr	7	10	14	19	8	0	10	11	0	13	6	20	32	22
Cu	5	5	5	0	2	5	0	4	4	8	4	0	2	0
Ga	22	19	13	15	13	11	16	21	14	19	20	16	17	19
Ni	4	0	3	3	3	28	6	6	4	3	3	8	8	8
Nb	22	16	14	18	19	17	19	23	22	26	25	21	20	23
Pb	30	20	10	6	44	14	17	15	9	11	15	25	35	12
Rb	77	77	87	40	147	201	68	179	176	130	154	53	48	146
Sc	16			12	7		11	13	18	14	14	12	11	17
Sn	6			3	3	3	3	0	3	0	0	4	0	0
Sr	148	6	50	125	24	34	98	51	102	51	28	79	99	56
Th	13	8	9	16	10	9	16	19	17	16	20	16	15	17
U	4	3	3	3	5	5	3	1	2	6	3	3	5	5
V	8	10	16	6	9	8	3	9	6	9	9	18	21	9
W	6			5	4	0	2	6	5	3	9	2	2	7
Y	78	53	58	69	56	67	79	86	79	88	83	65	67	83
Zn	92	58	63	41	50	57	29	77	41	41	46	100	115	57
Zr	161	139	135	153	130	159	171	323	211	372	354	338	339	346
La	34	9	17	30	22	30	30	41	38	45	35	37	51	51
Ce	60	35	49	70	72	62	58	99	75	111	94	93	121	111
S	0			0	6	0		0	458	26	350	279	620	7
Mo	0			0	0	0			0			1	0	0
Ag	0			0					0			0	0	0
Cd	0			0					0			0	0	0
Sb	0			2	10			4	14	0	0	2	0	0
Cs	4			0	4	8		3	2	6	6	0	0	2
Bi	0			0	0				0			0	0	0

TL 069	Gwern Gof tuff, Nant Gwynant valley	6728 6019	
KB 1219	Penamnen tuff, Crimea Pass	6978 4951	
KB 1204	Clogwyn Gottal tuff, Nant Gwynant valley	6383 4902	

Capel Curig Volcanic Formation

Member 1

TL 084	Tryfan	6677 6020
KB 112	Craig yr Ysfa	6920 6445
TL 009	Dyffryn Mymbyr	7086 5781
KB 569	Dolwyddelan, pod facies	6876 5009

Member 2

KB 118	Craig yr Ysfa	6920 6490
TL 012	Dyffryn Mymbyr	7087 5795

Member 3

KB 121	Craig yr Ysfa	6920 6490
KB 128	Llyn Dulyn	7025 6626

Member 4

TL 094	Tryfan	6638 6023
TL 097	Gallt yr Ogof	6938 5938
TL 090	Tryfan	6663 6030

Table 5   Intrusions, related to 1st Eruptive Cycle

	098	205	088	106	008	021	TL005	006	182	159	TL037
SiO_2	61.12	59.81	62.73	63.52	61.81	62.75	64.70	67.58	66.62	72.17	66.11
Al_2O_3	14.33	15.08	15.07	15.32	14.15	14.31	14.18	14.78	14.06	13.19	15.44
TiO_2	1.00	0.77	0.80	0.80	1.25	1.22	0.95	0.77	0.84	0.44	0.65
Fe_2O_3	7.00	6.92	5.90	5.83	6.60	6.64	5.37	3.94	4.70	3.62	3.98
MgO	2.98	4.50	2.30	2.42	2.17	1.82	1.23	1.41	0.99	0.53	0.76
CaO	4.54	6.17	5.05	4.55	3.24	3.46	2.54	1.85	2.39	0.67	2.48
Na_2O	3.66	3.02	3.35	3.75	3.59	4.28	3.43	4.56	4.64	4.21	4.92
K_2O	2.72	2.26	2.67	2.92	4.56	2.81	3.06	4.00	3.33	3.94	2.12
MnO	0.19	0.16	0.13	0.15	0.14	0.13	0.21	0.09	0.12	0.09	0.16
P_2O_5	0.17	0.11	0.11	0.12	0.29	0.29	0.25	0.19	0.22	0.04	0.19
LOI	2.21	1.03	1.20	1.07	3.73	1.80	3.48	1.26	3.31	1.35	2.64
Total	99.92	98.66	99.31	100.45	100.34	99.92	99.40	100.43	99.82	100.24	99.44
As	0	0	3	3	5	7	2	0	0	0	2
Ba	416	327	522	513	611	588	615	565	757	480	900
Co	0	27	15	5	11	8	7	17	5	9	3
Cr	73	142	57	60	15	18	13	18	7	10	0
Cu	42	53	20	18	11	11	6	0	6	1	8
Ga	15	16	17	19	20	21	20	15	20	19	23
Ni	33	149	12	17	10	8	64	9	1	3	3
Nb	16	10	13	14	12	13	15	20	24	22	20
Pb	19	16	19	19	16	16	13	13	24	6	16
Rb	103	74	110	114	133	86	120	114	80	104	67
Sc	21		19	18				9	18		11
Sn	0	2	5	0			5	5	4	2	0
Sr	86	116	130	128	110	183	73	119	194	45	180
Th	17	3	12	8	10	13	13	16	5.5	25	10
U	3	0	5	4	0	2	3	3	1	3	0
V	132	104	100	112	108	97	59	71	39	27	42
W	0	0	0	0			0	10	0	8	0
Y	55	34	56	53	59	56	67	41	49	60	46
Zn	75	62	68	65	75	77	89	49	90	17	62
Zr	187	145	176	162	282	306	339	263	372	305	438
La	18	17.70*	22	21	32.20*	32.80*		37.30*	41	40	47
Ce	54	37.70*	62	44	67.50*	69.30*		78.10*	87	91	83
Pr*		4.50			8.10	8.30		8.60			
Nd*		21.30			38.30	39.40		37.50			
Sm*		5.60			9.30	9.40		7.90			
Eu*		1.43			2.20	2.33		1.55			
Gd*		6.10			9.90	10.00		7.30			
Dy*		6.34			9.89	9.82		6.91			
Ho*		1.30			2.02	2.02		1.43			
Er*		4.01			6.00	6.02		4.22			
Yb*		3.85			5.68	5.92		4.17			
Lu*		0.62			0.90	0.95		0.69			
S	168	127	39	88	63	0		0	56	126	56
Sb	0	0	0	0				0	0		0
Cs	0	0	0	2				3			3
Mo								0			0
Ag								0			0
Cd								0			0
Bi								0			0
Cl									98		

Penmaenmawr, microdiorite
KB 098   6970 7540
KB 205   6985 7620

Carreg Fawr, microdiorite
KB 088   6910 7350

Dinas, microdiorite
KB 106   6980 7340

Aber-Drosgl, microtonalite
KB 008   6648 6781
KB 021   6691 6987

Carreg y Gatn, microgranodiorite
TL 005   6713 6594

Gyrn, microgranite
KB 006   6468 6865

Mynydd Perfedd, microgranite
KB 182   6250 6230

Bwlch y Cwyion, microgranite
KB 159   6380 6105

Talgau, rhyolite
TL 037   6001 5810

* ICP at King's College, University of London

Table 6 Pitts Head Tuff Formation

	518	630	556	660	553	TL034	TL061	663	666	516
SiO$_2$	75.13	73.48	76.00	75.13	74.55	79.04	74.81	76.06	73.94	77.70
Al$_2$O$_3$	12.74	13.43	12.14	10.85	12.79	10.79	12.20	10.59	10.47	11.29
TiO$_2$	0.26	0.32	0.22	0.20	0.24	0.17	0.23	0.18	0.15	0.13
FeO	2.11	3.10	1.17	5.23	1.61	1.16	1.90	4.09	6.33	1.54
MgO	0.18	0.08	0.85	0.62	0.92	0.40	0.38	0.47	0.67	0.09
CaO	0.36	0.01	0.27	0.02	0.07	0.40	1.31	0.03	0.03	0.07
Na$_2$O	3.13	3.11	1.47	0.24	0.34	1.96	4.43	0.20	0.24	3.17
K$_2$O	4.96	5.16	7.19	6.35	8.53	5.50	3.56	7.38	6.75	5.09
MnO	0.02	0.12	0.07	0.08	0.08	0.05	0.14	0.06	0.11	0.02
P$_2$O$_5$	0.03	0.03	0.01	0.01	0.01	0.03	0.03	0.02	0.02	0.00
LOI	1.09	1.17	0.78	1.47	1.13	0.66	1.13	1.04	1.52	0.98
Total	100.01	100.01	100.17	100.19	100.27	100.15	100.13	100.10	100.00	100.08
As	0	2	40	8	3	0	0	8	11	0
Ba	608	534	1329	1264	1236	1023	764	1424	1297	60
Co	0	0	0	5	0	0	0	0	12	0
Cr	10	11	0	0	3	3	26	0	0	10
Cu	4	10	0	12	5	2	5	7	5	0
Ga	20	23	15	15	16	12	9	17	15	24
Hf	8	5	0		7					19
Ni	4	8	7	0	7	10	2	0	2	0
Nb	26	30	23	23	24	20	24	31	27	63
Pb	12		16	8	10	26	11	9	10	32
Rb	156	163	164	178	236	117	66	194	158	136
Sc	6	6	4		2		6			3
Sn	4	17	0		0	6	0			10
Sr	34	21	71	29	43	48	90	22	19	38
Th	18	18	13	13	17	13	14	22	19	22
U	3	2	3	2	4	0	4	2	4	2
V	8	11	9	12	9	3	3	8	11	6
W	4		3		7	0	0			0
Y	54	64	52	31	54	41	49	104	60	41
Zn	51	58	30	54	62	24	50	83	88	136
Zr	317	294	249	237	277	267	276	293	256	507
La	56	57	66	11	49	30	59	45	19	8
Ce	112	105	103	31	82	65	122	101	58	22
S	6	356				0	0			148
Cl	9									18
Sb	2						0			0
Cs		10	17		4		1			
Nb				0		0	2			
Bi				0		0	0			
Ag							0			
Cd							0			

Llwyd Mawr, intracaldera facies
KB 518   Craig Cwm Silyn   5170 5010
KB 630   NW of Cae Amos   5150 4558

Lower outflow tuff
Subaerial
KB 556   Rhyd ddu   5833 5274
KB 660   Cwmystradlyn   5591 4564
KB 553   Moel ddu, raft   5860 4444

Subaqueous
TL 034   Pont y Gromlech   6294 5649
TL 061   Ogwen Cottage   6490 6058

Upper outflow tuff
KB 663   Cwmystradlyn   5592 4578
KB 666   Cwmystradlyn   5582 4571

Intracaldera rhyolite
KB 516   Craig Cwm dulyn   4970 4930

Table 7

	TL120	TL126	546	695	560	698	246	551	605	682	1214	1232	686	TL022	1236	690
SiO_2	47.44#	45.99#	71.47	73.08	72.99	75.04	74.24	72.77	77.52	73.72	70.24	77.21	62.49	75.29	74.41	62.49
Al_2O_3	16.32#	17.34#	14.35	13.50	12.37	12.11	12.29	12.36	12.08	13.00	14.54	9.31	14.28	12.20	13.20	14.28
TiO_2	0.96#	1.42#	0.26	0.31	0.21	0.22	0.19	0.21	0.18	0.29	0.30	0.23	0.78	0.26	0.23	0.78
Fe_2O_3	8.88#	10.03#	1.57	2.91	3.16	2.84	2.99	3.37	1.23	3.15	5.74	7.08	7.27	2.83	2.82	7.27
MgO	6.87#	10.54#	1.34	0.77	0.58	0.42	0.61	0.67	0.73	1.31	1.84	0.87	3.07	1.05	0.99	3.07
CaO	5.02#	5.49#	0.19	0.08	0.07	0.03	0.06	0.06	0.28	0.29	0.04	0.03	0.87	1.22	0.05	0.87
Na_2O	3.88#	3.96#	1.64	2.42	0.42	0.15	0.20	0.30	4.56	3.94	0.13	0.11	2.86	2.24	3.54	2.86
K_2O	2.88#	0.63#	7.69	5.26	9.23	7.89	8.13	8.65	2.80	2.80	4.47	2.70	4.93	2.47	3.13	4.93
MnO	0.69#	0.22#	0.07	0.08	0.07	0.09	0.07	0.12	0.03	0.08	0.14	0.12	0.10	0.07	0.11	0.10
P_2O_5	0.07#	0.08#	0.01	0.04	0.01	0.04	0.02	0.02	0.03	0.03	0.03	0.03	0.04	0.03	0.03	0.04
LOI	6.73#	4.65#	1.31	1.42	0.70	1.27	1.14	1.03	0.82	1.34	2.62	2.01	2.93	2.49	1.43	2.93
Total	99.58#	100.39#	99.90	99.89	99.80	100.09	99.94	99.57	100.26	99.95	100.08	99.70	99.62	100.13	99.95	99.62
As			0	5	2	3	2	4	6	3	7	31	3	2	3	2
Ba	1106#	94#	524	475	1052	1000	979	1113	826	596	305	670	525	314	708	288
Co	49#	58#	0	4	5	0	0	0	0	4	5	15	0	5	7	5
Cr	286#	304#	8	17	4	0	0	5	6	15	15	12	13	15	9	44
Cu	3#	21#	5	2	5	6	6	4	5	15	9	18	5	3	3	9
Ga			19	17	20	24	24	21	18	25	28	20	22	22	21	42
Hf			5			5		8	5							
Ni	114#	176#	7	0	9	0	19	3	3	0	0	0	0	6	0	8
Nb	2#	<2#	14	15	34	36	37	36	25	42	36	27	35	34	35	91
Pb	<5#	<5#	10	8	10	8	10	13	28	20	3	3	21	27	16	14
Rb	87#	28#	282	228	269	246	253	222	54	135	234	150	148	123	136	99
Sc			5		19			22	16					20		
Sn			8			8	11		5					5		
Sr	142#	377#	63	55	21	55	27	47	211	64	6	1	93	120	34	36
Th			34	28	23	24	24	21	17	24	22	12	25	22	20	43
U			8	5	5	4	4	4	3	6	6	3	5	4	4	8
V			16	24	6	10	6	6	8	18	17	18	15	12	8	61
W			2			5	0	8	6							
Y	29#	23#	49	65	90	93	88	84	62	101	84	61	95	99	92	210
Zn	65#	75#	36	24	16	28	27	47	26	59	27	35	93	94	23	211
Zr	69#	77#	201	209	344	355	339	346	255	372	426	278	317	295	325	1780
La	3.86*	9.04*	25	20	64	73	63	66	41	26	58	19	35	70	49	108
Ce	10.41*	10.15*	60	66	140	169	129	134	90	80	134	64	107	127	120	261
Pr*	1.11	1.53														
Nd*	7.15	8.75														
Sm*	2.26	3.05														
Eu*	1.13	1.19														
Gd*	2.88	3.93														
Dy*	3.83	3.90														
Ho*	0.70	0.88														
Er*	2.04	2.45														
Yb*	1.77	2.03														
Lu*	0.25	0.30														
S								0						332		
Mo			0		4		3							0		
Ag														0		
Cd														0		
Sb														0		
Cs			9		5		4		5					6		
Bi			0		0		0		0							

Basalt
TL 120 Cwm Llan 6015 5283
TL 126 Glyder Fawr 6450 5820

Yr Arddu tuffs
KB 546 Yr Arddu 6245 4554
KB 695 Moel y Dyniewyd, accretionary lapilli tuff 6462 4844

Lower Rhyolitic Tuff Formation
KB 560 Yr Aran, basal welded tuff 5905 7230
KB 698 Moel y Dyniewyd, basal welded tuff 6180 4802
KB 246 Cwm Llan, welded pod 6155 5243
KB 551 Moel ddu, basal welded tuff 5842 4254

Main caldera phase
KB 605 Dolwyddelan 6825 5065
KB 682 Pen y Pass 6462 5548
KB 1214 Aberglaslyn Pass 5934 4687
KB 1232 Cwm Tregalan 6146 5393
KB 686 Cwm Idwal 6454 5894
TL 022 Creigiau Gleision 7271 6168

Reworked facies
KB 1236 Cwm Tregalan 6139 5410
KB 690 Cwm Idwal 6410 5879

* ICP at Royal Holloway and Bedford New College. University of London
# XRF at Leeds University

Table 8 Rhyolites associated with Snowdon Centre

	440	445	451	465	472	476	257	261	450	459	109	258	471
SiO_2	73.97	75.31	69.39	73.02	71.77	71.50	75.76	74.56	72.06	73.54	75.62	74.35	76.75
Al_2O_3	12.27	12.47	14.35	12.78	12.69	13.06	12.25	13.05	12.87	12.04	10.78	10.76	9.46
TiO_2	0.20	0.18	0.19	0.28	0.21	0.26	0.40	0.28	0.25	0.25	0.28	0.34	0.24
Fe_2O_3	2.10	1.25	1.83	3.58	2.67	3.20	0.55	1.78	2.33	2.91	5.09	4.65	4.58
MgO	0.61	1.55	0.85	0.93	0.80	1.19	0.00	0.99	0.58	1.06	0.81	0.67	1.59
CaO	0.04	0.02	2.32	0.14	1.86	0.63	0.06	0.30	0.29	0.13	0.15	0.61	0.04
Na_2O	1.58	0.93	4.65	3.15	4.60	5.64	0.31	2.43	3.31	1.68	2.09	4.22	0.74
K_2O	6.85	5.29	2.16	3.98	1.61	1.65	10.16	6.74	5.84	5.52	4.72	2.91	4.79
MnO	0.11	0.05	0.05	0.17	0.05	0.17	0.03	0.02	0.09	0.07	0.05	0.10	0.08
P_2O_5	0.01	0.02	0.01	0.04	0.01	0.04	0.04	0.03	0.03	0.02	0.01	0.02	0.02
LOI	0.98	1.85	3.04	1.45	2.62	1.37	0.49	0.00	0.71	1.33	0.72	1.28	1.64
Total	98.72	98.94	98.85	99.51	98.89	98.71	100.04	100.18	98.37	98.56	100.32	99.90	99.91
As	2	0	0	0	0	0	46	0	0	0	1	0	0
Ba	546	336	321	511	91	211	1718	431	773	814	157	267	232
Co*	1.9	2.4	3.8	3.52	2.8	2.4	1.9	1.4	2.9	2.1	2.7	1.0	1.2
Cr*	5.6	4.2	7.2	10.0	7.4	7.0	12.2	6.4	3.4	2.2	13.7	14.0	5.3
Cu	5	1	3	2	0	3	13	5	3	2	6	1	2
Ga	18	17	14	21	24	22	15	20	16	21	42	39	32
Hf*	6.28	5.65	5.56	10.36	12.37	12.19	9.12	10.2	10.83	10.26	26.07	27.18	28.59
Ni	3	5	3	6	6	4	22	42	3	6	2	10	7
Nb	16	14	14	40	46	44	15	22	23	22	71	69	66
Pb	47	25	21	17	18	16	25	21	20	15	22	24	20
Rb	277	245	131	157	68	48	325	192	104	155	122	86	129
Sc*	3.37	2.86	2.97	3.6	2.05	2.60	6.75	4.9	5.27	5.49	0.39	0.10	0.33
Sn	5	3	4	11	9	11	10	9	5	6	11	14	24
Sr	61	20	543	26	94	44	69	79	86	75	44	85	29
Ta*	1.66	1.53	1.61	2.83	2.97	3.00	1.48	1.51	1.67	1.62	4.08	4.59	4.67
Th	34	29	33	26	31	28	38	27	23	24	34	37	32
U	5	4	6	4	4	5	6	5	5	4	3	6	7
V	12	4	6	12	3	3	46	5	0	3	2	9	0
W	6	12	2	9	7	6	4	0	6	5	7	5	13
Y	46	55	60	72	84	77	48	57	58	74	169	162	207
Zn	28	20	22	76	65	73	89	39	44	53	180	212	137
Zr	207	189	192	377	439	465	290	437	424	398	1129	1184	1292
La*	46.32	53.50	55.43	61.25	61.32	69.05	51.87	50.59	49.46	52.37	96.39	98.39	169.40
Ce*	98.78	105.90	111.80	90.76	130.50	137.70	103.10	102.40	115.60	105.80	236.20	239.40	253.10
Nd*	31.64	42.01	36.30	50.10	46.61	60.39	39.84	46.05	43.45	47.06	94.76	105.70	166.30
Sm*	7.29	8.39	7.84	10.53	10.80	12.18	8.12	9.47	9.58	10.95	24.54	25.75	41.06
Eu*	0.556	0.668	0.638	0.847	0.463	0.736	1.068	1.012	0.897	1.147	1.852	2.012	2.795
Gd*	10.12	9.15	10.33	11.75	13.45	13.69	9.67	11.05	10.30	12.24	26.37	28.89	34.94
Ho*	2.49	2.30	2.50	2.55	3.41	3.47	2.10	2.93	2.49	2.86	6.60	6.58	6.26
Tb*	1.90	1.52	1.89	1.99	2.38	2.31	1.68	1.90	1.70	2.06	5.35	4.87	5.13
Yb*	6.13	5.43	4.91	6.57	8.26	7.89	5.50	6.34	6.14	6.15	17.59	17.47	18.63
Lu*	0.921	0.762	0.752	1.062	1.213	1.206	0.863	0.919	1.051	1.050	2.600	2.908	2.962
S							762	0			462	0	
Sb												0	
Cs												3	

Group A1			Group B2		
KB 440	Yr Arddu	6278 4633	KB 261	Carnedd y Cribau	6708 5365
KB 445	Yr Arddu	6324 4683	KB 450	Cefn y cerrig	6645 5533
KB 451	Bwlch Ehediad	6683 5242	KB 459	Castell	6382 4780
Group A2			**Group B3**		
KB 465	Moel Hebog	5622 4740	KB 109	Twll du	6420 5870
KB 472	Moel Lefn	5533 4850	KB 258	Carnedd y Cribau	6657 5333
KB 476	Beddgelert	5653 4942	KB 471	Moel yr Ogof	5568 4793
Group B1					
KB 257	Bylchau Teyrn	6170 5076	* INAA at Goldsmith's College, University of London		

Table 9 Intrusions related to Snowdon Centre

	270	290	342	347	315	348	331	1221	038	039	TL105	TL107	362	365	480
SiO$_2$	74.13	65.27	71.43	72.78	73.53	72.04	75.92	77.99	71.14	72.48	76.74	76.14	75.52	73.74	65.88
Al$_2$O$_3$	12.36	13.17	12.44	13.52	13.28	13.68	10.74	11.13	13.40	13.60	11.79	12.06	12.12	12.58	15.01
TiO$_2$	0.30	0.84	0.27	0.33	0.28	0.23	0.28	0.28	0.38	0.35	0.11	0.09	0.21	0.25	0.78
Fe$_2$O$_3$	3.57	5.68	2.58	2.98	2.53	2.64	4.71	4.20	2.74	2.31	2.55	2.44	2.21	2.88	4.85
MgO	0.29	1.99	0.51	0.23	0.18	0.35	0.63	0.63	1.07	0.62	0.15	0.08	0.12	0.29	1.35
CaO	0.51	3.61	3.75	0.07	0.61	1.70	1.19	0.16	0.34	0.41	0.05	0.23	0.20	0.70	2.39
Na$_2$O	0.96	4.86	0.35	3.37	3.42	4.70	3.43	1.50	3.31	3.62	4.16	4.74	3.52	3.60	3.73
K$_2$O	6.31	0.46	3.12	6.13	5.84	1.48	0.92	2.61	5.42	5.60	4.13	4.12	5.22	4.87	3.02
MnO	0.15	0.16	0.29	0.05	0.06	0.11	0.24	0.09	0.08	0.06	0.05	0.10	0.04	0.05	0.17
P$_2$O$_5$	0.04	0.12	0.03	0.04	0.02	0.03	0.01	0.02	0.05	0.04	0.01	0.01	0.03	0.03	0.20
LOI	1.17	3.70	5.43	0.88	0.69	2.42	2.12	1.79	1.24	1.17	0.37	0.16	0.81	0.83	1.29
Total	99.80	99.85	100.19	100.39	100.42	99.40	100.19	100.40	99.19	100.23	100.10	100.17	100.01	99.80	98.65
As	4	12	0	3	0	3	75	59	2	0	0	2	2	0	0
Ba	1349	185	264	734	607	243	240	297	597	463	23	0	895	763	799
Co	0	13	0	0	0	0	0	19	5	3	0	0	0	0	9
Cr	0	30	11	3	0	7	6	8	12	4	18	6	3	5	9
Cu	6	10	3	3	5	2	2	2	6	0	0	0	2	2	8
Ga	16	19	20	22	21	17	31	29	21	20	28	33	19	17	19
Hf								30							9
Ni	6	14	3	2	7	5	30	5	4	88	7	2	5	4	6
Nb	20	16	23	26	22	14	66	66	25	22	21	127	18	20	18
Pb	14	15	11	22	27	17	27	23	17	18	74	44	15	15	18
Rb	211	23	163	209	237	61	44	123	170	217	187	344	152	166	77
Sc									7	24	0	0			
Sn	0	5			5		11	3	3	6	11	21			0
Sr	49	164	52	46	56	87	57	11	63	51	9	0	32	55	227
Th	34	20	24	28	29	7	28	33	30	29	23	30	12	15	10
U	6	3	4	6	4	3	5	6	5	6	3	7	3	3	2
V	11	91	11	13	12	10	5	7	29	15	0	0	8	14	46
W	0	2			0		6	16	5	0	6	7			5
Y	82	54	67	62	75	87	162	166	79	58	110	16	80	63	44
Zn	70	64	22	49	49	79	304	31	15	26	95	194	23	31	56
Zr	346	286	418	520	420	342	1146	1210	434	380	732	724	273	248	362
La	54	35	45	51	54	35	89	50	46	53.60*	14	28	48	50	42
Ce	83	78	101	117	122	87	205	131	132	102.80*	41	70	91	115	111
Pr*										11.60					
Nd*										48.90					
Sm*										10.10					
Eu*										1.11					
Gd*										9.40					
Dy*										9.00					
Ho*										1.86					
Er*										5.67					
Yb*										5.87					
Lu*										0.92					
S	0	0	178	0	0	0	0		0	0	0	0	0	0	
Sb									0	0	0	0			
Cs									2	1	0	2	4	0	
Mo									0	0	2	0			
Ag									0	0	0	0			
Cd									0	0	0	0			
Bi									0	2	3	1	0	2	
Cl			0	40		8					53	0	287	251	

Feldspar porphyries
KB 270 Nantmor/Croesor 6310 4523
KB 290 Nantmor/Croesor 6270 4533
KB 342 Llyn Croesor 6613 4576
KB 347 Cnicht 6836 4672

Rhyolites
KB 315 Llyn Cwm Corseog 6640 4740
KB 348 Ceseiliau Moelwyn 6665 4464

Microgranite
KB 331 Bwlch y battel 6380 4690
KB 1221 Cwmystradlyn 5522 4440

Ogwen microgranite
KB 038 Llyn Ogwen 6529 6058
KB 039 Llyn Ogwen 6537 6058

Mynydd Mawr microgranite
TL 105 Mynydd Mawr 5456 5510
TL 107 Mynydd Mawr 5506 5530

Tan y Grisiau granite
KB 362 Tan y Grisiau 6847 4453
KB 365 Tan y Grisiau 6752 4344

Y Garn feldspar porphyry
KB 480 Y Garn 5537 5277

* ICP at King's College, University of London

Table 10

	493	506	510	863	TL049	801	865	959	1260	1261	251	253	614
SiO_2	42.52	45.50	47.27	48.53	45.32	44.91	46.66	45.80	70.31	81.44	74.99	76.79	51.99
Al_2O_3	17.27	15.34	16.85	13.29	16.06	14.90	14.30	15.91	10.71	6.22	10.59	10.65	18.25
TiO_2	2.76	3.04	2.94	3.72	1.98	2.08	3.85	2.42	0.87	0.23	0.31	0.28	2.11
FeO	13.14	13.61	13.11	20.22	12.48	12.17	17.02	14.24	6.42	4.62	4.27	2.16	11.75
MgO	5.02	4.70	7.99	4.95	7.57	10.92	6.97	8.14	2.84	2.02	1.02	0.68	3.89
CaO	5.78	5.50	0.64	0.70	9.55	5.98	0.89	2.78	0.09	0.03	0.23	0.60	1.15
Na_2O	4.90	3.65	3.10	0.15	2.62	3.78	0.19	3.71	0.08	0.14	2.72	3.37	4.78
K_2O	0.88	0.65	2.64	2.32	1.05	0.53	3.53	0.05	5.64	2.95	4.30	3.50	1.84
MnO	0.18	0.15	0.09	1.25	0.17	0.18	1.73	0.20	0.79	0.42	0.01	0.03	0.29
P_2O_5	0.37	0.40	0.38	0.55	0.24	0.24	0.55	0.28	0.05	0.02	0.00	0.00	0.25
LOI	7.52	7.52	4.55	3.98	2.62	4.30	4.57	6.46	1.94	1.83	1.66	1.69	4.00
Total	100.35	100.07	99.57	99.66	99.64	99.99	100.24	99.99	99.74	99.82	100.10	99.74	100.31
As	0	8	27	121	2	0	27	0	9	10	0	0	5
Ba	472	53	1443	573	245	82	1050	49	621	354	337	361	83
Co	41.0 *	45.0 *	28.8 *	39.0 *	48.0 *	47	37	48	11	4	1.8 #	1.8 #	62
Cr	147*	172*	187*	91*	158*	175	120	215	54	3	9.8 #	3.9 #	330
Cu	28	38	21	35	61	51	15	10	6	6	3	1	74
Ga	24	21	23	26	18	19	23	21	26	19	47	43	20
Hf	4.76*	5.14*	5.13*	6.40*	3.40*						27.56#	22.96#	3
Ni	51	52	65	87	77	72	36	70	18	8	4	9	146
Nb	13	14	14	26	11	13	26	15	48	43	71	57	7
Pb	2	5	2	617	8	6	107	7	20	71	5	11	10
Rb	27	29	38	97	24	22	93	4	156	72	108	98	90
Sc	40.0 *	43.0 *	42.7 *	43.7 *	38.7 *						0.34 #	0.31 #	50
Sn	0	6	3						5	3	12	6	
Sr	218	129	78	40	431	315	54	65	76	16	40	51	78
Ta	1.26*	0.95*	0.85*	1.67*	0.66*						4.51 #	3.94 #	
Th	2.77*	2.22*	1.73*	1.97*	0.64*	0	0	0	21	16	29	33	2
U	0.8 *	0.7 *	0	0.8 *	0	2	2	0	4	2	4	2	0
V	328	399	381	460	271	284	474	314	81	11	1	3	304
W	0	6	17	33	4	0		5	3	3	5	0	
Y	36	40	42	50	38	31	44	30	121	87	169	148	31
Zn	102	105	106	1635	85	88	752	113	102	185	39	35	78
Zr	197	221	205	265	154	169	298	174	869	754	1120	1194	148
La	15.6 *	15.8 *	16.4 *	22.0 *	10.4 *	10	23	8	63	41	98.59 #	90.95 #	11
Ce	40.8 *	40.5 *	39.1 *	51.6 *	26.0 *	38	58	23	181	88	211.20 #	176.60 #	42
Nd*	25.3	28.1	27.6	34.3	19.8						107.80 #	88.46 #	
Sm*	6.42	6.33	6.94	8.92	4.54						26.98 #	19.74 #	
Eu*	2.00	1.84	2.07	2.66	1.82						2.024#	1.413#	
Gd#											30.02	20.97	
Ho#											6.27	5.34	
Tb*	1.15	1.20	1.31	1.88	0.97						5.04 #	3.44 #	
Yb*	3.33	3.49	3.69	4.19	2.46						18.98 #	16.30 #	
Lu*	0.52	0.56	0.57	0.33	0.39						2.89 #	2.53 #	
S	56	76	554	2326	0	0	0	0	1984	552			874
Cl	38	33	41						87	87			
Sb	2	0	0										
Cs	1.07*	0.89*	0.61*	2.26*	1.08*								
Mo									9	0			
Bi									2	0			

Bedded Pyroclastic Formation
Arc tholeiite affinity
KB 493   Gallt y Wenallt   6435 5340
KB 506   Clogwyn y Person   6174 5538
KB 510   Llyn Gwynant   6475 5270
Ocean-island basalt affinity
KB 863   Cwm Llan   6234 5250
TL 049   Twll du   6389 5881

Volcaniclastic sediments
KB 801   Cwm Idwal   6398 5868
KB 865   Cwm Llan   6232 5261
KB 959   Bwlch main   6064 5388

Upper Rhyolitic Tuff Formation
KB 1260   Crib y ddysgl, ash-flow tuff   6105 5516
KB 1261   Crib y ddysgl, bedded tuff   6119 5514
KB 251   Gallt y Wenallt, rhyolite   6380 5342
KB 253   Gallt y Wenallt, rhyolite   6335 5360
KB 614   Garnedd Ugain, basalt   6100 5516

* INAA at Open University
# INAA at Goldsmith's College, University of London

Table 11

	308	372	447	473	479	508	611	289	294	367	499	278	318	255	262
SiO_2	45.74	43.59	47.83	45.72	45.21	47.34	50.32	51.42	46.22	41.23	45.10	43.78	53.15	59.42	58.96
Al_2O_3	15.53	13.56	14.26	14.15	16.87	15.76	15.01	14.31	13.53	13.77	17.21	12.85	14.22	13.97	14.08
TiO_2	2.27	2.50	1.76	2.90	1.35	1.65	1.02	2.10	3.43	2.87	1.15	3.16	1.19	1.90	1.95
Fe_2O_3	12.47	12.76	12.31	13.77	10.40	10.67	9.35	12.82	15.27	17.84	9.73	13.75	8.35	9.04	8.95
MgO	7.30	5.99	6.55	6.90	9.11	7.76	8.60	5.86	5.36	6.05	9.89	5.07	4.07	2.30	2.59
CaO	10.07	9.90	9.89	9.22	9.60	8.28	8.65	5.87	9.90	7.58	10.30	10.37	7.03	4.53	4.08
Na_2O	3.22	0.71	4.19	2.83	3.23	3.53	2.64	4.01	1.89	1.38	2.10	1.19	4.19	3.52	3.96
K_2O	0.03	0.40	0.06	1.19	0.15	0.84	0.90	0.13	0.07	0.00	0.60	0.10	1.37	2.82	3.04
MnO	0.20	0.20	0.19	0.19	0.23	0.17	0.14	0.24	0.28	0.71	0.15	0.29	0.18	0.19	0.17
P_2O_5	0.30	0.40	0.15	0.40	0.14	0.18	0.09	0.24	0.56	0.30	0.14	0.52	0.26	0.50	0.51
LOI	3.36	9.92	2.22	3.03	3.86	3.29	3.39	2.70	3.06	8.68	3.43	8.10	6.28	1.81	1.63
Total	100.48	99.93	99.42	100.29	100.16	99.38	100.13	99.69	99.57	100.40	99.80	99.90	100.29	100.00	99.91
As	4	6	0	0	0	5	1	42	29	8	0	88	0	3	0
Ba	39	50	83	112	121	224	163	96	39	17	156	37	654	520	520
Co	52	37	51	37	48	37	40	39	42	39	46	39	27	12	16
Cr	189	158	318	161	233	233	235	88	75	205	252	60	99	8	10
Cu	53	35	86	56	53	37	45	42	49	6	37	37	20	4	3
Ga	22	20	17	23	17	18	17	23	26	23	15	23	24	21	22
Hf	6		0	3	2	0	0				0				
Ni	79	53	54	43	123	130	97	46	31	48	221	24	18	24	12
Nb	13	16	6	18	6	7	5	12	27	11	7	26	19	30	30
Pb	5	9	10	14	9	8	8	8	13	11	2	15	11	14	17
Rb	9	23	4	29	6	21	33	9	5	3	25	6	31	82	97
Sc	32					30	23				25				
Sn	0		0	4	4	4		5	7		0	3	3	3	5
Sr	258	100	137	444	373	268	213	381	509	86	265	186	150	323	244
Th	0	0	5	0	0	2	3	3	1	0	2	2	6	10	4
U	0	0	0	2	0	0	0	2	2	0	0	2	0	2	4
V	244	267	233	322	167	196	143	277	400	367	139	350	118	132	145
W	4		4	0	0	0		0	4		0	2	3	0	0
Y	30	38	34	38	23	26	24	44	39	38	21	39	60	54	54
Zn	100	97	62	133	83	80	66	101	131	223	64	119	73	110	112
Zr	145	222	129	216	104	127	104	214	269	190	98	255	395	380	377
La	15	21	10	16	8	6	0	15	19	9	0	18	18	44	40
Ce	45	42	33	54	33	17	31	52	65	37	9	62	55	102	87
S	1411	1199				409	53	672	3101	0	94	221	0	0	0
Cl		11				60					20	63			
Sb						0						2			
Cs							4								

Dolerites			Dyke		
KB 308	Croesor	6243 4510	KB 278	Croesor	6272 4514
KB 372	Croesor	6193 4428	**Microdiorite**		
KB 447	Yr Arddu	6717 5509	KB 318	Cwm Croesor, basaltic	6645 4690
KB 473	Moel yr Ogof	5537 4850	**Icelandite**		
KB 479	Beddgelert	5574 5097	KB 255	Llyn Teyrn	6431 5466
KB 508	Dinas Mot	6227 5635	KB 256	Llyn Teyrn	6431 5466
KB 611	Llyn yr Adair	6522 4768			

Dolerites of multiple intrusions

KB 289	Croesor	6270 4533
KB 294	Croesor	6252 4507
KB 367	Cnicht	6400 4600
KB 499	Rhyd ddu	5885 5400

Table 12

	TL027	TL021	TL018	TL030	TL042	240	214	307	229	231
SiO_2	73.06	67.84	74.80	78.44	73.59	76.43	51.46	42.23	41.91	39.95
Al_2O_3	12.70	14.51	9.88	9.87	11.79	0.17	13.44	14.16	16.72	16.76
TiO_2	0.31	0.35	0.33	0.28	0.41	12.41	1.61	2.71	2.66	3.28
FeO	3.55	3.83	4.76	4.49	4.55	0.28	10.34	13.09	14.45	15.28
MgO	1.44	1.61	1.75	1.06	1.80	0.04	7.03	9.80	8.16	9.86
CaO	0.34	0.40	1.30	0.13	1.11	0.00	8.97	6.52	5.49	5.29
Na_2O	4.93	0.86	2.30	1.50	0.05	0.09	4.03	2.63	1.29	1.11
K_2O	2.44	9.14	3.08	1.69	3.72	1.25	0.14	0.65	1.33	1.00
MnO	0.06	0.08	0.07	0.04	0.05	8.83	0.15	0.23	0.13	0.19
P_2O_5	0.04	0.04	0.03	0.01	0.01	0.04	0.16	0.27	0.27	0.33
LOI	1.15	1.37	1.93	2.15	2.88	0.71	2.71	6.34	7.51	7.45
Total	100.03	100.03	100.22	99.65	99.95	100.25	100.04	100.09	99.93	100.50
As	2	0	2	0	1	5	0	0	2	1
Ba	120	1753	1211	273	3501	1997	87	605	671	771
Co	4	3	6	7	7	0	31.6 *	47	52	59
Cr	11	6	14	5	0	0	137*	128	117	142
Cu	9	9	20	2	12	6	35	34	26	44
Ga	19	24	24	26	28	9	16	20	23	24
Hf							2.99*			
Ni	9	9	26	7	4	50	107	61	62	73
Nb	34	41	55	58	52	33	9	11	14	18
Pb	20	18	20	24	60	135	17	10	8	9
Rb	37	168	47	80	99	189	6	16	47	36
Sc	6	6	3	0	4		33.5 *			
Sn	4	4	3	7	11	0		6	2	0
Sr	132	134	71	112	149	106	181	154	114	76
Ta							0.57*			
Th	17	18	28	26	27	26	155*	0	1	0
U	6	5	3	4	6	4	0.8 *	2	3	0
V	20	19	24	6	12	7	211	360	375	428
W	6	6	6	9	9	0	0	0	9	4
Y	115	139	150	118	111	58	27	44	33	45
Zn	108	126	174	126	71	11	84	60	103	114
Zr	475	468	1065	1001	394	251	132	161	184	230
La	69	57	56	42	49	39	11.1 *	15	8	6
Ce	164	137	161	129	106	63	26.4 *	40	30	32
Pr*										
Nd*							18.1			
Sm*							449			
Eu*							1.51			
Gd*										
Tb*							0.84			
Ho*										
Er*										
Yb*							2.62			
Lu*							0.41			
S	636	1189	2028	0	742	0	30	0	12	0
Ag	0	0	0		0					
Cd	0	0	0		0					
Cs	2	3	0		5					
Sb	0	0	0		1					
Bi	0	0	0		0					
Mo	0	0	0		1					

Lower Crafnant Volcanic Formation
Ash-flow tuff, Unit 2
TL 027   Creigiau Gleision   7305 6171
TL 021   Capel Curig   7302 5882
Ash-flow tuff, Unit 3
TL 018   Capel Curig   7304 5874
TL 030   Creigiau Gleision   7322 6201

Upper Crafnant Volcanic Formation
TL 042   Ash-flow tuff
Betws y Coed   7978 5776
Cae Coch, rhyolite
KB 240   Cae Coch   7777 6588
Tal y Fan Volcanic Formation
KB 214   Tal y Fan, basalt   7460 7345
KB 307   Tal y Fan, basic tuff   7325 7210

Dolgarrog Volcanic Formation
KB 229   Afon Porthlwyd, hyaloclastite   7588 6720
KB 231   Afon Porthlwyd, hyaloclastite   7880 6720

* INAA at Open University

Table 13   Dolerite intrusions at Crafnant Centre

	034	134	150	209	329	301	304	284	287	285	TL077	TL082
SiO$_2$	45.67	46.66	46.52	46.36	44.60	45.46	44.60	44.01	48.37	47.18	47.19	48.60
Al$_2$O$_3$	14.10	16.79	16.59	13.98	15.19	14.98	15.04	15.84	15.25	14.71	14.73	15.31
TiO$_2$	2.14	1.29	1.53	2.67	2.76	2.67	2.10	1.45	1.80	2.53	2.07	1.91
Fe$_2$O$_3$	12.90	10.08	10.05	13.22	12.59	11.11	11.33	10.01	8.94	11.09	12.35	10.90
MgO	9.13	8.68	8.08	7.88	7.82	9.21	9.22	10.93	7.05	8.13	6.11	6.68
CaO	8.33	8.61	10.11	6.47	7.25	5.48	8.77	9.57	10.12	6.14	9.83	9.69
Na$_2$O	3.32	3.67	2.44	3.18	3.88	3.59	2.66	1.98	3.64	3.54	4.10	3.98
K$_2$O	0.34	0.23	0.77	1.27	0.35	1.28	0.88	0.55	0.59	1.14	0.15	0.52
MnO	0.23	0.16	0.16	0.21	0.18	0.23	0.25	0.19	0.20	0.21	0.26	0.24
P$_2$O$_5$	0.15	0.12	0.22	0.45	0.52	0.39	0.31	0.26	0.35	0.45	0.41	0.27
LOI	3.14	3.08	3.52	3.26	3.38	4.67	3.83	4.34	2.74	3.55	2.16	2.22
Total	99.44	99.37	99.98	98.96	98.53	100.31	100.25	100.24	100.03	99.90	99.36	100.33
As	0	0	0	0	2	0	0	0	0	0	2	4
Ba	196	185	567	437	224	555	379	235	167	1848	58	140
Co	48	41	46	42	42	48	59	62	46	43	48	41
Cr	190	195	252	175	136	135	103	63	380	92	150	239
Cu	63	52	51	51	23	26	128	27	26	41	38	94
Ga	16	16	16	18	19	20	19	16	18	15	20	18
Hf							3.60#	2.64#				
Ni	56	152	126	43	53	83	103	135	45	51	21	47
Nb	5	6	7	19	23	12	10	8	11	13	8	13
Pb	4	2	10	8	8	8	21	10	10	8	9	8
Rb	8	0	17	22	6	21	20	17	11	22	5	6
Sc	45	27										37
Sn			0	0	4	0	0	5	0	0	0	0
Sr	185	413	700	202	299	240	566	344	530	460	173	286
Ta							0.65#	0.45#				
Th	0	0	2	0	1	0	0.46#	0.30#	0	0	5.58#	4
U	0	0	2	0	3	0	0	0	0	3	3	0
V	343	172	181	325	322	331	246	145	238	309	276	220
W			6	2	0	5	0	0	0	0	3	0
Y	23	14	24	42	43	32	26	21	30	32	36	39
Zn	111	68	67	109	102	58	134	76	68	81	99	86
Zr	101	59	120	233	248	186	138	114	166	205	136	166
La	6.60*	5.20*	7.80*	18	20	13.35*	10.35*	8.49	12.29*	14.75*	10	26
Ce	16.90*	12.80*	19.20*	34	51	31.62*	24.79*	19.84	29.81*	35.54*	27	48
Pr*	2.70	1.80	2.80			4.35	3.66	2.75	4.41	5.13		
Nd*	14.30	9.80	13.90			21.61	17.46	13.49	20.51	25.20		
Sm*	4.30	2.80	4.00			5.52	4.35	3.40	5.19	6.00		
Eu*	1.40	1.32	1.52			1.95	1.69	1.34	1.86	2.20		
Gd*	5.10	3.30	4.40			6.19	4.80	3.74	5.77	6.76		
Dy*	5.12	3.14	4.21			5.82	4.66	3.57	5.53	6.36		
Ho*	1.03	0.64	0.86			1.16	1.05	0.74	1.22	1.35		
Er*	3.04	1.91	2.64			3.23	2.69	2.05	3.10	3.58		
Yb*	2.61	1.75	2.25			2.83	2.32	1.83	2.67	3.11		
Lu*	0.43	0.30	0.38			0.41	0.33	0.25	0.39	0.46		
S	0	0	0		812	31	0	0	0	0	0	0
Mo											0	0
Sb											0	2
Bi											0	0
Ag												0
Cd												0
Cs												3

KB 034	Nant Llyn y Foel	7168 5472	**Tal y Fan dolerite**			**Dolerite**	
KB 134	Cwm Eigiau	7215 6385	KB 301	Tal y Fan	7295 7270	TL 077	Foel Grach	6961 6534
KB 150	Pen y craig-gron	7430 6206	KB 304	Tal y Fan	7295 7270	TL 082	Ysgolion Duon	6735 6310
KB 209	Dolgarrog	7666 7630	KB 284	Tal y Fan	7293 7265		
KB 329	Pen y gadair	7365 6909	KB 287	Tal y Fan	7329 7208	# INAA at Goldsmith's College. University of London
			KB 285	Tal y Fan	7329 7208	* ICP at King's College. University of London

INDEX